Methods in Molecular Biology™

Series Editor
John M. Walker
School of Life Sciences
University of Hertfordshire
Hatfield, Hertfordshire, AL10 9AB, UK

For further volumes:
http://www.springer.com/series/7651

In Silico Models for Drug Discovery

Edited by

Sandhya Kortagere

Department of Microbiology and Immunology, Drexel University College of Medicine, Philadelphia, PA, USA

Humana Press

Editor
Sandhya Kortagere
Department of Microbiology and Immunology
Drexel University College of Medicine
Philadelphia, PA, USA

ISSN 1064-3745 ISSN 1940-6029 (electronic)
ISBN 978-1-4939-6297-6 ISBN 978-1-62703-342-8 (eBook)
DOI 10.1007/978-1-62703-342-8
Springer New York Heidelberg Dordrecht London

Printed on acid-free paper

Humana Press is a brand of Springer
Springer is part of Springer Science+Business Media (www.springer.com)

Preface

Infectious diseases caused by viruses, parasites, bacteria, and fungi are the number one cause of death worldwide. In addition to treatment costs, the social and economic challenges faced by governments, individuals, and families in managing these diseases and in preventing epidemics present an enormous burden to society. Although new technologies have improved diagnosis of infectious diseases, treatment remains a challenge. The efficacy of all known current anti-infective agents is threatened by the spread of drug-resistant forms of the pathogens. Hence the need remains urgent to develop anti-infective agents that target drug-resistant pathogens.

This book presents a comprehensive discussion of the role of in silico models in understanding infectious diseases and in developing novel therapeutics to treat them. This includes the role of in silico methods in vaccine development as well as small molecule development against known and new drug targets. Each chapter is written by a leading expert and addresses a unique aspect of in silico methods in drug design. The book is divided into two main sections, with the first ten chapters providing an overview of the methods and techniques used in drug design and the later six chapters detailing applications of these methods to real-world drug discovery problems. Chapter 1 provides an overview of the in silico models used in virtual screening, and Chapters 2 and 3 describe techniques to derive novel antimicrobial targets. One of the major problems associated with curing infectious diseases is the ability of the pathogen to acquire drug resistance. Chapter 4 provides an excellent overview and describes methods of predicting drug resistance using in silico models. Current trends in drug discovery have shifted towards developing novel therapeutics with systemic effects. Chapter 5 details methods of interpreting these polypharmacology based drug target networks. Chapter 6 addresses the modeling techniques involved in building structure–activity relationships of molecules. These techniques can be used for optimization of drug candidates. Chapter 7 highlights the molecular dynamics techniques used to compute binding energies of drugs to their target proteins. Chapters 8 and 9 address in silico immunology and describe the tools and databases that aid in B-cell epitope prediction and vaccine design. Chapter 10 explores a unique concept in drug design that features collaborative efforts between scientists in academia and in the biotechnology or pharmaceutical industries working under an integrated platform of drug design. Chapters 11 through 16 describe various applications of in silico models to real-world problems of drug design. These include using nontrivial homology models to derive functional relationships in *Helicobacter pylori* and designing novel inhibitors of anthrax toxin, HIV-1, malaria parasite kinase proteins, *Trypanosoma brucei*, and tuberculosis-causing agents, respectively.

The contribution of in silico models to vaccine development comprises algorithms for accelerated in silico identification of relevant protein candidates; in silico design of novel immunogens with improved expression, safety, and immunogenicity profiles; and in silico design of nucleic acid-based, vectored, or live attenuated vaccines. In small molecule development, in silico models play a major role in comparative genomics, whole genome analysis, pathway analysis, delineation of novel protein–protein interactions, explication of systems and network biology, target identification, virtual screening, and identification of multidrug targets for combination therapy.

The contribution of in silico models to the field of drug discovery for infectious diseases has not previously been comprehensively reviewed in the literature. Hence we think this book should be of interest to all those involved in the study and treatment of infectious diseases, including academic researchers, students, industrial and pharmaceutical scientists, and other healthcare professionals. In addition, each chapter in the book covers a unique in silico technique and hence the book could also be used in the microbiology and immunology curricula in medical and graduate schools.

Philadelphia, PA, USA ***Sandhya Kortagere***

Acknowledgments

I am extremely grateful to the series editor, Dr. John M. Walker, for inviting me to edit the book and for all his advice in choosing the chapters and authors. I would like to thank all the authors for their excellent cooperation, for sharing their enthusiasm, and for their timely efforts to make this project a success. I am grateful to the dedicated staff of Academic Publishing Services at Drexel University College of Medicine for formatting and editing all the chapters. Special note of thanks to Ms. Diana Winters for coordinating with authors in sourcing all the materials and editing the manuscripts. I would like to thank my family, friends and colleagues for their support and encouragement.

Contents

Contributors

EZEKIEL ADEBIYI • *Bio-informatic Research Group, Department of Computer and Information Sciences, Covenant University, Ota, Nigeria*
MARION ADEBIYI • *Bio-informatic Research Group, Department of Computer and Information Sciences, Covenant University, Ota, Nigeria*
ROMMIE E. AMARO • *Department of Chemistry and Biochemistry, University of California, San Diego, La Jolla, CA, USA*
ELIZABETH A. AMIN • *Department of Medicinal Chemistry, College of Pharmacy, University of Minnesota, Minneapolis, MN, USA*
HIFZUR RAHMAN ANSARI • *Bioinformatics Centre, Institute of Microbial Technology, Chandigarh, India*
KRISTEN M. BULLARD • *School of Biological Sciences, University of Northern Colorado, Greeley, CO, USA*
BARRY A. BUNIN • *Collaborative Drug Discovery, Burlingame, CA, USA*
HAI-FENG CHEN • *Shanghai Jiaotong University, Shanghai, China*
RUNSHENG CHEN • *Institute of Biophysics, Chinese Academy of Sciences, Chaoyang District, Beijing, China*
TING-LAN CHIU • *Department of Medicinal Chemistry, College of Pharmacy, University of Minnesota, Minneapolis, MN, USA*
ASIM K. DEBNATH • *Lindsley F. Kimball Research Institute, New York Blood Center, New York, NY, USA*
ROBERT KIRK DELISLE • *Scientific Computing/Bioinformatics, Array BioPharma Inc., Boulder, CO, USA*
ÖZLEM DEMIR • *Department of Chemistry and Biochemistry, University of California, San Diego, La Jolla, CA, USA*
SEAN EKINS • *Collaborations in Chemistry, Fuquay Varina, NC, USA; Collaborative Drug Discovery, Burlingame, CA, USA*
SEGUN FATUMO • *Bio-informatic Research Group, Department of Computer and Information Sciences, Covenant University, Ota, Nigeria; Institute of Structural and Molecular Biology, University College London, London, UK*
JOEL S. FREUNDLICH • *Department of Pharmacology and Physiology, Center for Emerging and Reemerging Pathogens UMDNJ – New Jersey Medical School, Newark, NJ, USA; Department of Medicine, Center for Emerging and Reemerging Pathogens UMDNJ – New Jersey Medical School, Newark, NJ, USA*
AMIT K. GALANDE • *Center for Advanced Drug Research (CADRE), SRI International, Harrisonburg, VA, USA*
RAJARSHI GUHA • *NIH Center for Advancing Translational Science, Rockville, MD, USA*
YONGQUN HE • *Unit for Laboratory Animal Medicine, Department of Microbiology and Immunology, Center for Computational Medicine and Bioinformatics, University of Michigan Medical School, Ann Arbor, MI, USA; Comprehensive Cancer Center, University of Michigan Medical School, Ann Arbor, MI, USA*

SUSAN M. KEENAN • *School of Biological Sciences, University of Northern Colorado, Greeley, CO, USA*
JOHN E. KERRIGAN • *The Cancer Institute of New Jersey, New Brunswick, NJ, USA*
MARKUS LILL • *Department of Medicinal Chemistry and Molecular Pharmacology, Purdue University, West Lafayette, IN, USA*
KIMBERLY M. MAIZE • *Department of Medicinal Chemistry, College of Pharmacy, University of Minnesota, Minneapolis, MN, USA*
HIROSHI MAMITSUKA • *Bioinformatics Center, Institute for Chemical Research, Kyoto University, Kyoto, Japan; School of Pharmaceutical Sciences, Kyoto University, Kyoto, Japan*
GAJENDRA P.S. RAGHAVA • *Bioinformatics Centre, Institute of Microbial Technology, Chandigarh, India*
MALABIKA SARKER • *Center for Advanced Drug Research (CADRE), SRI International, Harrisonburg, VA, USA*
NARAYANASWAMY SRINIVASAN • *Molecular Biophysics Unit, Indian Institute of Science, Bangalore, India*
HONG SUN • *Shanghai Center for Bioinformation Technology, Shanghai Institutes for Biological Sciences, Chinese Academy of Sciences, Shanghai, China; Key Laboratory of Systems Biology, Shanghai Institutes for Biological Sciences, Chinese Academy of Sciences, Shanghai, China*
ICHIGAKU TAKIGAWA • *Bioinformatics Center, Institute for Chemical Research, Kyoto University, Kyoto, Japan; School of Pharmaceutical Sciences, Kyoto University, Kyoto, Japan*
CAROLYN TALCOTT • *Center for Advanced Drug Research (CADRE), SRI International, Harrisonburg, VA, USA*
KOJI TSUDA • *Bioinformatics Center, Institute for Chemical Research, Kyoto University, Kyoto, Japan; School of Pharmaceutical Sciences, Kyoto University, Kyoto, Japan*
NIDHI TYAGI • *Molecular Biophysics Unit, Indian Institute of Science, Bangalore, India*
ZUOSHUANG XIANG • *Unit for Laboratory Animal Medicine, Department of Microbiology and Immunology, Center for Computational Medicine and Bioinformatics, University of Michigan Medical School, Ann Arbor, MI, USA; Comprehensive Cancer Center, University of Michigan Medical School, Ann Arbor, MI, USA*

Chapter 1

Virtual Screening in Drug Design

Markus Lill

Abstract

Virtual screening has become a standard tool in drug discovery to identify novel lead compounds that target a biomolecule of interest. I present several concepts in ligand-based and structure-based virtual screening and discuss some of the current shortcomings and new developments. I also highlight approaches that combine concepts from structure- and ligand-based design.

Key words Virtual screening, Structure-based methods, Ligand-based methods, Drug design, Docking and scoring, Protein flexibility, Similarity, Shape matching, Machine learning

1 Introduction

High-throughput screening (HTS) has become a standard method in the drug discovery toolbox to identify hits when screening large libraries of compounds against a protein target (1). Despite significant progress, HTS is not free from false negatives (actives that are not identified in the experimental screen) and false positives (inactive compounds predicted to be active in the HTS experiment). HTS requires a significant investment in infrastructure and has only recently become accessible to academic laboratories (2, 3). Its success can also be limited by the number and type of chemicals contained in the available compound library.

Virtual screening is a cost-effective alternative to HTS or can aid the HTS process. In the latter, virtual screening is used to filter large libraries of compounds for potential actives, reducing the size of the library before proceeding to more costly HTS. Virtual screening does not require the physical synthesis of compounds; therefore, unlike HTS, it is not limited by the experimentally accessible chemical space. However, virtual screening does require experimental information, either a protein structure for structure-based virtual screening or a set of known actives for ligand-based virtual screening.

Sandhya Kortagere (ed.), *In Silico Models for Drug Discovery*, Methods in Molecular Biology, vol. 993, DOI 10.1007/978-1-62703-342-8_1,

2 Structure-Based Virtual Screening

Docking methods are widely applied to virtual screening (4–8) in large part due to the efficiency of these calculations. Docking typically requires only a few minutes of computing time on a single core per ligand; thus, 10,000 to 100,000 compounds can be screened in a single day on a small- to medium-sized cluster. Docking methods involve a combined posing and scoring process, in which many different protein–ligand conformations are sampled and a scoring function is used to rank the estimated interaction energies of each conformation.

2.1 Search Algorithms

Different categories of search (or posing) algorithms have been developed or adapted for docking applications (9), often classified as systematic, stochastic, and deterministic simulation-based search methods. In brute-force systematic search algorithms, in addition to translational and rotational motion, stepwise rotations around the torsional bonds of the molecule typically are allowed. The number of conformations grows exponentially with the number of flexible dihedrals; therefore, this method can generate an impractically large number of ligand conformations. For example, sampling every dihedral of a ligand with ten torsional bonds in 10° steps would result in $36^{10} \approx 3.7 \times 10^{15}$ poses (not including translational and rotational motions). Scanning alternative dihedral values every 120° would reduce the number of conformations to approximately 60,000 but might bypass the native binding pose (Fig. 1).

Fragment-based incremental construction and place-and-join algorithms (Fig. 2) are alternatives to systematic searches that overcome the combinatorial explosion of ligand conformations. Such methods are used in programs such as DOCK (10), FlexX (11), Glide (12, 13), and Hammerhead (14). In incremental construction, the ligand is split into a relatively rigid core fragment and groups of side chains. The core fragment is first docked and the side chains are subsequently "grown" from the core fragment to reconstruct the original ligand. Throughout the reconstruction step, only the dihedrals of the attached side chains are sampled. In the place-and-join concept, the ligand is split into several fragments that are individually docked to the protein. The missing linker groups between the fragments are then added, and only the ligand poses with linkages that result in energetically low conformations are retained.

In stochastic search methods, such as the Monte Carlo or genetic algorithm methods, new poses are typically generated from previous ligand–protein conformations by random changes to the chosen dihedrals and translational and rotational degrees of motions. Using the Monte Carlo method, for example, a new pose is accepted if it is lower in energy than the previous pose. If the new

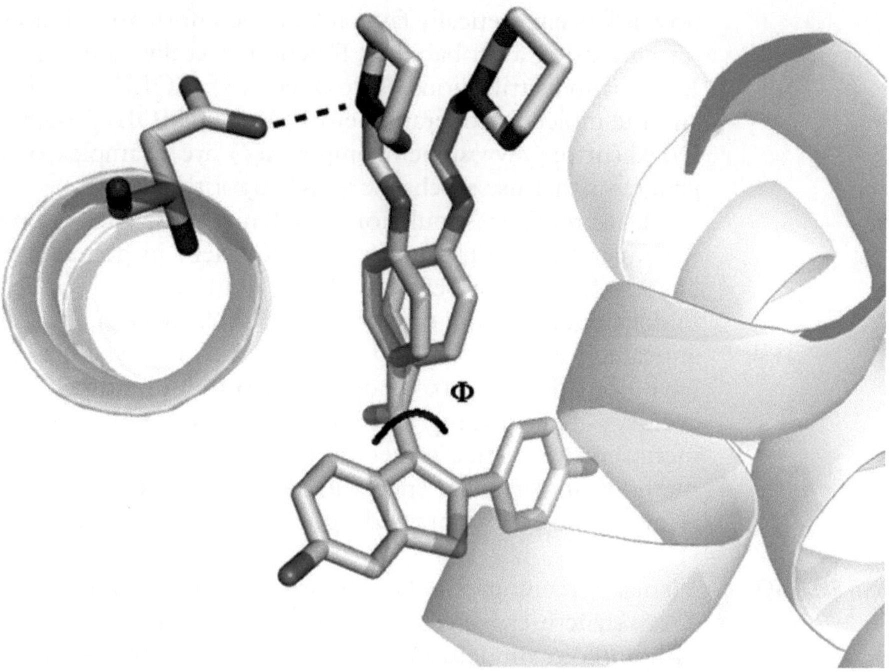

Fig. 1 A deviation of only 20° in the dihedral Φ of the selective estrogen receptor modulator raloxifene would result in a change of 2.5 Å in the tertiary amine position, missing the favorable hydrogen-bond interaction with Asp351

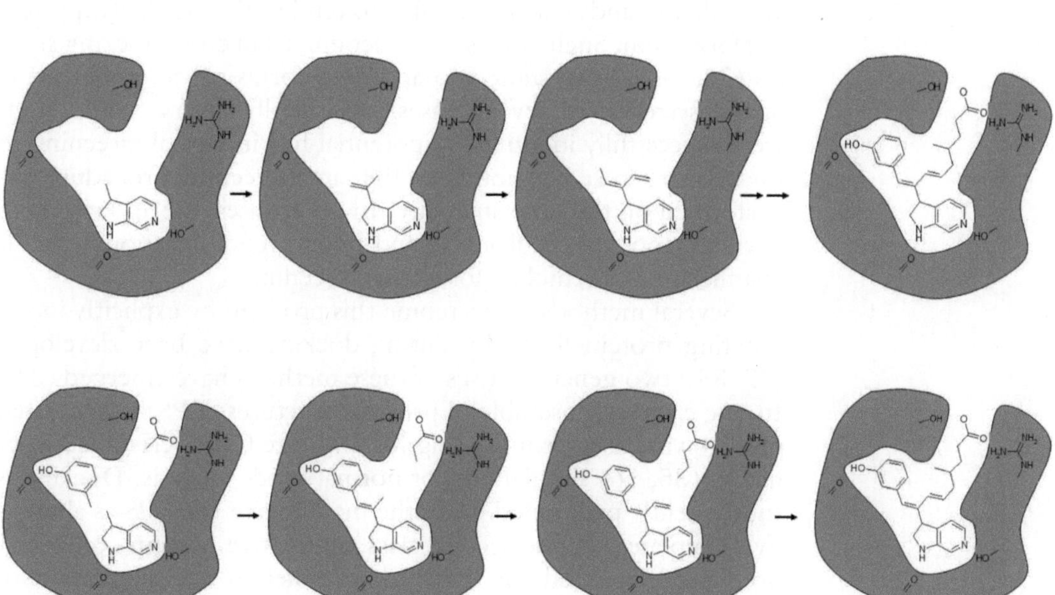

Fig. 2 Scheme for incremental construction (*top line*) and place-and-join (*bottom line*) systematic search algorithms

pose is less energetically favorable, the conformation is accepted or rejected using a probabilistic function typically based on statistical Boltzmann distributions. AutoDock (15), GOLD (16), ICM (17), and the molecular operating environment (MOE) program MOE-Dock (http://www.chemcomp.com/) are examples of docking programs that use stochastic search algorithms.

Deterministic simulation-based methods, such as molecular dynamics (MD) simulations, are inefficient in generating diverse ligand poses and therefore are rarely used as the sole search engine in docking. Although extensions of classical MD, such as simulated annealing, accelerated MD (18, 19), or metadynamics (20, 21), have been designed to study ligand–protein binding with higher accuracy than docking methods, these methods are still typically inefficient for virtual screening purposes. However, deterministic methods, in particular energy minimization, are frequently used to refine ligand poses obtained from other search methods.

2.2 Protein Flexibility

Virtual screening simulations are typically performed on static X-ray structures, selected nuclear magnetic resonance structures, or homology models. Proteins are not static objects, however, but are flexible and can dynamically adapt to different ligands. If the X-ray structure of a holo protein structure is used for docking, the ligand-bound form of the protein can be biased toward specific classes of ligands, e.g., related to the cocrystallized ligand (22, 23), and other ligands may not be able to bind to this particular protein conformation. Such effects were recognized in cross-docking studies (22, 24–26) to static apo and holo forms of various protein–ligand systems. In several cases, structurally diverse compounds were successfully identified as potential hits in virtual screening to the apo structure but not when the same screening procedure was performed on the holo structure. In other cases, the apo structure deviates too much from ligand-bound conformations and is insufficient as a template for virtual screening.

Several methods to overcome this problem by explicitly incorporating protein flexibility during docking have been developed (27–33); two general classes of these methods have emerged (34). In one class, an ensemble of protein structures (EPS) is generated prior to virtual screening using MD, Monte Carlo, elastic network model (35, 36) simulations, or normal-mode analysis. Docking is subsequently performed using the members of the EPS as alternative templates. In the second class, alternative protein conformations are generated on the fly in parallel to the generation of alternative ligand poses. In both of these concepts, selected degrees of freedom are added to the search algorithm to capture important contributions of protein flexibility.

Docking efficiency is critical for virtual HTS; therefore, the number of degrees of freedom associated with protein flexibility must be minimized. Consequently, protein flexibility is only partially incorporated in docking-based virtual screening, for example, by considering only side-chain flexibility in the binding site. Furthermore, docking studies (37) have shown that using an EPS with only a few protein conformations or limiting the number of additional degrees of freedom to the most essential ones can increase the virtual screening quality, whereas using a large EPS or many additional degrees of freedom might diminish virtual screening performance. This observation can be largely attributed to the inherent inaccuracy of the scoring methods used in docking (34). As the number of degrees of freedom increases, so does the potential to generate false-positive protein–ligand conformations. Several methods have recently emerged to generate the smallest number of alternative templates for docking that improve docking performance (37–41).

2.3 Scoring Functions

Scoring functions are designed to efficiently estimate protein–ligand interaction energies of all sampled poses and ligands but often must compromise accuracy to gain efficiency. Scoring functions can be grouped into force-field, empirical, and knowledge-based functions (4, 9, 42, 43). Force-field methods rely heavily on molecular mechanics energy terms to compute the interaction energy between protein and ligand as well as the internal energy of the ligand. They typically contain terms describing van der Waals and electrostatic interactions. Empirical terms, however, are added to include missing factors for modeling the protein–ligand association such as solvation and entropy. AutoDock (15) and GOLD (16), for example, use force-field-based scoring functions.

Empirical scoring functions use simplified functional forms such as stepwise linear functions to model the distance or angular dependence of protein–ligand interactions, for example, hydrogen bonds. Solvation effects are typically modeled empirically using the concept of hydrophobic contacts. The parameters of these empirical terms are fitted, often using regression analysis, on known experimental data such as binding affinities or binding poses for a number of experimental protein–ligand complexes (44). Examples of empirical scoring functions include ChemScore (45) and X-Score (46).

Knowledge-based statistical scoring functions are derived from a statistical analysis of a large number of experimentally determined protein–ligand complexes. Atom types for ligands and proteins are defined, and the relative frequency $p_{ij}(r)$ of pairwise contacts in all complexes is measured for all atom-type pairs i and j for distances r. The interaction energy between a pair of atoms at distance r is then computed by the Boltzmann inversion of the probability density

$$E_{ij}(r) = -k_B T \cdot \ln\left(\frac{p_{ij}(r)}{p(r)}\right)$$

with normalized reference probability $p(r)$. The underlying concept is that atom pairs that form favorable interactions are more frequently identified at shorter distances than less favorable interactions. Examples of knowledge-based scoring functions are DrugScore (47), Potentials of Mean Force (PMF) (48), and Small Molecule Growth (SMoG) (49).

Scoring functions are typically designed to efficiently screen large ligand libraries. As a consequence, they are not accurate enough to reliably predict the binding affinities of ligands for all protein systems. The best-performing scoring functions are often target dependent (50), and protocols have been developed to optimize scoring functions to the target of interest (51–53). Another method to increase the accuracy of scoring functions is consensus scoring. These schemes combine different scoring functions in hopes of offsetting the inherent errors of the various scoring measures (4, 9, 54). In addition to consensus scoring, postprocessing techniques such as molecular mechanics Poisson–Boltzmann surface area methods based on MD simulations (55, 56) can be applied on a small set of top-ranked binding poses and top-ranked compounds to more accurately estimate binding energies.

3 Ligand-Based Virtual Screening

Ligand-based virtual screening is based on the similarity principle, which states that similar compounds cause similar biological effects. These concepts rely on one or a few experimentally identified hits. Large ligand libraries can then be efficiently searched for compounds similar in chemical properties to the known actives, resulting in the identification of novel potentially active compounds (57). The main difference between the various ligand-based virtual screening methods is the measure of similarity, which ranges from two-dimensional descriptors, in particular fingerprints, to shape comparisons and three-dimensional descriptors, e.g., using pharmacophores.

3.1 Two-Dimensional Similarity-Based Screening

Descriptors derived from two-dimensional structure representations of a molecule are popular in ligand-based screening owing to their efficiency and simplicity. Privileged motifs, i.e., substructures that frequently bind to certain receptor types or families, can sometimes be identified. A substructure search can then be used to filter large ligand libraries and identify compounds that contain privileged motifs. This type of search is also used to filter out flagged compounds, e.g., ligand-containing reactive groups that are associated with toxicity or metabolic instability.

Among the most widely applied techniques for similarity searching are molecular fingerprints (58, 59). The assumption is

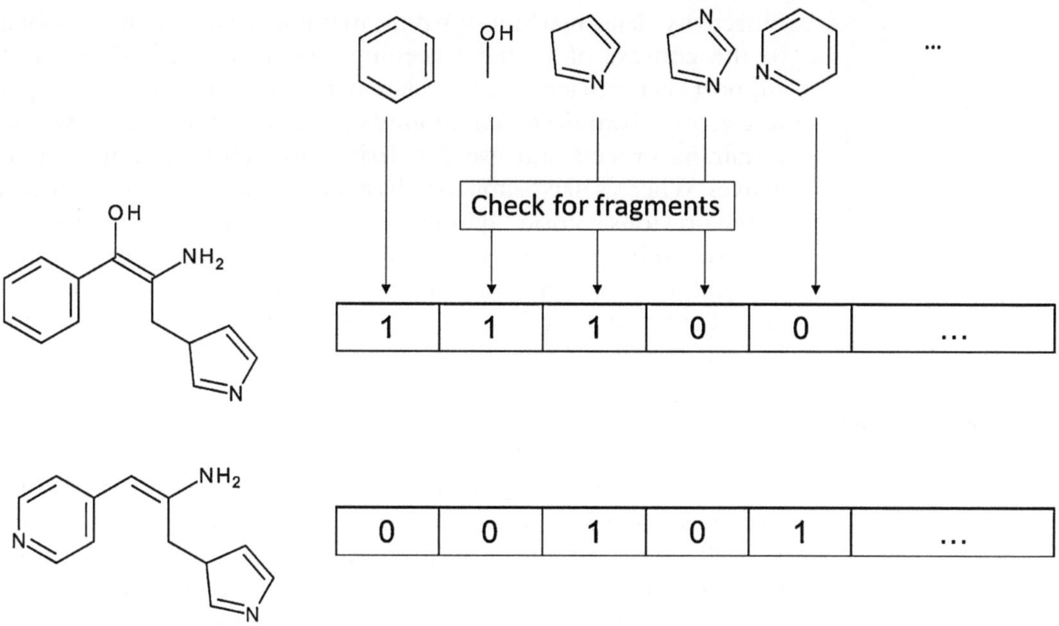

Fig. 3 Example of bit strings generated for two molecules using a small set of substructures. The Tanimoto coefficient for this simplified example would be $T_{AB} = 1/(3 + 2 - 1) = 0.25$

that two compounds are predicted to be similar if they contain similar substructures. Molecular fingerprints are stored as bit strings, wherein the absence ("0") or presence ("1") of a list of substructures is recorded for each ligand. The similarity between two molecules is computed by comparing the individual bits of their bit strings and applying similarity indices such as the Tanimoto coefficient (Fig. 3):

$$T_{AB} = \frac{n_{AB}}{n_A + n_B - n_{AB}}$$

where n_{AB} is the number of common bits set to "1" in both ligands A and B, and n_A and n_B are the numbers of bits set to "1" in ligands A and B individually.

The molecular fingerprints are calculated either on the basis of a predefined library of substructures, e.g., MACSS keys (60), or by an exhaustive enumeration of all fragments in the ligand library containing between N_{min} and N_{max} numbers of atoms. Fingerprints based on predefined libraries have the advantage of producing comparably short fingerprints but might neglect structural elements that are critical for differentiating ligands within the particular dataset.

3.2 Machine-Learning Methods

Because of the increasing quantities of publicly funded HTS data and annotated databases of protein–ligand binding affinity and bioactivity data, machine-learning methods have become popular in the context of virtual screening (61, 62). Methods such as support vector machines, Bayesian methods, or decision trees separate actives from decoy compounds using molecular descriptors in a training process and use the derived models to screen for new actives. Whereas naïve similarity-based screening methods described in the previous section "blindly" use the information of all descriptors (or substructures) in an unbiased form, machine-learning methods correlate the descriptors with biological activity. This process allows those methods to identify descriptors relevant for biological activity and positively biases the search for new actives.

3.3 Shape Matching

An important factor determining the strength of protein–ligand binding for most systems is the steric complementarity between the two entities. This complementarity led to the hypothesis that two molecules that share a similar shape will also have similar biological activity. In virtual screening, new potential actives can be identified on the basis of the similarity of their shapes to known actives. Several concepts have been developed to quantify the shape of a compound, ranging from surface descriptors (e.g., characteristic points defined by local optima of the surface curvature) to volume descriptors (63). The widely used method, Rapid Overlay of Chemical Structures (ROCS) (64), uses Gaussian functions centered on the atoms to describe the shape of a molecule. The maximum overlap between the Gaussian functions of two ligands is computed and used as a similarity measure. Although the molecular shape is an important descriptor for ligand binding to small and enclosed binding pockets, other physicochemical properties generally become more important for larger, solvent-exposed binding sites for which shape complementarity is only partially observed. Shape descriptors have been extended to add atom types (or "color") to the Gaussian functions, including other types of protein–ligand interactions such as hydrogen bonds and aromatic interactions, into the similarity analysis.

3.4 Pharmacophore-Based Screening

Another widely used three-dimensional descriptor for similarity-based virtual screening is pharmacophore models. Pharmacophore models are typically derived from a similarity analysis of several known actives. In addition to the manual definition of pharmacophores by experienced researchers, a number of methods (65–69) have emerged to deduce structural features common of biologically active ligands that are postulated to be important for biological activity. If experimental information about the three-dimensional structure of the binding pocket is known, these data can be used to guide development of the pharmacophore model. In the program LigandScout (70), for example, interactions between protein and ligand in an experimentally determined protein–ligand structure guide the pharmacophore selection process. All similarity-based methods, including pharmacophore models, are dependent on the chemical features present in the known actives. Features that are absent in the particular set of actives but are important for the binding of structurally different ligands may be neglected in the resulting pharmacophore model. Alternatively, the binding site of the target protein can also be used to generate a pharmacophore model without the inclusion of ligand information. Different strategies have been developed to select important pharmacophore elements. For example, an interaction map between chemical probes characterizing potential ligand features and binding pocket residues can be translated into potential locations of pharmacophore elements (71). In another approach, pharmacophores are selected using hydration-site information (72). The premise is that ligand functional groups that replace water molecules contribute to the overall binding affinity of the ligand owing to the additional free energy gained from releasing the water molecule into the bulk solvent.

4 Validation, Applications, and Current Trends

To validate or optimize a specific virtual screening procedure, retrospective screening is typically performed in which experimentally known actives are combined with a large number of decoys. Decoys are compounds that are nonactive against the target of interest, although their missing activity is often only assumed and not experimentally validated. The quality of a virtual screen is often assessed by plotting the percentage of identified actives as a function of the number of tested molecules or as a function of ranked decoys; the result is known as an enrichment plot. These simulations can demonstrate enrichment for known actives; retrospective screening, however, does not necessarily imply successful prospective screening of a novel ligand library. Similarity-based virtual screening studies can be biased toward previously identified active compounds and consequently might fail to identify novel active scaffolds as potential hits.

Virtual screening is not a replacement for experimental HTS and is perhaps best viewed as an aid to HTS. Using virtual screening as a prefilter can allow one to select subsets of compounds (focused library) from a larger library and reduces the cost and time required for subsequent experimental screening. Several success stories of virtual screening applications (73) demonstrate the utility of these computational methods for drug discovery, both in academia and industry.

There is no single method that performs best or performs well for all systems. Even if structural information about the protein is available, structure-based virtual screening methods will not necessarily outperform ligand-based approaches (74, 75). The different methods should not be viewed as exclusive; many concepts and applications have been published that combined structure-based and ligand-based approaches (76, 77). Methods have been devised to combine structure-based and ligand-based elements, or methods from both areas have been used sequentially or in parallel to achieve consensus predictions (77). Future research will likely address some of the shortcomings of current methods and will aim to develop new concepts that combine the advantages of both structure- and ligand-based approaches.

References

1. Macarron R, Banks MN, Bojanic D et al (2011) Impact of high-throughput screening in biomedical research. Nat Rev Drug Discov 10(3):188–195
2. NIH Center for Translational Therapeutics Web site (2012) http://nctt.nih.gov. Accessed
3. Academic Screening Facilities Directory. Society for Laboratory Automation and Screening Web site (2012) http://www.slas.org/screeningFacilities/facilityList.cfm. Accessed
4. Kitchen DB, Decornez H, Furr JR, Bajorath J (2004) Docking and scoring in virtual screening for drug discovery: methods and applications. Nat Rev Drug Discov 3(11):935–949
5. Villoutreix BO, Eudes R, Miteva MA (2009) Structure-based virtual ligand screening: recent success stories. Comb Chem High Throughput Screen 12(10):1000–1016
6. Waszkowycz B, Clark DE, Gancia E (2011) Outstanding challenges in protein-ligand docking and structure-based virtual screening. Wiley Interdiscip Rev Comput Mol Sci 1(2):229–259
7. McInnes C (2007) Virtual screening strategies in drug discovery. Curr Opin Chem Biol 11(5):494–502
8. Klebe G (2006) Virtual ligand screening: strategies, perspectives and limitations. Drug Discov Today 11(13–14):580–594
9. Halperin I, Ma BY, Wolfson H, Nussinov R (2002) Principles of docking: an overview of search algorithms and a guide to scoring functions. Proteins 47(4):409–443
10. Kuntz ID, Blaney JM, Oatley SJ et al (1982) A geometric approach to macromolecule-ligand interactions. J Mol Biol 161(2):269–288
11. Rarey M, Kramer B, Lengauer T, Klebe G (1996) A fast flexible docking method using an incremental construction algorithm. J Mol Biol 261(3):470–489
12. Friesner RA, Banks JL, Murphy RB et al (2004) Glide: a new approach for rapid, accurate docking and scoring. 1. Method and assessment of docking accuracy. J Med Chem 47(7):1739–1749
13. Halgren TA, Murphy RB, Friesner RA et al (2004) Glide: a new approach for rapid, accurate docking and scoring. 2. Enrichment factors in database screening. J Med Chem 47(7): 1750–1759
14. Welch W, Ruppert J, Jain AN (1996) Hammerhead: fast, fully automated docking of flexible ligands to protein binding sites. Chem Biol 3(6):449–462
15. Goodsell DS, Morris GM, Olson AJ (1996) Automated docking of flexible ligands: applications of AutoDock. J Mol Recognit 9(1): 1–5
16. Jones G, Willett P, Glen RC et al (1997) Development and validation of a genetic algorithm for flexible docking. J Mol Biol 267(3):727–748
17. Totrov M, Abagyan R (1997) Flexible protein-ligand docking by global energy optimization in internal coordinates. Proteins 1(Suppl 1): 215–220
18. Hamelberg D, Mongan J, McCammon JA (2004) Enhanced sampling of conformational transitions in proteins using full atomistic accelerated molecular dynamics simulations. Protein Sci 13:76–76
19. Hamelberg D, Mongan J, McCammon JA (2004) Accelerated molecular dynamics: a promising and efficient simulation method for biomolecules. J Chem Phys 120(24): 11919–11929
20. Gervasio FL, Laio A, Parrinello M (2005) Flexible docking in solution using metadynamics. J Am Chem Soc 127(8):2600–2607
21. Laio A, Parrinello M (2006) Computing free energies and accelerating rare events with metadynamics. In: Ferrario M, Ciccotti G, Binder K (eds) Computer simulations in condensed matter: from materials to chemical biology, vol 1, Springer. Berlin, Heidelberg, New York, pp 315–347
22. McGovern SL, Shoichet BK (2003) Information decay in molecular docking screens against

holo, apo, and modeled conformations of enzymes. J Med Chem 46(14):2895–2907
23. Xu M, Lill MA (2011) Significant enhancement of docking sensitivity using implicit ligand sampling. J Chem Inf Model 51: 693–706
24. Kua J, Zhang Y, McCammon JA (2002) Studying enzyme binding specificity in acetylcholinesterase using a combined molecular dynamics and multiple docking approach. J Am Chem Soc 124(28):8260–8267
25. Murray CW, Baxter CA, Frenkel AD (1999) The sensitivity of the results of molecular docking to induced fit effects: application to thrombin, thermolysin and neuraminidase. J Comput Aided Mol Des 13(6):547–562
26. Hoffmann D, Kramer B, Washio T et al (1999) Two-stage method for protein-ligand docking. J Med Chem 42(21):4422–4433
27. Carlson HA (2002) Protein flexibility and drug design: how to hit a moving target. Curr Opin Chem Biol 6(4):447–452
28. Teodoro ML, Kavraki LE (2003) Conformational flexibility models for the receptor in structure based drug design. Curr Pharm Des 9(20):1635–1648
29. Totrov M, Abagyan R (2008) Flexible ligand docking to multiple receptor conformations: a practical alternative. Curr Opin Struct Biol 18(2):178–184
30. Beier C, Zacharias M (2010) Tackling the challenges posed by target flexibility in drug design. Expert Opin Drug Discov 5(4): 347–359
31. Rao C, Subramanian J, Sharma SD (2009) Managing protein flexibility in docking and its applications. Drug Discov Today 14(7–8): 394–400
32. Sotriffer CA (2011) Accounting for induced-fit effects in docking: what is possible and what is not? Curr Top Med Chem 11(2):179–191
33. Lin JH (2011) Accommodating protein flexibility for structure-based drug design. Curr Top Med Chem 11(2):171–178
34. Lill MA (2011) Efficient incorporation of protein flexibility and dynamics into molecular docking simulations. Biochemistry 50(28): 6157–6169
35. Atilgan AR, Durell SR, Jernigan RL et al (2001) Anisotropy of fluctuation dynamics of proteins with an elastic network model. Biophys J 80(1):505–515
36. Bahar I, Atilgan AR, Erman B (1997) Direct evaluation of thermal fluctuations in proteins using a single-parameter harmonic potential. Fold Des 2(3):173–181
37. Armen RS, Chen J, Brooks CL (2009) An evaluation of explicit receptor flexibility in molecular docking using molecular dynamics and torsion angle molecular dynamics. J Chem Theory Comput 5(10):2909–2923
38. Barril X, Morley SD (2005) Unveiling the full potential of flexible receptor docking using multiple crystallographic structures. J Med Chem 48(13):4432–4443
39. Amaro RE, Baron R, McCammon JA (2008) An improved relaxed complex scheme for receptor flexibility in computer-aided drug design. J Comput Aided Mol Des 22(9): 693–705
40. Bolstad ES, Anderson AC (2009) In pursuit of virtual lead optimization: pruning ensembles of receptor structures for increased efficiency and accuracy during docking. Proteins 75(1):62–74
41. Xu M, Lill MA (2012) Utilizing experimental data for reducing ensemble size in flexible-protein docking. J Chem Inf Model 52(1): 187–198
42. Ferrara P, Gohlke H, Price DJ et al (2004) Assessing scoring functions for protein-ligand interactions. J Med Chem 47(12):3032–3047
43. Huang SY, Grinter SZ, Zou X (2010) Scoring functions and their evaluation methods for protein-ligand docking: recent advances and future directions. Phys Chem Chem Phys 12(40):12899–12908
44. Bohm HJ (1992) LUDI: rule-based automatic design of new substituents for enzyme inhibitor leads. J Comput Aided Mol Des 6(6):593–606
45. Eldridge MD, Murray CW, Auton TR et al (1997) Empirical scoring functions: I. The development of a fast empirical scoring function to estimate the binding affinity of ligands in receptor complexes. J Comput Aided Mol Des 11(5):425–445
46. Wang RX, Lai LH, Wang SM (2002) Further development and validation of empirical scoring functions for structure-based binding affinity prediction. J Comput Aided Mol Des 16(1):11–26
47. Gohlke H, Hendlich M, Klebe G (2000) Knowledge-based scoring function to predict protein-ligand interactions. J Mol Biol 295(2):337–356
48. Muegge I, Martin YC (1999) A general and fast scoring function for protein-ligand interactions: a simplified potential approach. J Med Chem 42(5):791–804
49. DeWitte RS, Shakhnovich EI (1996) SMoG: de novo design method based on simple, fast, and accurate free energy estimates. 1. Methodology and supporting evidence. J Am Chem Soc 118:11733–11744
50. Warren GL, Andrews CW, Capelli AM et al (2006) A critical assessment of docking programs and scoring functions. J Med Chem 49(20):5912–5931

51. Li L, Wang B, Merouch SO (2011) Support vector regression scoring of receptor-ligand complexes for rank-ordering and virtual screening of chemical libraries. J Chem Inf Model 51(9):2132–2138
52. Li LW, Khanna M, Jo IH et al (2011) Target-specific support vector machine scoring in structure-based virtual screening: computational validation, in vitro testing in kinases, and effects on lung cancer cell proliferation. J Chem Inf Model 51(4):755–759
53. Seifert MHJ (2009) Robust optimization of scoring functions for a target class. J Comput Aided Mol Des 23(9):633–644
54. Charifson PS, Corkery JJ, Murcko MA, Walters WP (1999) Consensus scoring: a method for obtaining improved hit rates from docking databases of three-dimensional structures into proteins. J Med Chem 42(25):5100–5109
55. Brown SP, Muchmore SW (2007) Rapid estimation of relative protein-ligand binding affinities using a high-throughput version of MM-PBSA. J Chem Inf Model 47(4): 1493–1503
56. Brown SP, Muchmore SW (2006) High-throughput calculation of protein-ligand binding affinities: Modification and adaptation of the MM-PBSA protocol to enterprise grid computing. J Chem Inf Model 46(3): 999–1005
57. Ripphausen P, Nisius B, Bajorath J (2011) State-of-the-art in ligand-based virtual screening. Drug Discov Today 16(9–10):372–376
58. Brown RD, Martin YC (1996) Use of structure–activity data to compare structure-based clustering methods and descriptors for use in compound selection. J Chem Inf Comput Sci 36:572–584
59. Brown RD, Martin YC (1997) The information content of 2D and 3D structural descriptors relevant to ligand-receptor binding. J Chem Inf Comput Sci 37:1–9
60. Durant JL, Leland BA, Henry DR, Nourse JG (2002) Reoptimization of MDL keys for use in drug discovery. J Chem Inf Comput Sci 42(6):1273–1280
61. Melville JL, Burke EK, Hirst JD (2009) Machine learning in virtual screening. Comb Chem High Throughput Screen 12(4): 332–343
62. Geppert H, Vogt M, Bajorath J (2010) Current trends in ligand-based virtual screening: molecular representations, data mining methods, new application areas, and performance evaluation. J Chem Inf Model 50(2):205–216
63. Nicholls A, McGaughey GB, Sheridan RP et al (2010) Molecular shape and medicinal chemistry: a perspective. J Med Chem 53(10): 3862–3886
64. Rush TS 3rd, Grant JA, Mosyak L, Nicholls A (2005) A shape-based 3-D scaffold hopping method and its application to a bacterial protein-protein interaction. J Med Chem 48(5): 1489–1495
65. Martin Y (1995) Distance comparisons (DISCO): a new strategy for examining 3D structure-activity relationships. American Chemical Society, Washington, DC
66. Barnum D, Greene J, Smellie A, Sprague P (1996) Identification of common functional configurations among molecules. J Chem Inf Comput Sci 36(3):563–571
67. Dixon SL, Smondyrev AM, Knoll EH et al (2006) PHASE: a new engine for pharmacophore perception, 3D QSAR model development, and 3D database screening: 1. Methodology and preliminary results. J Comput Aided Mol Des 20(10):647–671
68. Richmond NJ, Abrams CA, Wolohan PRN et al (2006) GALAHAD: 1. Pharmacophore identification by hypermolecular alignment of ligands in 3D. J Comput Aided Mol Des 20(9):567–587
69. Chen X, Rusinko A III, Tropsha A, Young SS (1999) Automated pharmacophore identification for large chemical data sets 1. J Chem Inf Comput Sci 39(5):887–896
70. Wolber G, Langer T (2005) LigandScout: 3-D pharmacophores derived from protein-bound ligands and their use as virtual screening filters. J Chem Inf Model 45(1):160–169
71. Kirchhoff PD, Brown R, Kahn S et al (2001) Application of structure-based focusing to the estrogen receptor. J Comput Chem 22(10): 993–1003
72. Hu B, Lill MA (2012) Protein pharmacophore selection using hydration-site analysis. J Chem Inf Model 52(4):1046–1060
73. Bollt EM, ben-Avraham D (2005) What is special about diffusion on scale-free nets? New J Phys 7:26
74. Hawkins PC, Skillman AG, Nicholls A (2007) Comparison of shape-matching and docking as virtual screening tools. J Med Chem 50(1): 74–82
75. McGaughey GB, Sheridan RP, Bayly CI et al (2007) Comparison of topological, shape, and docking methods in virtual screening. J Chem Inf Model 47(4):1504–1519
76. Tan L, Batista J, Bajorath J (2010) Computational methodologies for compound database searching that utilize experimental protein-ligand interaction information. Chem Biol Drug Des 76(3):191–200
77. Wilson GL, Lill MA (2011) Integrating structure-based and ligand-based approaches for computational drug design. Future Med Chem 3(6):735–750

Chapter 2

In Silico Systems Biology Approaches for the Identification of Antimicrobial Targets

Malabika Sarker, Carolyn Talcott, and Amit K. Galande

Abstract

Classical antibiotic discovery efforts have relied mainly on molecular library screening coupled with target-based lead optimization. The conventional approaches are unable to tackle the emergence of antibiotic resistance and are failing to provide understanding of multiple mechanisms behind drug actions and the off-target effects. These insufficiencies have prompted researchers to focus on a multidisciplinary approach of systems biology-based antibiotic discovery. Systems biology is capable of providing a big-picture view for therapeutic targets through interconnected networks of biochemical reactions derived from both experimental and computational techniques. In this chapter, we have compiled software tools and databases that are typically used for target identification through in silico analyses. We have also identified enzyme- and broad-spectrum metabolite-based drug targets that have emerged through in silico systems microbiology.

Key words Antimicrobials, Database, Metabolites, Omics, Software tools, Systems biology, Targets

1 Introduction

Antimicrobial research conducted in the second half of the twentieth century largely focused on two mainstream approaches. The first approach involved identifying leads for the next generation of antibiotics through screening diverse sets of molecular libraries (1), and the second approach consisted of identifying novel antimicrobial targets through reductionism. The reductionist approach, fueled by molecular biology methods, focuses on studying specific functions of individual genes, proteins, and cells separately to identify valid molecular targets for therapeutic intervention (2). While these two themes of "library screening" and "target-based discovery" are central to any drug research program, investigators in antimicrobial discovery were among the first to recognize the major deficit in these conventional approaches. Screening and target-based methods typically lack a "big picture" that shows molecular connectivity and provides global understanding of cellular physiology. This realization was mostly prompted by the rapid emergence of drug resistance along with the failure of lead candidates in the

Sandhya Kortagere (ed.), *In Silico Models for Drug Discovery*, Methods in Molecular Biology, vol. 993,
DOI 10.1007/978-1-62703-342-8_2, © Springer Science+Business Media, LLC 2013

late stages of drug development, owing especially to the off-target effects (3). One way to address the issues around these "isolated" drug discovery efforts is to obtain a global view of molecular mechanisms by studying cells as systems through the multidisciplinary and technology-driven approach of systems biology (4, 5).

Systems biology is the twenty-first century science that transformed the reductionist focus to a global view and added a new perspective to classical pharmaceutical research. In the context of drug discovery, systems biology involves integrating networks of biochemical reactions through experimental and computational approaches to provide a comprehensive understanding of the therapeutic target, including mechanisms of action. Systems biology emerged primarily as the result of the catalogue of genes provided by the multiple genome projects. The data from these genomics efforts were then utilized by other follow-up "omics" technologies—transcriptomics, proteomics, metabolomics, glycomics, among others—which were expected to provide a thorough understanding of the dynamics and interplay within biological systems. However, from the very beginning, the deluge of data produced by the experimental "omics" technologies proved overwhelming and created an immediate need to apply computational approaches for curation, pathway modeling, and bioinformatics (6, 7). Indeed, the past few years have seen monumental advances in this direction, and the in silico approaches have now taken center stage in systems biology (8, 9).

The availability of completed microbial genome sequences and the development of advanced microbial databases and high-performance software tools have opened up new opportunities for creating innovative computational methods for antimicrobial target identification. Traditionally, antimicrobial targets have been identified through knowledge of the function or essentiality of individual genes or proteins. Potential targets thus identified are generally taken through a validation process involving gene knockouts or site-directed mutagenesis experiments in whole cells or animals that lead to loss-of-function phenotypes. Experimental target validation can now be complemented with computational experiments such as in silico knockouts. In silico methods have the advantage of speed and low cost along with the ability to provide a systems view of the whole microbe at any given physiological stage. Consequently, in silico approaches are capable of generating hypotheses and questions that are unlikely to emerge through experimental methods (10).

In silico systems biology approaches are best used in combination with the experimental "omics" technologies. For example, when studying the proteome of *Mycobacterium tuberculosis* using mass spectrometry-based methods, we immediately realized that although the *M. tuberculosis* genome codes for about 4,000 proteins, several proteomics laboratories had identified only small subsets of this proteome. Also, because laboratories varied in sample preparation, chromatography, mass spectrometers, and bioinformatics, the types of proteins identified in these proteomics studies were significantly different (11–16). Moreover, when we combined the lists of *M. tuberculosis* proteins identified through multiple proteomics experiments by several different laboratories, a consolidated list of only about 2,000 nonredundant proteins was generated, which is only 50 % coverage of the *M. tuberculosis* proteome (unpublished observation). Thus, if proteomics is being used as a primary technology for target identification, many of the target proteins of interest are probably not "visible" to the mass spectrometers. Consequently, a logical next step would be to evaluate observed proteomics data in the context of genome-scale microbial in silico models to fill the gaps in the experimental data sets.

Our laboratory recently conducted a comprehensive study (17) that further underscores the utility of in silico approaches for novel target identification. We recently reported in silico analysis of metabolic networks of a panel of representative gram-positive and gram-negative bacteria and provided valuable findings on metabolites that could be used as antimicrobial targets. In-depth literature mining was performed to identify metabolites that are essential for the growth and survival of a broad spectrum of bacteria as determined by direct experimental evidence. To identify potential targets among these essential metabolites, in silico pathway analysis was performed through the BioCyc Pathway/Genome Database (PGDB). A PGDB was automatically generated in BioCyc from the annotated genome sequence of that organism using BioCyc Pathway Tools software (developed by Peter D. Karp and coworkers at the Bioinformatics Research Group at SRI International—http://www.ecocyc.org/download.shtml). BioCyc has 2038 available PGDBs till date, each containing the predicted metabolic network of an organism, including metabolic pathways, enzymes (and the genes encoding them), metabolites (with structural details), and reaction details. BioCyc Pathway Tools software further produces a pathway-based visualization of cellular biochemical networks, called the cellular overview diagram, which supports interrogation and systems biology analyses of the whole organism. BioCyc provides overview diagrams for more than 3000 organisms from bacteria to humans. We used the cellular overview diagram for comparative analyses of the complete metabolic networks of two or more organisms. In the display of an overview for one organism, the software can highlight all reactions that are either shared or not shared with other combinations of organisms. In our work, the entire human metabolic network from HumanCyc was compared with the networks of pathogens of interest to search for metabolites that are absent from humans. Moreover, the reactions around the selected metabolites were compared for their presence or absence among the bacteria of choice to determine whether the metabolites are shared and hence can have broad-spectrum action. The essential metabolites that were absent in humans and present in multiple pathogens were chosen as potential targets. Additionally, metabolites were selected for which no alternative compensatory pathway was present.

Based on these analyses, we identified ten metabolites as potential candidates for developing novel antibiotics. These are lipid II, meso-2, 6-diaminoheptanedioate (meso-DAP), pantothenate, biotin, shikimate, L-aspartyl-4-phosphate, deoxythymidine diphosphate (dTDP)-α-L-rhamnose, uridine diphosphate (UDP)-D-galacto-1,4-furanose, des-*N*-acetyl mycothiol, and siroheme. Previous identification of the first five metabolites as targets for antibiotic discovery validates our in silico approach and suggests that the latter five metabolites could be promising candidates as well. Identifying key metabolites and then developing metabolite scavengers as broad-spectrum antimicrobials is a relatively new approach that can benefit tremendously from in silico systems biology methods. Because conventional antimicrobial targets such as proteins and nucleic acids are amenable to mutations and antibiotic resistance, the relatively immutable metabolites provide a new frontier in antimicrobial drug discovery.

In addition to these efforts, here we have also performed in-depth data mining from publications that have reported antimicrobial targets derived through complementary in silico methods. We have classified these targets into two categories: metabolites (17–21) and metabolic enzymes (10, 19, 22–34). We observed that a few different in silico studies had independently arrived at the same metabolite or metabolic enzyme targets. Representative examples of such common targets are shown in Figs. 1 and 2. These targets not only present potential opportunities for developing broad-spectrum antimicrobials but also underscore the importance of conducting multiple and distinct in silico experiments for target identification.

Finally, whereas in silico approaches are showing great promise in antimicrobial drug discovery, the computational tools are scattered and lack standardization (35). One fundamental issue is the lack of a central resource that can provide comparative information on major microbial databases and software tools, which can be utilized for in silico systems biology, especially in the context of antimicrobial drug discovery. Accordingly, Tables 1, 2, and 3 in this chapter provide this information in a systematic tabular format to facilitate the selection of appropriate computational tools for researchers interested in in silico systems microbiology.

2 Microbial Databases

Systems-level investigation of genomic-scale information requires the development of integrated databases dealing with heterogeneous data, which can be queried for simple properties of genes or other database objects as well as for complex network-level properties, for the analysis and modeling of complex biological processes. Several database have been developed to provide valuable information from the bench chemist to biologist and from medical practitioner to pharmaceutical scientist, in a structured format. The advent of

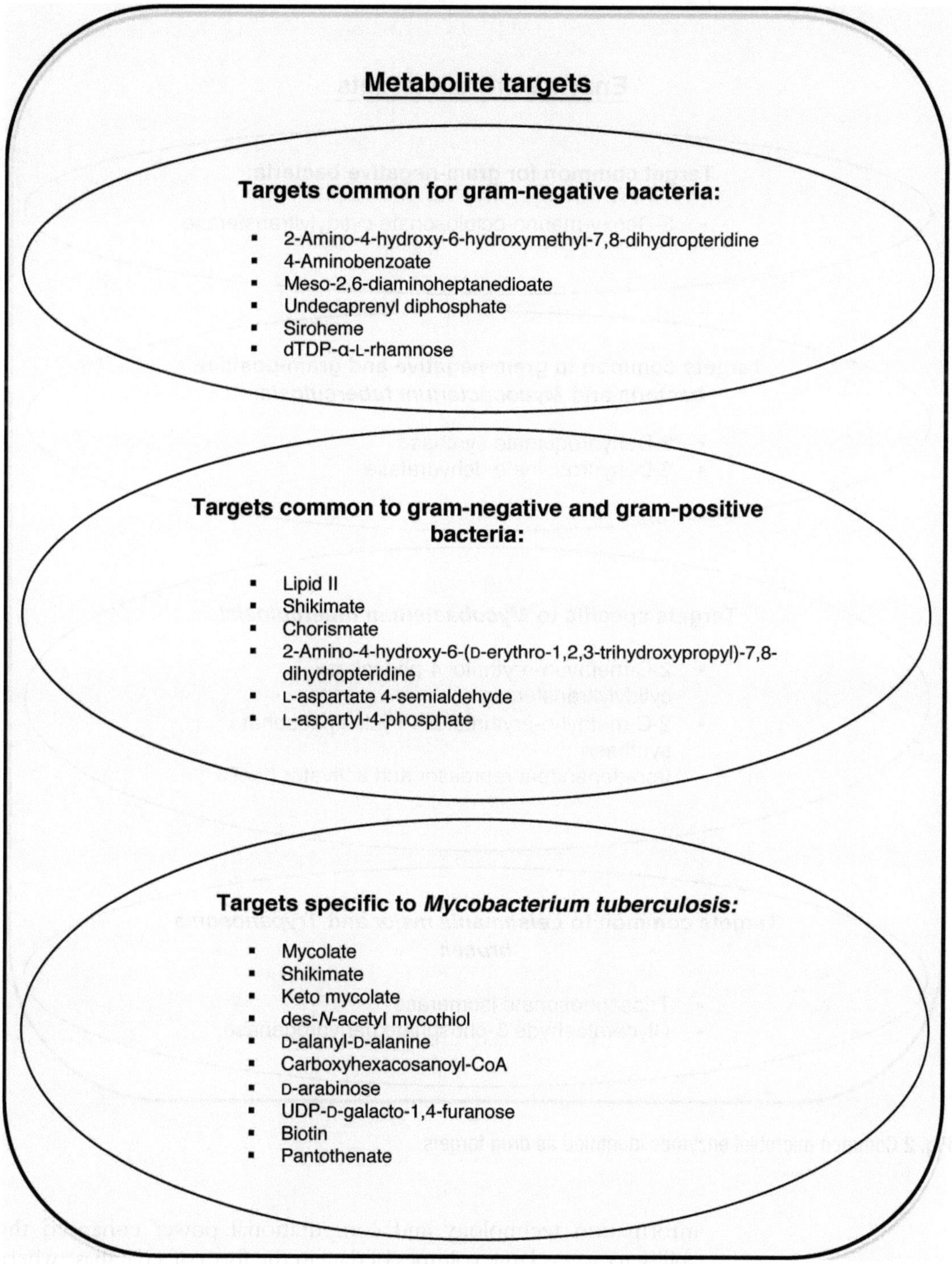

Fig. 1 Common microbial metabolites identified as drug targets. dTDP = deoxythymidine diphosphate, UDP = uridine diphosphate

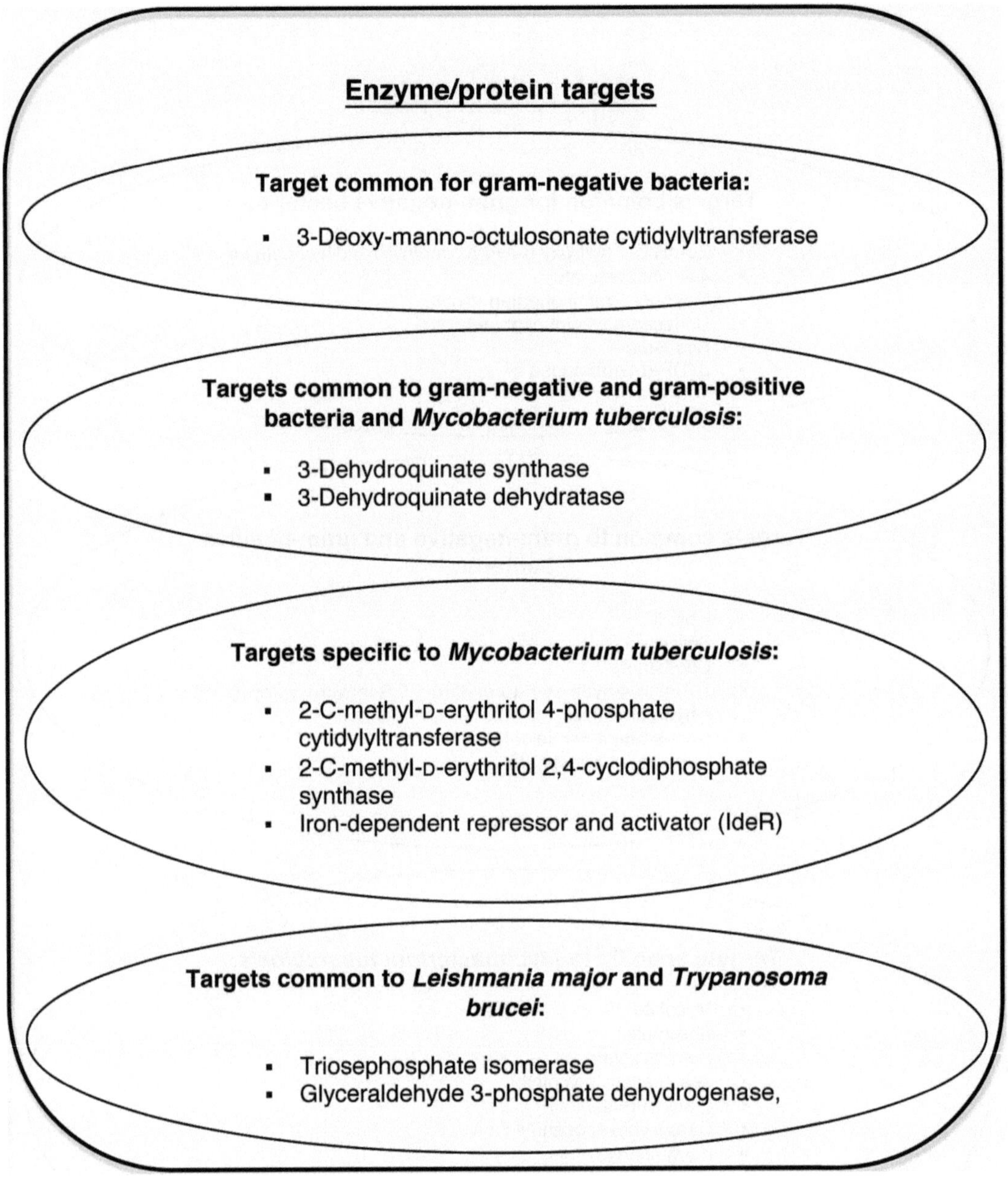

Fig. 2 Common microbial enzymes identified as drug targets

information technology and computational power enhanced the ability to access large volumes of data in the form of a database where one could do compilation, searching, archiving, analysis, and finally knowledge derivation (36).

Table 1 describes all the microbe-specific databases, and Table 2 lists databases for the microbial community. These databases provide genomic sequence data, gene and protein information, gene expression data, metabolic reactions and pathways, interaction network,

Table 1
Microbe-specific databases

Microbes	Database or Web resource	Description (URL)
Aspergillus spp.	*Aspergillus* Genome Database	Stanford University: Genomic sequence data and gene and protein information for aspergilli (http://www.aspgd.org/)
	Aspergillus Comparative Database	Broad Institute of MIT and Harvard: Comparative and functional genomics of seven aspergilli spp. (http://www.broadinstitute.org/annotation/genome/aspergillus_group/MultiHome.html)
Bacillus subtilis	SubtiList	Institut Pasteur: Genome annotation and analysis of bacterium *B. subtilis* 168 (http://genolist.pasteur.fr/SubtiList/)
	NRSub	University Lyon 1: Nonredundant, fully annotated database of sequences of *B. subtilis* 168 (http://pbil.univ-lyon1.fr/nrsub/nrsub.html)
Escherichia coli	EcoCyc	SRI International: Comprehensive literature-based curation of the entire genome and of transcriptional regulation, transporters, and metabolic pathways for bacterium *E. coli* K-12 MG1655 (http://ecocyc.org/)
	Ecogene	University of Miami School of Medicine: *E. coli* K-12 genome and proteome sequences, including extensive gene bibliographies (http://www.ecogene.org/3.0/)
	Colibri	Institut Pasteur: Genome analysis of *E. coli* (http://genolist.pasteur.fr/Colibri/)
Francisella tularensis	*Francisella tularensis* group Database	Broad Institute of MIT and Harvard: Comparative genomics analysis and virulence mechanisms of bacteria *F. tularensis* (http://www.broadinstitute.org/annotation/genome/francisella_tularensis_group/MultiHome.html)
Helicobacter pylori	PyloriGene	Institut Pasteur: Annotation and comparative analysis of bacteria *H. pylori* strains: 26695 and J99 (http://genolist.pasteur.fr/PyloriGene/)
Mycobacterium leprae	Leproma	Institut Pasteur: Genome analysis of the leprosy (Hansen disease) bacillus *M. leprae* (http://genolist.pasteur.fr/Leproma/)
Mycobacterium tuberculosis	TB Database	Stanford University: Provides genomic data (for 28 annotated genomes) and several thousand microarray datasets from in vitro experiments and *M. tuberculosis*-infected tissues (http://www.tbdb.org/)

(continued)

Table 1
(continued)

Microbes	Database or Web resource	Description (URL)
	TubercuList	Institut Pasteur: Complete dataset of DNA and protein sequences derived from *M. tuberculosis* H37Rv, linked to annotations and functional assignments (http://genolist.pasteur.fr/TubercuList/)
	*web*TB	TB Structural Genomics Consortium: Provides *M. tuberculosis* genome, structure summary for all known tuberculosis proteins, the *M. tuberculosis* regulatory database of proteins up- or downregulated in TB, top 100 persistence targets in TB (http://www.webtb.org/)
	TBrowse	Institute of Genomics and Integrative Biology—India: Resource for the integrative analysis of the *M. tuberculosis* genome (http://tbrowse.osdd.net/)
	TB Drug Resistance Mutation Database	Harvard School of Public Health: Provides mutations associated with *M. tuberculosis* drug resistance (http://www.tbdreamdb.com/)
Mycoplasma pulmonis	MypuList	Institut Pasteur: Genome analysis of the bacterium *M. pulmonis* (http://genolist.pasteur.fr/MypuList/)
Plasmodium falciparum	PlasmoDB	University of Georgia: Functional genomic database for malaria parasites, *P. falciparum*, *P. vivax*, *P. yoelii*, *P. berghei*, *P. chabaudi*, and *P. knowlesi* (http://plasmodb.org/plasmo/)
	Plasmodium falciparum database	Broad Institute of MIT and Harvard: Comparative genomics analysis of *Plasmodium* spp. (http://www.broadinstitute.org/annotation/genome/plasmodium_falciparum_spp/MultiHome.html)
Pseudomonas aeruginosa	*Pseudomonas* Genome Database	Simon Fraser University: Comparative genomics of all *Pseudomonas* spp. (http://www.pseudomonas.com/)
Saccharomyces cerevisiae	YeastCyc	SRI International: Comprehensive literature-based curation of the entire genome and of transcriptional regulation, transporters, and metabolic pathways of the budding yeast *S. cerevisiae* (http://biocyc.org/YEAST/organism-summary?object=YEAST)
	Saccharomyces Genome Database	Stanford University: Complete *S. cerevisiae* genomic sequence, its genes and their products, the phenotypes of its mutants, and the literature supporting these data (www.yeastgenome.org)
	Comprehensive Yeast Genome Database	Max-Planck-Institut für Biochemie: Molecular structure and functional network of *S. cerevisiae* and comparative analysis for related yeasts (http://mips.helmholtz-muenchen.de/genre/proj/yeast/)

(continued)

Table 1
(continued)

Microbes	Database or Web resource	Description (URL)
Vibrio cholerae	*Vibrio cholerae* Database	Broad Institute of MIT and Harvard: Comparative genomic studies of the different strains of *V. cholerae* (http://www.broadinstitute.org/annotation/genome/vibrio_cholerae/MultiHome.html)
Influenza virus	Influenza Research Database	University of Texas Southwestern Medical Center: Comprehensive, integrated data about influenza virus genome sequences, virus phenotypic characteristics, and results from surveillance activities for the discovery and development of influenza virus vaccines, diagnostics, and therapeutics (http://www.fludb.org/brc/home.do?decorator=influenza)
HIV	HIV Databases	Los Alamos National Laboratory: Contain data on HIV genetic sequences, immunological epitopes, drug resistance-associated mutations, and vaccine trials (http://www.hiv.lanl.gov/content/index)

Table 2
Databases and web resources for in silico systems microbiology

Database or Web resource	Description (URL)
ARDB	University of Maryland: Information on antibiotic resistance genes in sequenced bacteria (http://ardb.cbcb.umd.edu/)
BiGG	University of California San Diego: Knowledge base of large-scale metabolic reconstructions and high-quality curated metabolic models (http://bigg.ucsd.edu)
BioCyc	SRI International: Describes the genome and metabolic pathways of a single organism—total 1690 pathway/genome databases (http://biocyc.org/)
BioModels Database	EMBL-EBI: Repository of peer-reviewed, published, computational models (http://www.ebi.ac.uk/biomodels-main/)
CDD	Collaborative Drug Discovery: Repository of small-molecule libraries of more than 300,000 compounds derived from patents, literature, and high-throughput screening data shared by academic and pharmaceutical laboratories tested against *M. tuberculosis*; preliminary public antimalarial database from multiple sources on 30,000 public compounds (https://www.collaborativedrug.com/pages/public_access)

(continued)

Table 2 (continued)

Database or Web resource	Description (URL)
CSB.DB	Max Planck Institute of Molecular Plant Physiology: Presents the results of biostatistical analyses on gene expression data in association with additional biochemical and physiological knowledge (http://csbdb.mpimp-golm.mpg.de/csbdb/home/databases.html)
DOE Systems Biology Knowledgebase	U.S. Department of Energy: Community-driven cyberinfrastructure for sharing and integrating data and analytical tools for experimental design as well as modeling and simulation (http://genomicscience.energy.gov/compbio/index.shtml#page=news)
ERGO Light	Integrated Genomics: Curated database of public and proprietary genomic DNA with connected similarities, functions, pathways, functional models, and clusters (http://www.ergo-light.com/)
EuPathDB	Bioinformatics Resource Center: Provides genomic-scale datasets associated with the eukaryotic pathogens (http://eupathdb.org/eupathdb/)
GeneDB	The Wellcome Trust Sanger Institute: Genome database for prokaryotic and eukaryotic organisms (http://www.genedb.org/Homepage)
GTD	Institute of Integrative Omics and Applied Biotechnology—India: Provides putative genomic drug targets of most common human bacterial pathogens (http://iioab-dgd.webs.com/)
HAMAP	Swiss Institute of Bioinformatics: Database of completely sequenced microbial proteome sets and manually curated microbial protein families in UniProtKB/Swiss-Prot (http://hamap.expasy.org/)
HOGENOM	Université Claude Bernard : Database of homologous genes from fully sequenced organisms including bacteria (http://pbil.univ-lyon1.fr/databases/hogenom/acceuil.php)
KEGG	Kyoto University: Comprehensive database of biological systems, including genes, enzymes, metabolites, reactions, and pathways (http://www.genome.jp/kegg/)
MetaCyc	SRI International: Database of nonredundant, experimentally elucidated metabolic pathways—1,790 pathways from more than 2,216 different organisms (http://metacyc.org/)
MicrobesOnline	Virtual Institute for Microbial Stress and Survival: Community resource for comparative and functional genome analysis for over 1,000 complete genomes of bacteria, archaea, and fungi and thousands of expression microarrays from diverse organisms (http://www.microbesonline.org/)
MicroScope	LABGeM—Genoscope: Platform for microbial genome annotation and comparative genomics for 640 organisms (https://www.genoscope.cns.fr/agc/microscope/home/index.php)

(continued)

Table 2 (continued)

Database or Web resource	Description (URL)
MPIDB	J Craig Venter Institute: Provides all known physical microbial interactions (24,295 experimentally determined interactions) among proteins of 250 bacterial species/strains (http://www.jcvi.org/mpidb/about.php)
NIAID Systems Biology for Infectious Diseases Research	National Institute of Allergy and Infectious Diseases: Systems Biology Program for Infectious Disease Research comprising four centers: The TB Systems Biology Center, The Systems Virology Center, The Center for Systems Influenza, and The Center for Systems Biology for EnteroPathogens (http://www.niaid.nih.gov/labsandresources/resources/dmid/sb/Pages/default.aspx)
Pathema	J Craig Venter Institute: Core resource that supports basic research for a set of six target NIAID category A-C pathogens—genome sequencing and intergenomic comparisons (http://pathema.jcvi.org/Pathema/)
PATRIC	Virginia Bioinformatics Institute: Provides rich data and analysis tools for all bacterial species in the selected NIAID category A–C priority pathogens list (http://www.patricbrc.org/portal/portal/patric/Home)
RCBPR	Resource Center for Biodefense Proteomics Research: Proteomics and host-pathogen interactions for biodefense-related microorganisms (http://pir.georgetown.edu/pirwww/proteomics/)
TargetTrack	Protein Structure Initiative: Provides information on the experimental progress and status of target amino acid sequences selected for structural determination (http://sbkb.org/tt/)
TDR Targets Database	Universidad Nacional de General San Martín: Provides diverse datasets to facilitate the identification and prioritization of drugs and drug targets in neglected disease pathogens as both a Web site and a tool (http://tdrtargets.org/)
TransportDB	The Institute for Genomic Research: Describes the predicted cytoplasmic membrane transport protein complement for organisms whose complete genome sequence is available—includes 288 bacteria (http://www.membranetransport.org/)
VIDA	Virus Database at University College London: Contains a complete collection of homologous protein families derived from open reading frames from complete and partial virus genomes (http://www.biochem.ucl.ac.uk/bsm/virus_database/VIDA.html)
ViPR	NIAID Virus Pathogen Database and Analysis Resource: Provides a comprehensive data repository for all types of data related to 13 families of human pathogenic category A–C viral pathogens (http://www.viprbrc.org/brc/home.do?decorator=toga)
xBASE	University of Birmingham: Comprehensive resource for comparative bacterial genomics of 191 genomes (http://www.xbase.ac.uk/)

Table 3
Software tools for in silico systems biology

Software Tools	Description (URL)
Automated metabolic network reconstruction tools	
The SEED	University of Chicago: Develop comparative genomics environment and curated genomic data (http://theseed.uchicago.edu/FIG/index.cgi)
YANAvergence	Universität Würzburg: Provides a software framework for rapid network assembly, network overview, and network performance analysis (http://www.bioinfo.biozentrum.uni-wuerzburg.de/computing/yana)
Pathway tools software	SRI International: Pathway Tools software supports creation, editing, querying, visualization, analysis, and publishing of Pathway/Genome Database (http://bioinformatics.ai.sri.com/ptools/)
Metabolic network reconstruction software	
ERGO	Integrated Genomics, Inc.: Supports both automatic and manual genome-wide curation (https://ergo.integratedgenomics.com/)
SimPheny	Intrexon Corporation: Enables the development of predictive computer models of organisms, from bacteria to humans (http://g6g-softwaredirectory.com/bio/cross-omics/agent-based/20629-GT-Life-Sci-Genomatica-SimPheny.php)
Metabolic network analysis tools	
BioSDP	Universität Stuttgart: Matlab component specially designed for the analysis of uncertain biochemical networks via semidefinite programming (http://biosdp.sourceforge.net/)
geWorkbench	MAGNet: A Java-based open-source platform for integrated genomics (http://wiki.c2b2.columbia.edu/workbench/index.php/Home)
Machine learning tool	Technische Universität Wien: Does latent class analysis, short-time Fourier transform, fuzzy clustering, support vector machines, shortest path computation, bagged clustering, naive Bayes classifier, etc. (http://cran.r-project.org/web/packages/e1071/index.html)
NeAT (Network Analysis Tools)	Université Libre de Bruxelles: Toolbox for the analysis of biological networks, clusters, classes, and pathways (http://rsat.bigre.ulb.ac.be/rsat/index_neat.html)
Modeling, simulation, and analysis software	
CellDesigner	Systems Biology Institute—Japan: Structured diagram editor for drawing gene regulatory and biochemical networks with links to simulation and other analysis packages (http://www.celldesigner.org/)
Cellware	Bioinformatics Institute—Singapore: Grid-based modeling and simulation tool that conducts modeling and simulation of gene regulatory and metabolic pathways (http://www.bii.a-star.edu.sg/achievements/applications/cellware/index.php)
COPASI	Virginia Bioinformatics Institute & EML Research: Software application for simulation and analysis of biochemical networks and their dynamics (http://www.copasi.org/tiki-index.php)
Dizzy	Stephen Ramsey—Institute for Systems Biology: Chemical kinetics stochastic simulation software package written in Java (http://magnet.systemsbiology.net/software/Dizzy/)

(continued)

Table 3
(continued)

Software Tools	Description (URL)
Dynetica	California Institute of Technology: Simulator of dynamic networks written in Java that does model building for systems expressed as reaction networks (http://www.duke.edu/~you/Dynetica_page.htm)
E-Cell	Keio University: Object-oriented software suite for modeling, simulation, and analysis of large-scale complex systems such as biological cells (http://www.e-cell.org/models/)
Simmune	NIAID Laboratory of Systems Biology: Suite of software tools tool for simulating and analyzing immune system behavior (http://www.niaid.nih.gov/LabsAndResources/labs/aboutlabs/lsb/Pages/simmuneproject.aspx)
VCell	University of Connecticut Health Center: Supports complex models with a Web-based Java interface to specify compartmental topology and geometry, molecular characteristics, and relevant interaction parameters (http://www.nrcam.uchc.edu/)
Simulation software (flux balance analysis)	
Clp	Coin-or-linear programming: Open-source linear programming solver written in C++ to find solutions of mathematical optimization (https://projects.coin-or.org/Clp)
COBRA Toolbox	University of California San Diego: MATLAB package for constraint-based reconstruction and analysis methods to simulate, analyze, and predict a variety of metabolic phenotypes using genome-scale models (http://opencobra.sourceforge.net/openCOBRA/Welcome.html)
Fluxor	Jeremy Zucker—Harvard Medical School: Python command-line tool that takes a metabolic network specified in Systems Biology Markup Language (SBML) and performs flux balance analysis using the GNU Linear Programming Kit (GLPK) and SWIG (http://fluxor.sourceforge.net/)
GAMS	GAMS Development Corporation: High-level modeling system for mathematical programming and optimization (http://www.gams.com/)
GPLK	GNU Project: Set of routines written in ANSI C for solving large-scale linear programming (http://www.gnu.org/s/glpk/)
GLPKMEX	GNU Project: Matlab MEX interface for the GLPK library for solving linear programming (http://glpkmex.sourceforge.net/)
ILOG CPLEX 8.100	IBM: Provides flexible, high-performance mathematical programming solvers for linear programming (http://www-01.ibm.com/software/integration/optimization/cplex-optimizer/)
Matlab	MathWorks: High-level language and interactive environment that enables simulation of biochemical networks using integrated flux balance analysis, regulatory flux balance analysis, and ordinary differential equations (http://www.mathworks.com/products/matlab/)
MetaFluxNet	Korea Advanced Institute of Science and Technology: Program package for quantitatively analyzing metabolic fluxes (http://mbel.kaist.ac.kr/lab/mfaml/main.html?page=metafluxnet.html)
Yices	SRI International: Constraint solver that can handle flux balance analysis (http://yices.csl.sri.com/)

(continued)

Table 3 (continued)

Software Tools	Description (URL)
Visualization software	
Cytoscape	Institute of Systems Biology: Open-source software platform for network data integration, analysis, and visualization (http://cytoscape.org/)
GraphViz	Stephen C. North—AT&T Labs Research: Open-source graph visualization software (http://www.graphviz.org/)
Paintomics	Centro de Investigaciones Príncipe Felipe: Web tool for the integration and visualization of transcriptomics and metabolomics data (http://www.paintomics.org/)
VisANT	Boston University: Integrative visual analysis tool for biological networks and pathways (http://visant.bu.edu/)
Network layout tools	
EPE	University of Edinburgh: Visual editor designed for annotation, visualization, and presentation of wide variety of biological networks, including metabolic, genetic, and signal transduction pathways (http://epe.sourceforge.net/SourceForge/EPE.html)
JDesigner	ERATO project—Caltech: Win32 application, which allows one to draw a biochemical network and export the network in the form of SBML (http://sbw.kgi.edu/software/jdesigner.htm)
Pathway Projector	Keio University: Provides integrated pathway maps that are based upon the KEGG Atlas, with the addition of nodes for genes and enzymes, implemented as a scalable, zoomable map utilizing the Google Maps API (http://www.g-language.org/PathwayProjector/)
yEd	yWorks: Powerful diagram editors that can be used to quickly and effectively generate high-quality drawings of diagrams (http://www.yworks.com/en/products_yed_about.html)
Microarray data analysis tools	
GO.tools	The Gene Ontology: Tools for analysis of microarray data (http://www.geneontology.org/GO.tools.microarray.shtml)
Mayday	University of Tübingen: Graphical user interface for visualization, analysis, and storage of microarray data (http://www.microarray-analysis.org/mayday)
Microarray DB	Keio University: Tool for mapping transcriptome data onto KEGG pathways and for creating a Web-based database with an overview of the entire pathway (http://www.g-language.org/data/marray/)
SAM	Stanford University: Supervised learning software for genomic expression data mining (http://www-stat.stanford.edu/~tibs/SAM/)
General "omics" data analysis tools	
BL-SOM	Platform for Riken Metabolomics: An integrated analytical tool for a range of "omics" data (http://prime.psc.riken.jp/?action=blsom_index)
DAnTE	Pacific Northwest National Laboratory: Allows users to perform various downstream data analysis, normalization, data reduction, and hypothesis testing steps (http://omics.pnl.gov/software/DAnTE.php)

(continued)

Table 3
(continued)

Software Tools	Description (URL)
Pathway Tools Omics Viewer	EcoCyc—SRI International: Paints data values from the user's high-throughput and other experiments onto the cellular overview diagram for an organism (http://biocyc.org/expression.html)
VANTED	Leibniz Institute of Plant Genetics and Crop Plant Research—Germany: A tool for the visualization and analysis of networks with related experimental data (http://vanted.ipk-gatersleben.de/)
Metabolomics-specific data analysis tools	
MathDAMP	Keio University: Allows visualization of differences between metabolite profiles acquired by hyphenated mass spectrometry techniques (http://mathdamp.iab.keio.ac.jp/)
metAlign	Wageningen UR: Computer software tool for the analysis, alignment, and comparison of full-scan mass spectrometry datasets (http://www.metalign.wur.nl/UK/Download+and+publications/)
MetATT	University of Alberta: A web-based tool for time-series and two-factor metabolomic data analysis (http://metatt.metabolomics.ca/MetATT/)
MetaboAnalyst	University of Alberta: A web-based analytical pipeline for high-throughput metabolomics studies (http://www.metaboanalyst.ca/MetaboAnalyst/faces/Home.jsp)
metaP-Server	Helmholtz Zentrum München: Automates data analysis for the processing of metabolomics experiments (http://metabolomics.helmholtz-muenchen.de/metap2/)
MSFACTs	The Samuel Roberts Noble Foundation: Metabolomics spectral formatting, alignment, and conversion tools (http://www.noble.org/PlantBio/Sumner/msfacts/index.html)
MZmine 2	VTT Technical Research Centre of Finland: Toolbox for processing and visualization of mass spectrometry-based molecular profile data (http://mzmine.sourceforge.net/download.shtml)
SpectConnect	Massachusetts Institute of Technology: Systematic identification of conserved metabolites in gas chromatography/mass spectrometry data for metabolomics (http://spectconnect.mit.edu/)
SpinAssign	Platform for Riken Metabolomics: Provides batch annotations of a large number of metabolites against user nuclear magnetic resonance peaks (http://prime.psc.riken.jp/?action=nmr_search)
XCMS	Scripps Center for Metabolomics: Software for processing liquid chromatography–mass spectrometry-based metabolomics data (http://metlin.scripps.edu/xcms/)
Statistical computing software	
SAS software	SAS: An integrated system of software products for statistical analysis (http://www.sas.com/technologies/analytics/statistics/stat/)
SPSS	IBM: A computer program used for data mining and statistical analysis (http://www-01.ibm.com/software/analytics/spss/)

(continued)

Table 3 (continued)

Software Tools	Description (URL)
STATISTICA	StatSoft: Provides a comprehensive and integrated set of tools and solutions for data visualization, graphical data analysis, visual data mining, visual querying (http://www.statsoft.com/unique-features/statistica-general-overview/)
R Project	The R Foundation: A free software environment for statistical computing and graphics (http://www.r-project.org/)

comparative and functional genomics, information on mutation, virulence and drug resistance, libraries of small-molecule lead compounds, high-throughput experimental data based on transcriptomics, proteomics, and metabolomics. The microbes include gram-negative and gram-positive bacteria, protozoa, and viruses.

3 Software Tools for Microbial Systems Analysis

System-level studies are often built on molecular and genetic findings and "omics" studies including genomics, proteomics, and metabolomics. The main challenges in systems biology are the complexity of the systems, the vast quantities of data, and the scattered pieces of knowledge, all of which must be integrated; therefore, systematic computational tools are crucially important. Understanding complex biological systems requires extensive support from software tools. Such tools are needed at each step of a systems biology computational workflow, which typically consists of data handling, network inference, deep curation, dynamical simulation, and model analysis (37).

Table 3 lists the major advanced computational software tools that are currently used for data analysis, visualization, modeling, simulation, and statistical computing, especially for microbial metabolic networks, models, and "omics" experiments. The given selection while intended to cover currently available software in this field is subjective, and the reader should consider available literature to focus on the specialized aspects and specific applications of the listed databases and software tools.

References

1. Bevan P, Ryder H, Shaw I (1995) Identifying small-molecule lead compounds: the screening approach to drug discovery. Trends Biotechnol 13:115–121
2. Kitano H (2002) Systems biology: a brief overview. Science 295:1662–1664
3. Gwynn MN, Portnoy A, Rittenhouse SF et al (2010) Challenges of antibacterial discovery revisited. Ann N Y Acad Sci 1213:5–19
4. Westerhoff HV, Palsson BO (2004) The evolution of molecular biology into systems biology. Nat Biotechnol 22:1249–1252

5. Butcher EC, Berg EL, Kunkel EJ (2004) Systems biology in drug discovery. Nat Biotechnol 22:1253–1259
6. Davidov E, Holland J, Marple E et al (2003) Advancing drug discovery through systems biology. Drug Discov Today 8:175–183
7. Aderem A (2005) Systems biology: its practice and challenges. Cell 121:511–513
8. Palsson B (2000) The challenges of in silico biology. Nat Biotechnol 18:1147–1150
9. Kitano H (2002) Computational systems biology. Nature 420:206–210
10. Raman K, Yeturu K, Chandra N (2008) TargetTB: a target identification pipeline for *Mycobacterium tuberculosis* through an interactome, reactome and genome-scale structural analysis. BMC Syst Biol 2:109
11. Gu S, Chen J, Dobos KM et al (2003) Comprehensive proteomic profiling of the membrane constituents of a *Mycobacterium tuberculosis* strain. Mol Cell Proteomics 2:1284–1296
12. Bahk YY, Kim SA, Kim JS et al (2004) Antigens secreted from *Mycobacterium tuberculosis*: identification by proteomics approach and test for diagnostic marker. Proteomics 4: 3299–3307
13. Mawuenyega KG, Forst CV, Dobos KM et al (2005) *Mycobacterium tuberculosis* functional network analysis by global subcellular protein profiling. Mol Biol Cell 16:396–404
14. Mattow J, Siejak F, Hagens K et al (2007) An improved strategy for selective and efficient enrichment of integral plasma membrane proteins of mycobacteria. Proteomics 7:1687–1701
15. Malen H, Berven FS, Fladmark KE et al (2007) Comprehensive analysis of exported proteins from *Mycobacterium tuberculosis* H37Rv. Proteomics 7:1702–1718
16. Gonzalez-Zamorano M, Mendoza-Hernandez G, Xolalpa W et al (2009) *Mycobacterium tuberculosis* glycoproteomics based on ConA-lectin affinity capture of mannosylated proteins. J Proteome Res 8:721–733
17. Sarker M, Chopra S, Mortelmans K et al (2011) In silico pathway analysis predicts metabolites that are potential antimicrobial targets. J Comput Sci Syst Biol 4:021–026
18. Munger J, Bennett BD, Parikh A et al (2008) Systems-level metabolic flux profiling identifies fatty acid synthesis as a target for antiviral therapy. Nat Biotechnol 26:1179–1186
19. Kim HU, Kim TY, Lee SY (2010) Genome-scale metabolic network analysis and drug targeting of multi-drug resistant pathogen Acinetobacter baumannii AYE. Mol Biosyst 6:339–348
20. Kim TY, Kim HU, Lee SY (2010) Metabolite-centric approaches for the discovery of antibacterials using genome-scale metabolic networks. Metab Eng 12:105–111
21. Kim HU, Kim SY, Jeong H et al (2011) Integrative genome-scale metabolic analysis of Vibrio vulnificus for drug targeting and discovery. Mol Syst Biol 7:460
22. Schilling CH, Palsson BO (2000) Assessment of the metabolic capabilities of Haemophilus influenzae Rd through a genome-scale pathway analysis. J Theor Biol 203:249–283
23. Yeh I, Hanekamp T, Tsoka S et al (2004) Computational analysis of *Plasmodium falciparum* metabolism: organizing genomic information to facilitate drug discovery. Genome Res 14:917–924
24. Rahman SA, Schomburg D (2006) Observing local and global properties of metabolic pathways: 'load points' and 'choke points' in the metabolic networks. Bioinformatics 22:1767–1774
25. Jamshidi N, Palsson BO (2007) Investigating the metabolic capabilities of Mycobacterium tuberculosis H37Rv using the in silico strain iNJ661 and proposing alternative drug targets. BMC Syst Biol 1:26
26. Chavali AK, Whittemore JD, Eddy JA et al (2008) Systems analysis of metabolism in the pathogenic trypanosomatid Leishmania major. Mol Syst Biol 4:177
27. Mazumdar V, Snitkin ES, Amar S et al (2009) Metabolic network model of a human oral pathogen. J Bacteriol 191:74–90
28. Oberhardt MA, Goldberg JB, Hogardt M et al (2010) Metabolic network analysis of *Pseudomonas aeruginosa* during chronic cystic fibrosis lung infection. J Bacteriol 192:5534–5548
29. Raghunathan A, Shin S, Daefler S (2010) Systems approach to investigating host-pathogen interactions in infections with the biothreat agent Francisella. Constraints-based model of *Francisella tularensis*. BMC Syst Biol 4:118
30. Crowther GJ, Shanmugam D, Carmona SJ et al (2010) Identification of attractive drug targets in neglected-disease pathogens using an in silico approach. PLoS Negl Trop Dis 4:e804
31. Plata G, Hsiao TL, Olszewski KL et al (2010) Reconstruction and flux-balance analysis of the Plasmodium falciparum metabolic network. Mol Syst Biol 6:408
32. Navratil V, De Chassey B, Combe CR et al (2011) When the human viral infectome and diseasome networks collide: towards a systems biology platform for the aetiology of human diseases. BMC Syst Biol 5:13
33. Fatumo S, Plaimas K, Adebiyi E et al (2011) Comparing metabolic network models based on genomic and automatically inferred enzyme information from Plasmodium and its human host to define drug targets in silico. Infect Genet Evol 11:708–715

34. Fang K, Zhao H, Sun C et al (2011) Exploring the metabolic network of the epidemic pathogen *Burkholderia cenocepacia* J2315 via genome-scale reconstruction. BMC Syst Biol 5:83
35. Ng A, Bursteinas B, Gao Q et al (2006) Resources for integrative systems biology: from data through databases to networks and dynamic system models. Brief Bioinform 7:318–330
36. Jagarlapudi SA, Kishan KV (2009) Database systems for knowledge-based discovery. Methods Mol Biol 575:159–172
37. Ghosh S, Matsuoka Y, Asai Y et al (2011) Software for systems biology: from tools to integrated platforms. Nat Rev Genet 12:821–832

Chapter 3

Genome Comparisons as a Tool for Antimicrobial Target Discovery

Hong Sun, Hai-Feng Chen, and Runsheng Chen

Abstract

Essential genes are frequently conserved among bacterial species and thus microbial and eukaryote genome comparisons can be used to compile datasets of homologous proteins and families that can be utilized to identify attractive targets for the design of antimicrobial agents and other drugs. These searches can now often be conducted using Web tools. A number of such resources that provide sequence information and comparative software as well as computational tools for convenient analysis of the data are summarized here and their step-by-step use explained.

Key words Computational drug target discovery, Bacterial gene families, Microbial genome comparison, Eukaryote genome comparison, Antimicrobial agents

1 Introduction

Early analyses of the genome sequences of pathogenic bacteria provided the first steps toward computational drug target discovery (1). Essential genes are important targets for the development of broad-spectrum antibiotics, and such genes are frequently conserved among different bacterial species. Similarly, bacterial gene families that are not found in human or other eukaryotes but are conserved in prokaryotes may also constitute important targets for potential novel antibiotics (1, 2). Combined microbial and eukaryote genome comparisons can thus be used to compile datasets of homologous proteins and families that can be utilized to identify attractive targets for the subsequent design of antimicrobial agents and other drugs (1).

Sandhya Kortagere (ed.), *In Silico Models for Drug Discovery*, Methods in Molecular Biology, vol. 993, DOI 10.1007/978-1-62703-342-8_3, © Springer Science+Business Media, LLC 2013

2 Materials

A number of resources are now available for conducting microbial genome comparisons. Most of these are accessible over the Internet through convenient Web browser interfaces. These resources generally provide sequence information and comparative software as well as computational tools for convenient analysis of the data.

2.1 GenomeComp

GenomeComp is a DNA sequence analysis tool (3) that enables visualization of results of genome-wide sequence comparisons, highlighting cross-genomic variation in such elements as repeat regions, insertions, deletions, and sequence rearrangements.

2.2 Microbial Genome Database

The Microbial Genome Database (MBGD) (4) facilitates the comparative analysis of complete microbial genome sequences, emphasizing such aspects as gene order and paralogue clustering, identification of orthologous genes, and comparison of specific sequence motifs.

2.3 Microbiota-Associated Databases (Integrate Functional Annotation Data with Comparative Genome Analysis)

1. MicrobesOnline: a phylogenetic approach to the analysis of microbial genes and genomes (5); currently (January 2012) includes 1,750 bacterial, 94 archaean, and 119 eukaryotic genomes as well as data from several thousand gene expression experiments (i.e., microarray data).
2. Comprehensive Microbial Resource (CMR): contains information about all publicly available microbial genomes and provides tools for cross-genome analysis (6).
3. eggNOG: a database of orthologous groups of genes that currently contains 721,801 functionally annotated orthologous groups covering 1,133 species at 41 different taxonomic ranges (7).

2.4 The Integrated Microbial Genomes System

1. Use as a tool (see Note 1) for genome search and annotation that integrates microbial genome sequence data with most eukaryotic sequence data as well as viral and plasmid sequences (8).
2. Use to facilitate comprehensive extraction of gene and functional information displayed as lists from which adequate information can be selected and saved for further analysis.
3. Use its "Phylogenetic Profiler" tool to identify genes that are either specifically present *or* absent in one or more genomes of interest. For instance, use the tool to find genes that are present in the genome of a parasitic bacterium but absent in the genome of a closely related free-living bacterium. Such genes are likely to have a higher probability of being important for pathogenicity, and the proteins translated from these genes can be used for structure-based drug research.

4. Search for genes that are unique to a particular organism to identify candidates for narrow-spectrum antibiotic development.

2.5 Functional Analysis Tools

1. Use functional annotation databases to provide insights into the putative functions of the selected genes and gene families and analyze the target structures for possible binding sites, active sites, and other functionally important domains.
2. Use the SWISS-MODEL (9) to build possible protein structures for these genes; dock potential broad-spectrum antibiotics candidates to the active sites with AutoDock (10) and use molecular dynamics simulations to research possible binding modes between antibiotics and target proteins (11).

3 Methods

3.1 Genome Comparison and Antimicrobial Target Discovery: Practice 1 (see Notes 2 and 3)

1. Go to the IMG home page (http://img.jgi.doe.gov/cgi-bin/w/main.cgi); click on the "Find Genes" link at the top middle of the page.
2. Hover on the "Phylogenetic Profiler" in the drop down menu and then select the "Single Genes" button.
3. Find genes that are conserved in a wider group of bacteria (i.e., *Bacillus*); select all of the genomes belonging to "*Bacillus*" for homologue comparisons by using the associated radio buttons in the "With Homologs In" column (see Note 4).
4. Select a specific query genome within *Bacillus*, "*Bacillus subtilis* BSn5," by using the associated radio button in the "Find Genes In" column.
5. Avoid selecting genes that have homologues in vertebrate genomes. Select the "*Chordata*" for homologue comparisons with the target genomes by using the associated radio buttons in the "Without Homologs In" column.
6. Set parameters: Go to the bottom of the Phylogenetic Profiler page. Under "Similarity Cutoffs," set the Maximum E-value and Minimum Percent Identity for which results are reported. Select "Yes" to exclude pseudogenes during the search. Click "Go" to find the genes in the target genome that satisfy the homologue presence/absence condition (see Note 5).
7. Click on "Gene Object ID" for details on the individual genes, e.g., "649937608."
8. Click on "285 bp" in the "DNA Coordinates" row in the "Gene Information" table to get the DNA sequence.
9. Click on "94 aa" in the "Amino Acid Sequence Length" row in the "Protein Information" table to get the amino acid sequence.

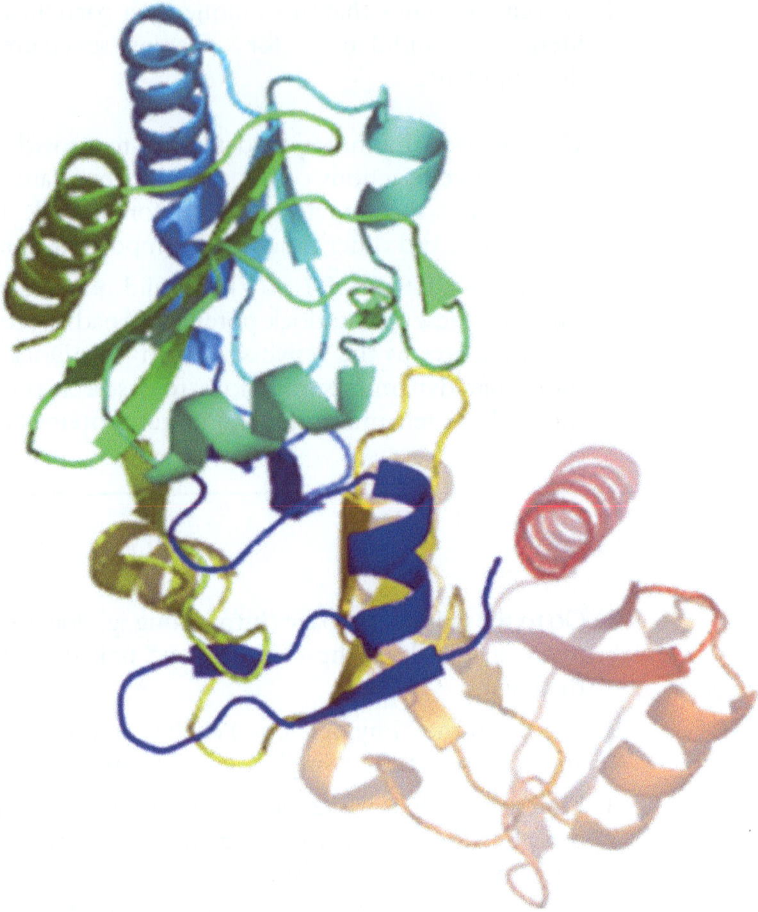

Fig. 1 Structure of the *Bacillus subtilis SpoVT_AbrB*

10. Go to the MicrobesOnline database (http://www.microbesonline.org), enter the selected gene Pfam name (e.g., "*SpoVT_AbrB*"), click the "Search" button, and click "E" for gene expression information.
11. Go to the eggNOG database (http://www.eggnog.embl.de/version_3.0), enter the selected gene Pfam name (e.g.,"*SpoVT_AbrB*"), and click on the "Search" button to get functional description lines.
12. Go to the SWISS-MODEL Web site (http://www.swissmodel.expasy.org), select "Automated Mode," input the e-mail address and the amino acid sequence of the protein retrieved from the Protein Information table, and click "Submit Modeling Request" to build target protein structure. (See the crystal structure of *SpoVT_AbrB* (pdb code: 2W1T) for *Bacillus subtilis* in Fig. 1.)

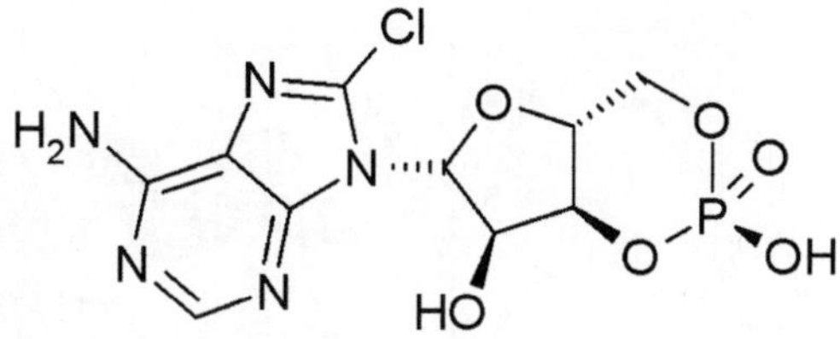

Fig. 2 Structure of 8-chloro-cAMP

13. Download the AutoDock 4.2 package from http://www.autodock.scripps.edu/ to dock the crystal structure (e.g., 2W1T) and small molecules against the active site. This step constitutes a virtual screening for potential antibiotics. The residues of F151, V120, V123, L102, and V135 for *SpoVT_AbrB* form a potential binding site for the ligands. The 8-chloro-adenosine-3p,5p-cyclic-monophosphate (8-chloro-cAMP) structure can be found from the virtual screen (Fig. 2).
14. Perform molecular dynamics simulations and energy minimizations for the target protein–small molecule complex (2W1T-8-chloro-cAMP) using AMBER 11 (downloaded from http://www.ambermd.org/) simulation package.

3.2 Genome Comparison and Antimicrobial Target Discovery: Practice 2 (see Notes 2 and 6)

1. Go to the IMG home page (http://www.img.jgi.doe.gov/cgi-bin/w/main.cgi) and click the "Find Genes" link at the top middle of the page.
2. Hover on "Phylogenetic Profiler" in the drop down menu and then select the "Single Genes" button.
3. Select the genomes in which homologues of the targets genes should not exist (i.e., *Streptomyces*, *Enterobacter*, and *Chordata*) by using the associated radio buttons in the "Without Homologs In" columns. In the default setting, all genomes are "ignored."
4. Select the query genome "*Streptomyces violaceusniger* Tu 4113" by using the associated radio button in the "Find Genes In" column.
5. Set parameters. Go to the bottom of the Phylogenetic Profiler page. Under "Similarity Cutoffs," set the "Maximum E-value" and "Minimum Percent Identity" for which results are reported. Select "yes" to exclude pseudogenes during the search. Click "Go" to find the genes in the target genome that satisfy the homologue presence/absence condition.
6. The result shows the genes in the query genome along with their functional characterization. As of January 2012, this list contains 777 actual and/or putative genes.

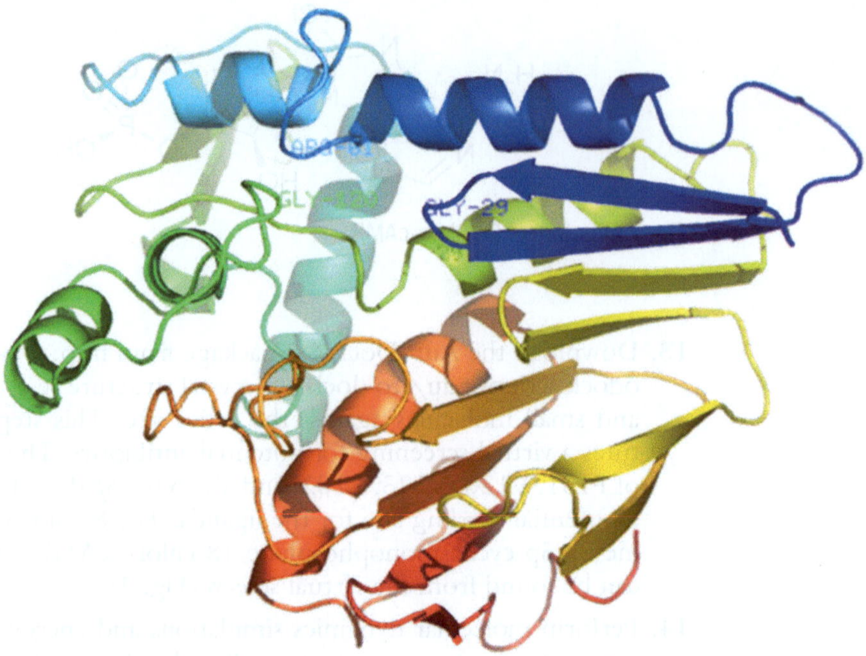

Fig. 3 Structure of a model of protein ApbE

7. Click on the "Gene Object ID" number for details on individual genes. Click "648749380" to find a membrane-associated lipoprotein involved in thiamine biosynthesis.
8. Click "957 bp" in the "DNA Coordinates" row in the "Gene Information" table to get the DNA sequence.
9. Click on "318 aa" in the "Amino Acid Sequence Length" row in the "Protein Information" table to get the amino acid sequence.
10. Enter the selected gene name (e.g., "*ApbE*") at http://www.microbesonline.org/ (MicrobesOnline database); click "Search" button; click "E" for gene expression information.
11. Enter selected gene name (e.g., "*ApbE*)"at http://www.eggnog.embl.de/version_3.0/ (eggNog database); click "Search" button to get functional description lines.
12. Select "Automated Mode" at http://www.swissmodel.expasy.org/. Input the e-mail address and the amino acid sequence of the protein retrieved from the Gene Information table; click "Submit Modeling Request" to build the target protein structure. The structure is shown in Fig. 3. The sequence identity is 23.5%. R61, D117, and I28 form a potential binding site for a ligand.

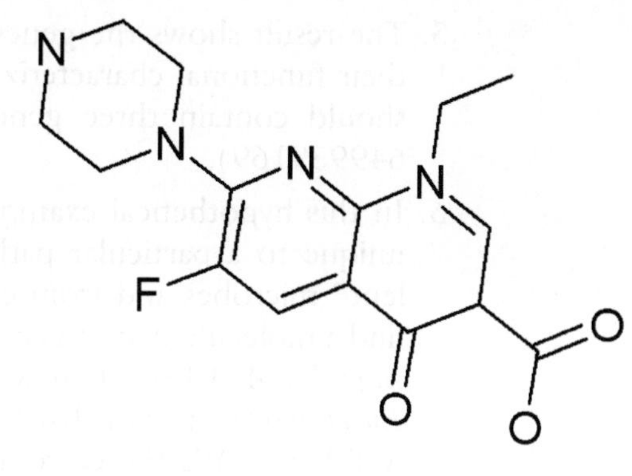

Fig. 4 Structure of enoxacin

13. Download the AutoDock 4.2 package from http://www.autodock.scripps.edu/ to dock the modeling protein structure and small molecules for virtual screening for potential antibiotics binding the active site. The structure of one such potential antibiotic (enoxacin) is shown in Fig. 4. If we assume that a given pocket is important for the function of the protein, we screen for small molecules that can dock to that site.
14. Perform molecular dynamics simulations and energy minimizations for the target protein–small molecule complex using the AMBER 11 (download from http://www.ambermd.org/) simulation package.

4 Notes

1. We use the IMG system to find gene families that are conserved among bacteria but are missing from eukaryotes. These gene families then constitute a pool of potential targets for broad-spectrum antibiotic development.
2. The specific procedures and examples highlighted in the two Practices apply to Web sites and databases as of January 2012. Because both are subject to frequent modifications and updates, we do not expect that every specific instruction and outcome in the described procedures will be valid indefinitely; however, the general instructions should be applicable for some time to come.
3. In this practical example, we want to find genes and gene families that are conserved among bacteria but missing from eukaryotes. These genes and gene families will constitute a pool of potential targets for broad-spectrum antibiotic development.
4. In the default setting, all genomes are "ignored."

5. The result shows the genes in the query genome along with their functional characterization. As of January 2012, the list should contain three genes (649937608, 649937627, and 649939169).
6. In this hypothetical example, we want to find genes that are unique to a particular pathogen but missing from "benevolent" microbes and from eukaryotes. Specifically, we want to find a molecule that can act as an antibiotic against *S. violaceusniger* Tu 4113 but that will not harm other *Streptomyces* and *Enterobacter* species that may be important for digestion and well-being. We therefore look for genes that are found in *S. violaceusniger* Tu 4113 but that do not have homologues in other *Streptomyces*, *Enterobacter*, and *Chordata* species. Such genes are potential targets for narrow-spectrum antibiotics.

References

1. Tatusov RL, Koonin EV, Lipman DJ (1997) A genomic perspective on protein families. Science 278:631–637
2. Galperin MY, Koonin EV (1999) Searching for drug targets in microbial genomes. Curr Opin Biotechnol 10:571–578
3. Yang J, Wang J, Yao ZJ et al (2003) GenomeComp: a visualization tool for microbial genome comparison. J Microbiol Methods 54:423–426
4. Uchiyama I, Higuchi T, Kawai M (2010) MBGD update 2010: toward a comprehensive resource for exploring microbial genome diversity. Nucleic Acids Res 38:D361–D365
5. Dehal PS, Joachimiak MP, Price MN et al (2010) MicrobesOnline: an integrated portal for comparative and functional genomics. Nucleic Acids Res 38:D396–D400
6. Peterson JD, Umayam LA, Dickinson T et al (2001) The comprehensive microbial resource. Nucleic Acids Res 29:123–125
7. Powell S, Szklarczyk D, Trachana K et al (2012) eggNOG v3.0: orthologous groups covering 1,133 organisms at 41 different taxonomic ranges. Nucleic Acids Res 40: D284–D289
8. Markowitz VM, Chen IM, Palaniappan K et al (2010) The integrated microbial genomes system: an expanding comparative analysis resource. Nucleic Acids Res 38:D382–D390
9. Arnold K, Bordoli L, Kopp J et al (2006) The SWISS-MODEL workspace: a web-based environment for protein structure homology modelling. Bioinformatics 22:195–201
10. Morris GM, Huey R, Lindstrom W et al (2009) AutoDock4 and AutoDockTools4: automated docking with selective receptor flexibility. J Comput Chem 30:2785–2791
11. Case DA, Cheatham TE 3rd, Darden T et al (2005) The Amber biomolecular simulation programs. J Comput Chem 26:1668–1688

Chapter 4

In Silico Models for Drug Resistance

Segun Fatumo, Marion Adebiyi, and Ezekiel Adebiyi

Abstract

Resistance to drugs that treat infectious disease is a major problem worldwide. The rapid emergence of drug resistance is not well understood. We present two in silico models for the discovery of drug resistance mechanisms and for combating the evolution of resistance, respectively. In the first model, we computationally investigated subgraphs of a biological interaction network that show substantial adaptations when cells transcriptionally respond to a changing environment or treatment. As a case study, we investigated the response of the malaria parasite *Plasmodium falciparum* to chloroquine and tetracycline treatments. The second model involves a machine learning technique that combines clustering, common distance similarity measurements, and hierarchical clustering to propose new combinations of drug targets.

Key words In silico, Drug, Resistance, Model, Mechanism

1 Introduction

Controlling infectious diseases is becoming more difficult as a result of the emergence of resistance to available drugs on the market. Drug resistance has emerged in the most dangerous diseases affecting humans, including malaria, tuberculosis, and HIV infection. These diseases have increased the disease burden particularly in developing countries, especially in Africa.

In this report, we present two in silico models, one for the discovery of drug resistance mechanisms and another for combating the evolution of drug resistance. Although we have adapted and developed these models for malaria research, they can be employed in the study of other infectious diseases. The first model has not been previously published. A model similar to our second model has been developed for the treatment of gastrointestinal stromal tumor (GIST) (1). With tumors as heterogeneous as GIST, up to five different types of secondary mutations can occur in the same patient. The aim is not only to wait for mutations to emerge before selecting the right compound but also to predict and group mutations according to likelihood, enabling clinicians to prescribe an appropriate drug as soon as a patient displays a particular mutation.

Sandhya Kortagere (ed.), *In Silico Models for Drug Discovery*, Methods in Molecular Biology, vol. 993,
DOI 10.1007/978-1-62703-342-8_4, © Springer Science+Business Media, LLC 2013

This dynamic multidrug-targeted prevention technique has been proposed in the treatment of chronic myeloid leukemia and the positive results obtained with the newly introduced drugs nilotinib and dasatinib suggested that a combination of two or three kinase inhibitors, when carefully selected to cover all known resistant mutations, could shut off all mechanisms of escape.

1.1 In Silico Modeling

An improved knowledge of genomics and of the structure of individual proteins has helped us increase our understanding of biological systems. However, insight into functional interactions between the key components of cells, organs, and systems helps us in understanding their physiology. Perturbations in these interactions lead to various diseases. We therefore must compute these interactions to determine the characteristics of the system when it changes from the healthy to the diseased state. With the development of powerful computing hardware and algorithms and an increasing number of pathway databases and models of cells, tissues, and organs, we can now explore functionality in a mathematical manner from the level of genes to the physiological function of whole organs and regulatory systems (2). The simplified mathematical representation of the dynamics of a system is called modeling (3). Modeling has become an important research area in biology and bioinformatics. We use models to explain experimental observations. Hence, we can also use them to test a hypothesis about biological function. We also use models for storing experimental data on biological molecules and processes in databases so as to analyze them (4). While modeling of individual reactions has been under way for a long time, we have only recently begun to appreciate the importance of modeling complex reactions, biochemical pathways, and networks (5). Because experimental data on biochemical reactions are insufficient and difficult, expensive, and time consuming to obtain, computational models of biological networks help in filling this data gap. We use computational models both for simulation and for metabolic engineering (2). Using computational simulation of complex biological networks, we can not only validate the conclusions drawn by experimental studies but also propound fresh hypotheses for further experimental validation. This iterative process of experimental studies and computational simulation has helped us develop highly sophisticated and realistic models, e.g., models of heart cells (6).

2 Materials and Concepts

2.1 DNA Microarray

The advent of DNA microarray high-throughput profiling experiments has allowed us to explore a major subset or all genes of an organism under a variety of conditions such as alternative treatments (drug-influenced condition vs. condition influenced by factors considered normal), mutants, developmental stages, and time points. For example, the technique enables us to classify tumor samples (7), to define small sets of potential marker genes to distinguish leukemia (8), and to discover regulatory mechanisms (9, 10). For example, without prior information, the structure and function of the network that regulates the SOS pathway in *Escherichia coli* was elucidated via transcription profiles (11).

2.2 Biochemical Metabolic Network

Biochemical investigations especially in the past 40 years have revealed an increasingly consistent image of cellular metabolism; see, for example, Berg et al. (12). This is especially true for less complex organisms such as *E. coli* (13). However, this approach used alone provides a rather static image of the cell and thus investigations have been performed to discover cellular adaptation programs in response to changing environments such as nutrient excess, starvation, and other stresses (14). These observations originally followed linear interaction and reaction cascades; studies investigated single knockouts and tediously tracked transcripts for single genes, compounds, and proteins that might be influenced; see, for example, Neidhardt (15). By combining metabolic network data and microarray data, data on the physical and chemical interactions of proteins can be integrated. For example, knowledge of protein–protein interaction gained from the use of high-throughput techniques (16) applied to the analysis of gene expression data revealed novel regulatory circuits (17). Moreover, knowledge of biochemical network interactions has been used to support the clustering procedure for gene expression profiles of yeast (18, 19).

2.3 BioCyc: A Collection of Biochemical Pathway Databases

BioCyc (20) is a collection of more than 200 pathway/genome databases, containing whole databases dedicated to certain organisms. For example, EcoCyc, which falls under the giant umbrella of BioCyc, is a highly detailed bioinformatics database on the genome and metabolic reconstruction of *E. coli*, including thorough descriptions of various signaling pathways. The EcoCyc database can serve as a paradigm and model for any reconstruction. Additionally, MetaCyc, an encyclopedia of metabolic pathways, contains a wealth of information on metabolic reactions derived from more than 600 different organisms, including *Plasmodium* and *Homo sapiens*.

2.4 Pathway Tools

Pathway Tools is a bioinformatics package that assists in the construction of pathway/genome databases such as EcoCyc (21). Developed by Peter Karp and his associates at the SRI International Bioinformatics Group, Pathway Tools comprises several separate units that work together to generate new pathway/genome databases (20). First, PathoLogic takes an annotated genome of an organism and infers probable metabolic pathways, allowing the creation of a pathway/genome database for the organism.

Pathway Hole Filler can then be applied to predict likely genes to fill "holes" (missing steps) in predicted pathways. Thereafter, the Pathway Tools Navigator and Editor functions let users visualize, analyze, access, and update the database. Thus, by using PathoLogic and encyclopedias such as MetaCyc, an initial fast reconstruction can be developed automatically, and then, using the other units of Pathway Tools, a detailed manual update, curation, and verification step is possible.

3 First Model: In Silico Model for Deducing Drug Resistance Mechanisms

In the first model, we sought to reveal subgraphs of a biological interaction network that show substantial adaptations when cells transcriptionally respond to a changing environment or treatment. As a case study, we investigated the response of the malaria parasite *Plasmodium falciparum* to chloroquine and tetracycline treatments.

This work was designed to unveil the mechanisms that culminated in the widespread resistance of this parasite to these drugs. We hope that our results will be useful in developing combinations of antiresistance drugs for malaria patients. Simple clustering of gene expression on the metabolic network of *P. falciparum* can yield subgraphs (clusters or features) that are either stimulated or repressed when the organism attempts to resist a particular treatment given to a malaria patient. König and Eils (22) and König et al. (23) demonstrated a similar mechanism with tryptophan-treated cells and in the heterofermentative bacterium *E. coli* in response to oxygen deprivation (24).

Following this line of work, we made the discoveries reported here. Before we indicate these, we note that the microarray datasets that we analyzed for tetracycline and chloroquine do not contain many differentially regulated reactions. The possibility remains that studying drug resistance mechanisms of the malaria parasites at the transcriptional level of their proteins is not reliable (Karine Le Roch, personal communication).

Using the tetracycline microarray data, Dahl et al. (25) indicated that tetracyclines specifically block expression of the apicoplast genome and concluded that the loss of apicoplast function in the progeny of treated parasites leads to a slow but potent antimalarial effect. From the clusters we extracted, we show that this slow antimalarial effect is due in particular to excess glucose that is being made available. The fatty acid production is upregulated (beta oxidation, starting at acetyl-coA) together with the farnesyl pathway, which is needed for cholesterol and also leads to fatty acids and membrane components. We also discovered important genes and reactions that participated in the resistance mechanism of *P. falciparum* to tetracycline.

From the chloroquine microarray data, we found that tryptophanyl-tRNA synthetase production in the apicoplast is upregulated. Others have hypothesized that resistant *P. falciparum* parasites have a mechanism for releasing chloroquine via an efflux process (26, 27). We prove in this work that the upregulated tryptophanyl-tRNA synthetase production in the apicoplast suggests that this efflux process may have been made possible (caused) by the apicoplast, the mini bacterium living inside the malaria parasite. We hypothesize that when our results are experimentally proved, in particular for the case of chloroquine, our findings may lead to better and more cost-effective agents for eradication of the parasite from the human blood stream.

3.1 Gene Expression Data Used

Serial Analysis of Gene Expression (SAGE) tags of chloroquine-treated cells were obtained from the work of Gunasekera et al. (24) and the microarray data were obtained from Gunasekera et al. (28). Data from the microarray response to tetracycline treatment were taken from Dahl et al. (25). Chloroquine is designed to inhibit the parasitic enzyme heme polymerase and tetracycline is designed to inhibit the cytosolic ribosomes. Additionally, Dahl et al. (25) showed the antimalarial effect of chloroquine against the apicoplast genome of *P. falciparum*. In the following discussion, "chloroquine drug influence" refers to the microarray data on chloroquine treatment vs. control (in cases when this is not so, we will explicitly state this).

In Gunasekera et al. (28), the parasite culture preparation and RNA preparation/hybridization were done as follows. Blood-stage *P. falciparum* parasites were maintained in vitro at 37°C in RPMI 1640 (Roswell Park Memorial Institute) medium (Invitrogen, Carlsbad, CA) containing 25 mM HEPES, 0.2% sodium bicarbonate, 50 μg/mL hypoxanthine, 25 μg/mL gentamicin, 5% heat-inactivated human O^+ serum, 5% bovine serum albumin (Albumax II, Invitrogen), and 5% human O^+ blood, following standard protocols (29). 3D7 strain parasites were used for all experiments. Mixed-stage 3D7 parasites were treated with 120 and 400 nM chloroquine for 30 min and 6 h, alongside matched controls, yielding six samples (0 nM—30 min, 120 nM—30 min, 400 nM—30 min, 0 nM—6 h, 120 nM—6 h, and 400 nM—6 h). Two separate starting cultures at 8% parasitemia but with different stage profiles were subjected to each of the six treatments. The first consisted of approximately 1.7% rings, 2.5% early trophozoites, 3.4% late trophozoites, and 0.15% schizonts, and the second contained 1.7% rings, 5.4% early trophozoites, 0.6% late trophozoites, and 0.15% schizonts. Hence a total of 12 different cell states, representing parasites under varying drug concentrations (three), drug exposures (two), and staging profiles (two), were assayed. Total RNA was harvested at the end of each time point using the Tri-Reagent BD protocol (Molecular Research Center, Cincinnati, OH), labeled by a strand-specific protocol and hybridized to a custom-made high-density oligonucleotide array containing 260,596 25-mer probes from a predicted coding sequence of the parasite genome and 106,630 probes from a noncoding sequence (30). Probes mapping to coding sequences were used to compute gene expression levels by means of the match-only integral distribution algorithm (MOID) (31). We normalized the expression data using an established variance normalization method (32).

The malaria parasites preparation, culture, and microarray analysis by Dahl et al. (25) were performed using the following setup. *P. falciparum* parasites were cultured in human erythrocytes maintained at 2% hematocrit in RPMI 1640 medium with 0.5% (wt/vol) bovine serum albumin in 92% N_2, 5% CO_2, and 3% O_2. Synchrony was maintained by serial sorbitol treatments. Strain 3D7 was used here. Parasites stably expressing green fluorescent protein fused to an acyl carrier protein apicoplast-targeting sequence (ACP_1-GFP), kindly provided by Geoff McFadden (33), were maintained in medium containing 100 nM pyrimethamine. Dually transfected parasites stably expressing a red fluorescent protein fused to an acyl carrier protein apicoplast-targeting signal and a yellow fluorescent protein fused to a citrate synthetase mitochondrial targeting signal (ACP_1-DsRed and CS_1-YFP), also kindly provided by Geoff McFadden (34), were maintained in medium containing 5 nM WR99210.

Synchronized parasites were treated at the late ring/early trophozoite stage (approximately 20 h postinvasion) with 1 μM doxycycline or an equivalent volume of dimethyl sulfoxide for 24 h, until they reached the late schizont stage. The parasites were then subcultured and maintained in drug-free medium for an additional 35 h. Infected erythrocytes were collected every 5 h, lysed with 0.1% saponin for 5 min, centrifuged at 12,000 × *g* at 4°C, flash-frozen in an ethanol–dry ice bath, and stored at −80°C. Total parasite RNA was harvested using TRIzol reagent (Invitrogen). For each sample, 12 μg of total parasite RNA was reverse transcribed into cDNA containing amino-allyl-dUTP (Ambion, Invitrogen) using SuperScript II RNase H-Reverse Transcriptase (Invitrogen) and then coupled to succidimyl ester Cy5 dye (Amersham, GE Healthcare, Chalfont St. Giles, UK), as described previously (35). Cy5-labeled sample cDNA and a reference pool of Cy3-labeled cDNA representing all life cycle stages were competitively hybridized to a *P. falciparum* 70-mer microarray as described by Bozdech et al. (36). The microarrays were scanned using a GenePix 4000B scanner, and images were analyzed using GenePix3 Software (Molecular Devices, Sunnyvale, CA), stored, and normalized using the NOMAD database (http://ucsf-nomad.sourceforge.net). Expression data were log transformed and mean centered.

3.2 Mapping of SAGE Tags to Genes

We mapped all SAGE tags to the genes they represented as follows. We used the standalone Blast from NCBI (ftp://ftp.ncbi.nlm.nih.

gov/blast/executables) and the databases of coded regions of *P. falciparum* and blasted all SAGE tags against all open reading frames selecting only the perfect matches.

3.3 Model for Analyzing Gene Expression Data on Metabolic Networks

To analyze the gene expression data above on the metabolic network of *P. falciparum*, we used the following computational pipelines:

1. Construction of the metabolic network from PlasmoCyc obtained from BioCyc.
2. Network clustering using a simulated annealing and a Kernighan–Li clustering procedure.
3. Mapping gene expression data onto the reactions.
4. Feature extraction using a combinatorial approach.
5. Analysis of stimulated and repressed pathways.

We elaborate on these in the following sections. The pipelines explored here have been used by König et al. (23), but these investigators used another feature extraction technique, the Haar wavelet transform. We explored a novel feature extraction technique based on a combinatorial approach. We confirm further the results obtained via the pipelines above using the Haar wavelet transform. This transform was done using the clusters due to the consecutive-ones clustering technique (23).

3.4 Construction of the Metabolic Network from PlasmoCyc (BioCyc)

We constructed our network from the metabolic reaction database PlasmoCyc. The metabolites were taken as nodes. Two metabolites were connected by an edge if an enzymatic reaction existed that had them as an educt or product, respectively (23). We discarded highly connected metabolites such as water, CO_2, and adenosine triphosphate. These metabolites are needed in many reactions and are therefore unspecific in the metabolic network.

3.5 Network Clustering Using Kernighan–Li and Simulated Annealing Algorithms

Here we describe how the network given above will be clustered to group enzymes into parts of the network with their major connections. Formally, given the metabolic network as graph G(V,E) with node set V (metabolites) and edge set E (reactions), the goal of our clustering here is to identify clusters of G where each cluster was given by the node set of a highly connected subgraph. Note that the clusters are not required to be mutually disjoint.

For the network clustering problem, we used both the simulated annealing and Kernighan–Li (37–39) algorithms. We then applied our feature extraction technique, the combinatory approach on both clusters obtained via these algorithms. The idea is that if the clusters are similar (similarly ranked) in both results, this will help confirm our findings.

In the following, we explain each clustering technique briefly. We adapted these algorithms to cluster the metabolic network described above. For more details, readers should see Kernighan and Li (38), Dutt (39), and Brown and Huntley (37). The Kernighan–Li algorithm was designed to solve the following combinatorial problem: given a graph G with costs on its edges, partition the nodes of G into subsets no larger than a given maximum size, so as to minimize the total cost of the edges cut. We explain the two-way uniform partition of G using Kernighan and Li; its application in performing multiple-way partitions (as we did in this work) is achieved using the two-way procedure that allows us to partition into unequal-sized sets. Formally, let $G(V,E)$ be a graph with node set $V(G)$ and edge set $E(G)$, where there is a positive cost $c(\{v_i,v_j\})$ associated with every edge $\{v_i,v_j\} \in E(G)$ that may, for example, represent the width of the corresponding link. The problem is to partition $V(G)$ into partitions P_1 and P_2 so that $-1 \leq |P_1| - |P_2| \leq 1$, and the cost of the cut-set $\sum c(\{v_i,v_j\})$ is minimized, where v_i and v_j belong to different partitions. The resulting effect of this partitioning is that nodes that are densely connected to each other are placed near each other.

Let us take the following notations. Given a partition P_1 and P_2 of $V(G)$, for each $u \in V(G)$, let us define the external cost E_u and the internal cost I_u of u as follows:

$E_u = \Sigma_v \varepsilon_{Pi}\, c(\{u,v\})$, where $i=1, 2$, u does *not* belong to P_i, and $\{u,v\} \in E(G)$

$I_u = \Sigma_{v\,Pi}\, c(\{u,v\})$, where $i=1, 2$, u does belong to P_i, and $\{u,v\} \in E(G)$

We define the D value of node u as $D_u = E_u - I_u$. This is the gain (reduction in the cost of the cut-set) obtained by moving u from its current partition. Thus if $u \in P_1$ and $v \in P_2$, then it is easy to see that the gain $G_{u,v}$ associated with swapping the pair of nodes (u,v) is $D_u + D_v - 2c(\{u,v\})$ if $\{u,v\} \in E(G)$ and $D_u + D_v$ otherwise.

Assume that there are $n = 2\,m$ nodes in G, and the initial partitions are P_1 and P_2, with $|P_1| = |P_2| = m$. Let $P_1 = \{u_1, u_2, \ldots, u_m\}$ and $P_2 = \{v_1, v_2, \ldots, v_m\}$. A main data structure used in the Kernighan and Li algorithm is the symmetric cost matrix C, where $C_{u,v} = c(\{u,v\})$ if $\{u,v\}$ $E(G)$ and $C_{u,v} = 0$ otherwise. First the D value of each node u is computed using C. Then, that pair of nodes (u_{i1},v_{j1}) is chosen for swapping that has the maximum value of $G_{ui,vj}$. Node u_{i1} is removed from P_1, v_{j1} is removed from P_2, (u_{i1},v_{j1}), and inserted in an ordered set S of node pairs, and the D value of each node u is updated to reflect the fact that the pair (u_{i1},v_{j1}) has been swapped between the partitions. This procedure is iterated m times until P_1 and P_2 become empty, with the node pair (u_{ik}, v_{jk}) inserted in S in the kth iteration, $1 \leq k \leq m$ to give $S = [(u_{i1},v_{j1}), (u_{i2},v_{j2}), \ldots, (u_{im},v_{jm})]$. All partial sums $S_k = \Sigma_{t=1,\ldots,k}\, G_{uit,vjt}$ are computed, and p is chosen such that the partial sum S_p is the maximum. The sets of node pairs that are actually swapped are then $\{(u_{i1},v_{j1}), \ldots, (u_{ip}, v_{jp})\}$, such that the maximum gain $G = S_p$ is obtained. This whole process is called a pass. A number of passes are made until the maximum gain G

obtained is 0. This is a local maxima with respect to the initial partitions P_1 and P_2. Empirical evidence shows that the number of passes required to achieve a local maxima is 2–4.

The simulated annealing algorithm (37) for the partitional clustering as we required in this work was designed based on the following problem formation.

Let

Q be the set of all objects to be clustered (here, metabolites),

$n = |Q|$ be the number of objects in Q,

$k \leq n$ be the maximum number of clusters,

$P = \{p$: for every $i \in \{1, \ldots, n\}$, $p_i \in \{1, \ldots, k\}\}$ be the set of all partitionings,

$J: P \rightarrow R$ be the internal clustering criterion;

Then

Minimize $J(p)$ (here, based on minimizing the total cost of the edges cut)

Subject to

$p \in P$

The algorithm requires the perturbation operator δ and the annealing schedule (MaxIt, T_o, α, T_f). The perturbation operator for partitional clustering switches a randomly chosen object i in Q from one cluster to another randomly chosen cluster. A set L contains the cluster labels used in **p**. Similarly, L^c contains the labels not used in **p**. The switching procedure first selects an integer m in the range $[0,|L|]$. If m equals 0 and there exists an unused cluster label (i.e., $|L| < k$), then object i is placed in its own singleton cluster. Otherwise, i switches to another, existing cluster. The computational effort is made fair by allowing each run a fixed number of trial perturbations. The total number of perturbations tried in any run is MaxIt.NumTemp, where MaxIt is a fixed multiple of the number of objects to be clustered and NumTemp is a user-defined constant. The solution is made accurate using a very conservative annealing schedule (40, 41).

3.6 Mapping Gene Expression Data onto Reactions and Feature Extraction

To do the mapping, each reaction documented in the metabolic network created above is linked to the gene(s) that produced the enzymes that catalyze it. This way, the gene expression values obtained from the microarray experiment (for each time point) replace the corresponding reaction of the metabolic network. We take the average reading for the case where a reaction is catalyzed by more than one gene.

The features extraction, i.e., the discovery of clusters whose genes are differentially expressed, in particular within different time points of the microarray under the control condition (no drug)

and the drug-influenced condition, is carried out using a combinatorial approach. We explain this below.

In the combinatorial approach we developed here, all possible combinations of sums and differences of expression values in each cluster are calculated. Note that we do not need all combinations, only half, because the other half can be obtained by the multiplication of −1 with one-half. We explain the process further using a small example: If we have a cluster with three reactions and we have already mapped the expression values to the corresponding genes of the reactions, let these expression values be 1, 2, and 3. Then we will have four possible combinations, namely +1 +2 +3, +1 +2 −3, +1 −2 +3, and +1 −2 −3. Next, if we compared (subtracted) all combinations, the largest difference would be taken as *P* value and the clusters are ranked according to their *P* values. In the actual sense, the rationale behind choosing the largest *P* value is that it indicates the best probability that exists for the group of genes in the cluster in question not to be differentially expressed.

For each cluster, all combinations are calculated as described above. However, this is done for each experiment (time point) separately. Once all combinations are calculated, a Wilcoxon test is done to distinguish differences between the two different states (in our case control vs. drug). For each cluster, this is done for every calculated combination. Once all Wilcoxon tests for all clusters and all combinations are done, the *P* values are corrected for multiple testing. The clusters are then ranked according to the lowest *P* value that was achieved for the respective clusters.

3.7 Analysis of Stimulated or Repressed Pathways

The analysis of stimulated or repressed pathways was done manually and included an in-depth literature search. First, per cluster, we identified the product/function of each gene and the metabolic pathways in which each is functionally active. We did this using plasmoDB. For each drug, we identified the genes functionally active in the pathways it targeted, expecting our pattern extraction tool to capture the distinct differential expression of these genes between the drug-induced and the control samples. We looked for cases that did not show this format (such cases have been found) to give us hints on what collection of genes differentially coexpressed might deactivate the effectiveness of the drug on the targeted pathway.

3.8 Results and Discussion

Figure 1 shows the histograms of the Wilcoxon *P* value of each gene expression under tetracycline (A) and chloroquine (B) treatment conditions compared with its gene expression under no drug influence. There are many more discriminative coexpression patterns in the chloroquine data than in the tetracycline data. Using these data, we list in Tables 1 and 2 the genes that are at least 95% significantly differentially expressed under tetracycline influence and genes that are at least 99.999% significantly differentially

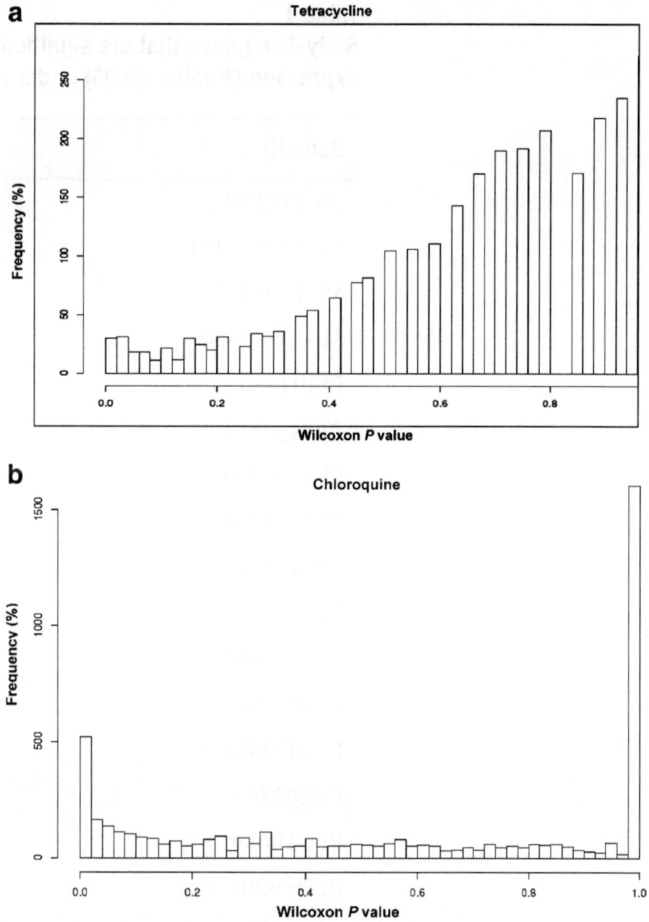

Fig. 1 Histograms of the Wilcoxon *P* values of each gene's expression under tetracycline treatment (**a**) and chloroquine treatment (**b**) compared with its expression under no drug influence. The *x*-axis lists ranges of all the *P* values estimated and the *y*-axis shows the frequency of each

expressed under chloroquine influence, respectively, with their corresponding Wilcoxon *P* values.

Currently from PlasmoCyc, 691 reactions of the malaria parasite have been curated and documented. We consider here top-ranking reactions of the parasite whose enzymes were significantly differentially expressed under drug treatment conditions compared with their expression under no drug influence for tetracycline and chloroquine (Tables 3 and 4, respectively). The first and second columns give the reaction's common name and unique ID in PlasmoCyc. The second column gives their Wilcoxon test *P* values. The Wilcoxon test is applicable here because we do not have the requirement for normally distributed data. The lower the *P* values of a reaction under the drug influence vs. control, the more highly significant the possibility that the reaction may have contributed to

Table 1
Sixty-five genes that are significantly differentially expressed (*P* value ≤0.05) under tetracycline treatment

Gene ID	Wilcoxon *P* values
pla_ORF78	0.003636625
MAL13P1.271	0.04490200
MAL13P1.312	0.02048920
MAL13P1.304	0.001829776
PFI0495w	0.01004454
PF14_0114	0.0006560272
PF10_0319	0.0284207
PF10_0026	0.02048920
PF14_0582	0.02048920
PF14_0294	0.0004955335
PF14_0695	0.02048920
PFB0425c	0.01004454
PF10_0313	0.01449251
PFE0230w	0.01209398
PFC1065w	0.0284207
PFD1090c	0.01209398
pla_tufA1	0.00555959
PF14_0278	0.0008579387
PFE0755c	0.00829316
PFA0430c	0.03872114
PFI0990c	0.03872114
MAL6P1.93	0.01727119
wPF10_0061	0.01727119
pla_ORF470	0.03872114
pla_rps11	0.02418426
PFL0635c	0.02418426
pla_rps17	0.03872114
MAL8P1.71	0.0284207
PF11_0086	0.03324143
PF14_0175	0.02418426
MAL6P1.104	0.04490200

(continued)

Table 1 (continued)

Gene ID	Wilcoxon *P* values
PF14_0409	0.01449251
PFD0260c	0.003636625
PFL2335w	0.01727119
PFL0835w	0.003636625
PFL1125w	0.03872114
PFD0400w	0.04490200
PF13_0332	0.01209398
PFE1455w	0.004513053
MAL13P1.33	0.0284207
PFC0260w	0.01727119
PFI1500w	0.03324143
PFC0750w	0.002316434
PF10_0213	0.00555959
PF14_0529	0.004513053
PFD0845w	0.001829776
PF13_0332	0.00829316
pla_rps7	0.02418426
PFD0885c	0.03324143
pla_tRNA-Gln	0.02418426
pla_tRNA-Gly	0.03872114
PFE1375c	0.0284207
pla_tRNA-Trp	0.03872114
PFL2325c	0.03872114
PFD0970c	0.03324143
PF14_0093	0.0284207
MAL6P1.105	0.0284207
MAL13P1.261	0.003636625
PF11_0433	0.01727119
PF10_0336	0.03872114
PF13_0210	0.04490200
PF11_0289	0.03872114
PFL0290w	0.02418426
PFA0430c	0.00555959

Table 2
Ninety genes that are significantly differentially expressed (*P* value ≤1.0e–5) under chloroquine

Gene ID	Wilcoxon *P* value
MAL13P1.245	7.396023e–07
MAL13P1.25	7.396023e–07
MAL6P1.181	7.396023e–07
MAL6P1.4	7.396023e–07
MAL6P1.60	7.396023e–07
MAL6P1.79	8.875228e–06
MAL7P1.104	5.177216e–06
MAL7P1.50	8.875228e–06
MAL8P1.22	7.396023e–07
MAL8P1.24	1.479205e–06
MAL8P1.97	7.396023e–07
PF07_0050	2.958409e–06
PF07_0055	7.396023e–07
PF07_0056	2.958409e–06
PF07_0111	7.396023e–07
PF07_0115	1.479205e–06
PF08_0008	7.396023e–07
PF08_0018	8.875228e–06
PF08_0021	7.396023e–07
PF08_0073	8.875228e–06
PF10_0002	7.396023e–07
PF10_0082	7.396023e–07
PF10_0132	8.875228e–06
PF10_0167	5.177216e–06
PF10_0177	2.958409e–06
PF10_0198	1.479205e–06
PF11_0021	2.958409e–06
PF11_0098	7.396023e–07
PF11_0127	7.396023e–07
PF11_0164	2.958409e–06

(continued)

Table 2
(continued)

Gene ID	Wilcoxon *P* value
PF11_0236	2.958409e–06
PF11_0289	7.396023e–07
PF13_0295	7.396023e–07
PF13_0317	7.396023e–07
PF14_0061	8.875228e–06
PF14_0161	5.177216e–06
PF14_0212	1.479205e–06
PF14_0217	2.958409e–06
PF14_0231	7.396023e–07
PF14_0303	1.479205e–06
PF14_0336	7.396023e–07
PF14_0481	7.396023e–07
PF14_0497	2.958409e–06
PF14_0512	5.177216e–06
PF14_0611	2.958409e–06
PF14_0701	5.177216e–06
PF14_0715	8.875228e–06
PFA0290w	8.875228e–06
PFA0460c	9.61483e–06
PFB0470w	1.479205e–06
PFB0820c	7.396023e–07
PFB0845w	1.479205e–06
PFC0195w	2.958409e–06
PFC0370w	5.177216e–06
PFC0470w	5.177216e–06
PFC0495w	7.396023e–07
PFC0575w	7.396023e–07
PFC0785c	1.479205e–06
PFD0035c	7.396023e–07
PFD0215c	8.875228e–06
PFD0430c	7.396023e–07

(continued)

Table 2
(continued)

Gene ID	Wilcoxon *P* value
PFD0490c	7.396023e–07
PFD0520c	7.396023e–07
PFD0820w	7.396023e–07
PFE0820c	7.396023e–07
PFE0890c	7.396023e–07
PFE0950c	1.479205e–06
PFE1300w	7.396023e–07
PFE1595c	1.479205e–06
PFE1605w	5.177216e–06
PFI0300w	8.875228e–06
PFI0315c	8.875228e–06
PFI0860c	8.875228e–06
PFI1080w	1.479205e–06
PFI1225w	5.916818e–06
PFI1420w	7.396023e–07
PFI1485c	7.396023e–07
PFL0370w	5.177216e–06
PFL0410w	1.479205e–06
PFL0920c	7.396023e–07
PFL1045w	7.396023e–07
PFL1150c	1.479205e–06
PFL1195w	2.958409e–06
PFL1270w	1.479205e–06
PFL1970w	2.958409e–06
PFL1980c	7.396023e–07
PFL2190c	8.875228e–06
PFL2390c	7.396023e–07
PFL2415w	7.396023e–07
PFL2555w	5.177216e–06

Table 3
Twenty-two top-ranking reactions of the parasite whose enzymes were significantly differentially expressed under drug (tetracycline) influence vs. their expression under no drug influence

Common reaction name	Unique ID in PlasmoCyc	Wilcoxon *P* value
Threonine–tRNA ligase	THREONINE–TRNA-LIGASE-RXN	0.2189208
Phenylalanine–tRNA ligase	ALANINE–TRNA-LIGASE-RXN	0.1781820
Ferrochelatase	PROTOHEMEFERROCHELAT-RXN	0.2591973
Adenylosuccinate lyase	AMPSYN-RXN	0.2320216
Adenylosuccinate lyase	AICARSYN-RXN	0.2320216
Fructose-bisphosphate aldolase	F16ALDOLASE-RXN	0.2415238
Lysine decarboxylase	LYSDECARBOX-RXN	0.2591973
Copper-exporting ATPase	3.6.3.4-RXN	0.1472773
Inositol-1,4,5-trisphosphate 5-phosphatase	3.1.3.56-RXN	0.2092226
Thiosulfate sulfurtransferase	THIOSULFATE-SULFURTRANSFERASE-RXN	0.2581537
UDP-*N*-acetylglucosamine–dolichyl-phosphate *N*-acetylglucosamine phosphotransferase	2.7.8.15-RXN	0.1718688
Adenylate kinase	ADENYL-KIN-RXN	0.0786467
Aromatic amino acid transferase	TYRAMINOTRANS-RXN	0.2415238
Aromatic amino acid transferase	PHEAMINOTRANS-RXN	0.2415238
Aspartate aminotransferase	ASPAMINOTRANS-RXN	0.2415238
Phenylalanine(histidine) aminotransferase	3-SULFINOALANINE-AMINOTRANSFERASE-RXN	0.2415238
Dihydrolipoamide *S*-acetyltransferase	RXN0-1133	0.1848905
Acetyl-coA C-acyltransferase	METHYLACETOACETYLCOATHIOL-RXN	0.1118256
Acetyl-coA C-acyltransferase	KETOACYLCOATHIOL-RXN	0.1118256
Histone acetyltransferase	HISTONE-ACETYLTRANSFERASE-RXN	0.07230222
Acetyl-coA C-acetyltransferase	ACETYL-COA-ACETYLTRANSFER-RXN	0.1118256
Pyruvate dehydrogenase (lipoamide)	RXN0-1134	0.2179721

Table 4
Fifty-two top-ranking reactions of the parasite whose enzymes were significantly differentially expressed under drug (chloroquine) influence vs. their expression under no drug influence

Reaction common name	Unique ID in PlasmoCyc	Wilcoxon *P* value
Acetyl-coA carboxylase	RXN0-5055	0.01262057
Acetyl-coA carboxylase	ACETYL-COA-CARBOXYLTRANSFER-RXN	0.01262057
Biotin carboxylase	BIOTIN-CARBOXYL-RXN	0.01262057
Phosphopantothenate–cysteine ligase	P-PANTOCYSLIG-RXN	0.05755659
Long-chain-fatty-acid–coA ligase	RXN-7904	0.008890324
Long-chain-fatty-acid–coA ligase	R223-RXN	0.008890324
Long-chain-fatty-acid–coA ligase	ACYLCOASYN-RXN	0.008890324
Tyrosine–tRNA ligase	TYROSINE–TRNA-LIGASE-RXN	0.000641938
Methionine–tRNA ligase	METHIONINE–TRNA-LIGASE-RXN	0.03842444
Lysine–tRNA ligase	LYSINE–TRNA-LIGASE-RXN	0.0015584469
Leucine–tRNA ligase	LEUCINE–TRNA-LIGASE-RXN	0.01209398
Isoleucine–tRNA ligase	ISOLEUCINE–TRNA-LIGASE-RXN	0.0341075
Histidine–tRNA ligase	HISTIDINE–TRNA-LIGASE-RXN	0.003636625
Phosphoacetylglucosamine mutase	PHOSACETYLGLUCOSAMINEMUT-RXN	0.007259989
Mannose-6-phosphate isomerase	MANNPISOM-RXN	0.05335639
Ferrochelatase	PROTOHEMEFERROCHELAT-RXN	0.07540837
Guanylate cyclase	GUANYLCYC-RXN	0.04010272
Pseudouridylate synthase	PSEUDOURIDYLATE-SYNTHASE-RXN	0.01381947
GDP-mannose 4,6-dehydratase	GDPMANDEHYDRA-RXN	0.09140153
1-Phosphatidylinositol-4,5-bisphosphate phosphodiesterase	3.1.4.11-RXN	8.875228e–06
Inositol-1,4,5-trisphosphate 5-phosphatase	3.1.3.56-RXN	0.006815353
Pyruvate, water dikinase	RXN0-308	0.09261214
Pantetheine-phosphate adenylyltransferase	PANTEPADENYLYLTRAN-RXN	0.0004955335
Adenylyltransferase	FADSYN-RXN	0.03872114
Mannose-1-phosphate guanylyltransferase	2.7.7.13-RXN	0.00829316

(continued)

Table 4
(continued)

Reaction common name	Unique ID in PlasmoCyc	Wilcoxon *P* value
Ribose-phosphate diphosphokinase	PRPPSYN-RXN	0.00250444
Pyruvate kinase	PEPDEPHOS-RXN	0.06123981
Ethanolamine kinase	ETHANOLAMINE-KINASE-RXN	0.08042361
Choline kinase	CHOLINE-KINASE-RXN	0.007112016
6-Phosphofructokinase	6PFRUCTPHOS-RXN	0.005852473
Diphosphate–fructose-6-phosphate 1-phosphotransferase	2.7.1.90-RXN	0.005852473
Glutathione transferase	GST-RXN	0.002898356
Glutathione transferase	GSHTRAN-RXN	0.002898356
Farnesyltranstransferase	FARNESYLTRANSTRANSFERASE-RXN	0.0002514648
Protein farnesyltranstransferase	2.5.1.58-RXN	0.0002514648
Formate C-acetyltransferase	RXN-1381	0.09165743
Histone acetyltransferase	HISTONE-ACETYLTRANSFERASE-RXN	0.08575171
Glycylpeptide *N*-tetradecanoyltransferase	2.3.1.97-RXN	0.002914033
Aminomethyltransferase	GCVT-RXN	0.002914033
Site-specific DNA-methyltransferase (cytosine-specific)	2.1.1.73-RXN	0.007795191
Cytochrome-b5 reductase	CYTOCHROME-B5-REDUCTASE-RXN	0.001432610
Sarcosine dehydrogenase	SARCOSINE-DEHYDROGENASE-RXN	0.002914033
Dimethylglycine dehydrogenase	DIMETHYLGLYCINE-DEHYDROGENASE-RXN	0.002914033
Pyridoxamine-phosphate oxidase	PMPOXI-RXN	0.003350211
Protoporphyrinogen oxidase	PROTOPORGENOXI-RXN	0.07529077
Pyruvate dehydrogenase (lipoamide)	RXN0-1134	0.05377831
Ferredoxin–NADP(+) reductase	FLAVONADPREDUCT-RXN	0.05004103
Ferredoxin–NADP(+) reductase	1.18.1.2-RXN	0.05004103
None	GDPREDUCT-RXN	0.08158513
None	CDPREDUCT-RXN	0.08158513
None	ADPREDUCT-RXN	0.08158513
L-Lactate dehydrogenase	L-LACTATE-DEHYDROGENASE-RXN	0.00250444

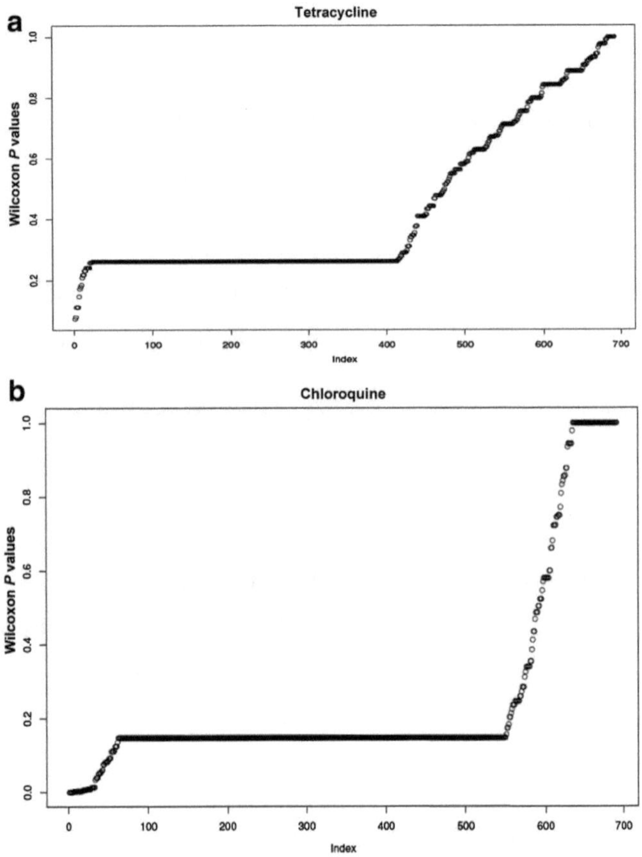

Fig. 2 Distribution of the sorted version of the *P* values for all reactions of the parasite for the tetracycline treatment (**a**) and the chloroquine treatment (**b**) compared with control. Each gene indexed is plotted on the *x*-axis and its corresponding Wilcoxon *P* value is plotted on the *y*-axis

the ability of the malaria parasite to resist these drugs. Figure 2 shows the distributions of the sorted versions of the *P* values for all reactions of the parasite for the tetracycline vs. control condition (A) and the chloroquine vs. control condition (B). Based on these findings, we listed all reactions whose *P* values are ≤0.25 for tetracycline and ≤0.1 for chloroquine.

Data in Table 1 suggested the following. Although a number of the genes in Table 1 are conserved 0.01004454 protein of unknown function, we were able to get important interpretation of the kind of results deducible from Table 1 via PFI0990. The gene PFI0990 is said to interact with the following genes: PF08_0026 (conserved *Plasmodium* protein of unknown function), PFL1385C (a merozoite surface protein 9), and PFL1315W (a potassium channel protein). It was found that these genes are inhibited PFI0990 (www.plasmodb.org), which is heavily expressed (by our results in Table 1) under tetracycline treatment compared with its normal expression in the absence of tetracycline treatment. This means that these genes must have been silenced for PFI0990 to be heavily expressed. First, PFL1385C (coding for a merozoite surface protein 9) confirm this statement by Dahl et al. (25): "Our results demonstrate that tetracyclines specifically block expression of the apicoplast genome, resulting in the distribution of nonfunctional apicoplasts into daughter merozoites." And second, it is known potassium channels are found in most cell types and control a wide variety of cell functions. Therefore, the inhibition of PFL1315W looks to have contributed to the negative effect of tetracycline on the parasite. We also found PF10_0061 (an apical membrane antigen 1) to be heavily expressed under tetracycline treatment compared with its normal expression in the absence of tetracycline treatment. Knowing the genes it interacts with can give us more insight into the biological mode of action of tetracycline.

In Table 2, little is known of the genes therein, interacted with. Information on gene PF07_0056 obtained from plasmoDB also gives us further information that can deduced from Table 2, if more information about the genes therein are available, PF07_0056, which is heavily expressed under chloroquine treatment, activates MAL8P1.23 which in turn activates PFF1300w (a pyruvate kinase). It is known that the enzyme pyruvate kinase affects the survival of red blood cells. In our prediction, via the chloroquine treatment, positively.

We did not find any significant differential expression between any clusters in the chloroquine SAGE and control samples. Le Roch also reached this conclusion (personal communication).

From the chloroquine microarray data (obtained using the combinatorial technique based on the Kernighan–Li clustering technique; second extracted subgraph), confirmed using the wavelets technique based on the consecutive-ones clustering technique (eighth extracted subgraph), we observed that tryptophanyl-tRNA synthetase production in the apicoplast is upregulated. Wellems and Plowe (42) state that "chloroquine's efficacy is thought to lie in its ability to interrupt hematin detoxification in malaria parasites as they grow within their host's red blood cells. Hematin is released in large amounts as the parasite consumes and digests hemoglobin in its digestive food vacuole. Hematin normally is detoxified by polymerization into innocuous crystals of hemozoin pigment and perhaps also by a glutathione-mediated process of destruction. Chloroquine binds with hematin in its μ-oxodimer form and also adsorbs to the growing faces of the hemozoin crystal, disrupting detoxification and poisoning the parasite. Chloroquine-resistance *P. falciparum* survives by reducing accumulation of the drug in the digestive vacuole; however, the mechanism by which this happens has not been determined. Leading proposals include mechanisms that involve alterations of digestive vacuole pH or changes in the flux of chloroquine across the parasite's cytoplasmic or digestive vacuole membrane." The second mechanism of flux of chloroquine was summarized by Krogstad et al. (43), who write that "... chloroquine-resistance *P. falciparum* accumulates less chloroquine than susceptible parasites. This observation suggests that chloroquine resistance in *P. falciparum* results from either decreased uptake or increased excretion of the drug by the resistant parasite ... resistance *P. falciparum* parasites have a mechanism for releasing chloroquine (an efflux process) (44). This efflux is either absent or greatly reduced in the susceptible parasite."

Therefore, that tryptophanyl-tRNA synthetase production in the apicoplast is upregulated (in the chloroquine-induced microarray data) may suggest that this efflux process was made possible (caused) by the apicoplast, the mini bacterium living inside the malaria parasite. Ralph et al. (45) state that "it is not yet clear what the key function of the apicoplast is but the organelle is clearly indispensable. Curiously though, parasites cured of their apicoplasts do not die immediately. Rather, they fail to invade new host cells successfully. This suggests that apicoplasts provide some component essential to invasion and or [sic] establishment of the parasitophorous vacuole in the host cell" (46, 47).

Thus a combination of chloroquine with the agents that cured *P. falciparum* of its apicoplast may be helpful in preventing the parasite from invading new host cells, and this combination may also kill the parasite, because it could not then flux out accumulated chloroquine in its digestive food vacuole.

Analyzing the two sets of microarray data together here provides the opportunity to identify reactions that may be upregulated via treatment with both drugs. In this line, we found the following reactions: FARNESYLTRANSTRANSFERASE-RXN, TRYPTOPHAN-TRNA-LIGASE, THREONINE-TRNA-LIGASE-RXN, and ALANINE-TRNA-LIGASE-RXN. These reactions appear very important in the parasite quest to resist the two antimalaria drugs we have considered in this paper (tetracycline and chloroquine). Our study represents the first attempt to unveil this.

We also observed that many (19 of 22) of the enzymes encoded by the genes active in the pathway targeted by chloroquine have not been identified. We are following these leads and we believe that further findings will be possible when such information is available.

4 Second Model: In Silico Model to Combat Resistance

In the second model, we extended our algorithm (48, 49) using a machine learning approach. The resulting algorithm is able to identify novel combinable drug targets from the metabolic network of *P. falciparum*. Using this approach we identified, among others, 19 drug targets confirmed from the literature. The machine learning approach combines clustering, common distance similarity measurements, and hierarchical clustering to propose new combinations of drug targets, see details in Fatumo et al. (50). Our result suggests that two or more enzymatic reactions from our list of drug targets that span across pathways could be combined to form an efficient malaria drug target, targeting distinct time points in the parasite's intraerythrocytic developmental cycle.

The metabolic network of *P. falciparum* was set up using the BioCyc database (http://biocyc.org) as described recently for *E. coli* (23). The metabolites were the nodes and the enzymatic reactions were the edges of the network. Our network yielded 554 metabolites and 575 reactions. Each compound can be substrate and product.

We set up a graph-based algorithm analyzing the structure of biochemical networks to infer differences (such as different paths) when exposed to changing nutrients and environmental conditions. Raymond and Segrè (51) showed that the access for metabolites changes drastically when oxygen is available. Following this strategy, we chose several sets of metabolites as sets of products. Then the investigated reaction was deleted from the network. The mutated network (the network with the deleted reaction) was investigated to determine whether the chosen products in each set could still be produced. We compared the number of products that could be produced in the wild-type network and the mutated network. The difference in the numbers gave an insight into whether the investigated reaction is essential or not.

4.1 Verifying the Essentiality of a Knockout Reaction

The algorithm investigates a reaction by deleting the reaction from the metabolic network and checking whether a chosen product can be produced without the deleted reaction.

4.2 Creating the Variety of Products

We assigned a list of all reactions in the neighborhood of compounds of the reaction under investigation. Thirty percent of all compounds of these reactions were set as a product to be produced by the remaining compounds. A total of 1,000 different combinations of the chosen product were assembled.

4.3 Minimizing the Number of Reactants and Reactions to Produce the Products

The algorithm checked every investigated reaction for a minimum number of needed reactions and reactants needed to produce the products. A "greedy" approach was employed for minimizing the number of reactants and reactions needed to produce the products.

4.4 Comparing the Results of Wild-Type and the Mutated Network to Obtain the Essentiality of the Investigated Reaction

We computed the average minimum sets of substrates for a knockout reaction in the mutated network vs. wild-type. Similarly, we computed the average minimum sets of reactions. We then compared the number of successful productions for the wild-type and knockout reactions. A total of 1,000 different sets of products were used.

4.5 Gene Expression Analysis

We identified 46 essential enzymatic reactions as reported by our algorithm. We used GENESIS (52), a sophisticated tool for analyzing gene expression data including clustering techniques, motif search, and visualization utilities, to analyze the essential reactions. Our DNA microarray data, which were obtained from Bozdech et al. (53) with 48 individual 1 h time points from the intraerythrocytic developmental cycle of *P. falciparum*, were organized by hierarchical clustering.

We clustered the 46 expressed genes into 6 groups according to their expression levels. Groups 1 and 2 had 4 enzymatic reactions, group 3 had 14 reactions, group 4 had 8 reactions, group 5 had 10 reactions, and group 6 had 6 reactions. We noticed that all the reactions in group 3 are responsible for transport and all coded for one gene; two reactions in group 6 also coded for the same gene. This left us with only 30 essential reactions as possible targets.

4.6 Comparative Screening Analysis of Possible Drug Targets

We found from the drug banks SIGMA and TDR targets for inhibitors or drugs for most of the possible drug targets we identified. We have at least one inhibitor/drug for 19 possible drug targets. We further did gene expression analysis of the 19 possible drug targets to determine whether two or more enzymatic reactions in the initial groups overlap with the new groups. It seems reasonable to combine the inhibitors for such possible drug targets because the resulting drug might attack the parasite at the same time point in its life cycle during its stay in the human red blood cells.

We clustered the gene expression data analysis using GENESIS (50). This analysis resulted in two new groups, with the initial groups 1, 5, and 6 now belonging to new group A and the initial groups 2 and 4 now belonging to group B. We hypothesize that it is beneficial to combine inhibitors/drugs for targets within each group.

5 Conclusion

With the first in silico model, we were able to use the biochemical network of *P. falciparum* to deduce its drug resistance mechanism(s) using two sets of gene expression data obtained from treatment of the parasite with chloroquine and tetracycline. Our work is the first to develop and apply computational means toward the elucidation of these mechanisms in *P. falciparum*. Our work suggests viable mechanisms for the resistance of the malaria parasite to chloroquine and tetracycline. When these results are experimentally tested they may provide useful weapons to efficiently cleanse malaria parasites from the blood stream.

With the second in silico model, we established a machine learning tool that identified drug targets confirmed from the literature, which we then further analyzed using a sophisticated gene expression analysis tool. Our data were organized using common distance similarity measurements and hierarchical clustering. Our results suggest that two or more enzymatic reactions from the list of our drug targets, which span about ten pathways, could be combinable if targeted at distinct pathways to produce an efficient malaria drug.

Acknowledgments

Many thanks go to Karine Le Roch, Svetlana Bulashesva, Benedikt Brors, Gunnar Schramm, Anna-Lena Kranz, Roland Eils, and Rainer Koenig for many useful discussions and contributions.

References

1. Pierotti MA, Tamborini E, Negri T et al (2011) Targeted therapy in GIST: in silico modeling for prediction of resistance. Nat Rev Clin Oncol 8:161–170. doi:10.1038/nrclinonc.2011.3
2. Noble D (2002) Modelling the heart – from genes to cells to the whole organ. Science 295(5560):1678–1682
3. Hammer GL, Sinclair TR, Chapman SC, Oosterom EV (2004) Scientific correspondence on systems thinking, systems biology and the in silico plant. Plant Physiol 134:909–911
4. Deville Y, Gilbert D, Helden JV, Wodak SJ (2003) An overview of data models for the analysis of biochemical pathways. Brief Bioinform 4(3):246–259
5. Crampin EJ, Schnell S (2004) New approaches to modelling and analysis of biochemical reactions, pathways and networks. Prog Biophys Mol Biol 86(1):1–4
6. Noble D, Rudy Y (2001) Models of cardiac ventricular action potentials: iterative interaction between experiment and simulation. Phil Trans R Soc Lond A 359:1127–1142
7. Van't Veer LJ, Dai H, van de Vijver MJ et al (2002) Gene expression profiling predicts clinical outcome of breast cancer. Nature 415:530–536
8. Stephanopoulos G, Hwang D, Schmitt WA, Mistra J (2002) Mapping physiological states from microarray expression measurements. Bioinformatics 18:1054–1063
9. Gasch AP, Spellman PT, Kao CM et al (2000) Genomic expression programs in the response of yeast cells to environmental changes. Mol Biol Cell 11:4241–4257
10. Spellman PT, Sherlock G, Zhang MQ et al (1998) Comprehensive identification of cell cycle-regulated genes of the yeast *Saccharomyces cerevisiae* by microarray hybridisation. Mol Biol Cell 9:3273–3297
11. Gardner TS, di Bernardo D, Lorenz D, Collins JJ (2003) Inferring genetic networks and identifying compound mode of action via expression profiling. Science 301:102–105
12. Berg JM, Tymoczko JL, Stryer L (2002) Biochemistry, 5th edn. W.H. Freeman, New York, p 1050
13. Karp PD, Riley M, Pellegrini-Toole A (2002) The MetaCyc database. Nucleic Acids Res 30:59–61
14. Khodursky AB, Peter BJ, Cozzarelli NR et al (2000) DNA microarray analysis of gene expression in response to physiological and genetic changes that affect tryptophan metabolism in *Escherichia coli*. Proc Natl Acad Sci USA 97:12170–12175
15. Neidhardt FC (1996) *Escherichia coli* and Salmonella: cellular and molecular biology. American Society for Microbiology, Washington, DC
16. Uetz P, Giot L, Cagney G et al (2000) A comprehensive analysis of protein-protein interactions in *Saccharomyces cerevisiae*. Nature 403:623–627
17. Ideker T, Ozier O, Schwikowski B, Siegel AF (2002) Discovering regulatory and signalling circuits in molecular interaction networks. Bioinformatics l8(Suppl 1):S233–S240
18. Hanisch D, Zien A, Zimmer R, Lengauer T (2002) Co-clustering of biological networks and gene expression data. Bioinformatics 8(Suppl 1):S145–S154
19. Zien A, Küffner R, Zimmer R, Lengauer T (2000) Analysis of gene expression data with pathway scores. Proc Int Conf Intell Syst Mol Biol 8:407–417
20. Karp PD, Ouzounis CA, Moore-Kochlacs C et al (2005) Expansion of the BioCyc collection of pathway/genome databases to 160 genomes. Nucleic Acids Res 33:6083–6089

21. Francke C, Siezen RJ, Teusink B (2005) Reconstructing the metabolic network of a bacterium from its genome. Trends Microbiol 13(11):550–558
22. König R, Eils R (2004) Gene expression analysis on biochemical networks using the Potts spin model. Bioinformatics 20:1500–1505
23. König R, Schramm G, Oswald M et al (2006) Discovering functional gene expression pattern in the metabolic network of *Escherichia coli* with wavelets transforms. BMC Bioinformatics 7:119
24. Gunasekera AM, Patankar S, Schug J et al (2003) Drug-induced alterations in gene expression of the asexual blood forms of *Plasmodium falciparum*. Mol Microbiol 50(4):1229–1239
25. Dahl EL, Shock JL, Shenai BR et al (2006) Tetracyclines specifically target the apicoplast of the malaria parasite *Plasmodium falciparum*. Antimicrob Agents Chemother 50(9): 3124–3131
26. Booth KS, Lueker GS (1976) Testing for the consecutive ones property, interval graphs, and graph planarity using *PQ*-Tree algorithms. J Comput Syst Sci 13:335–379
27. Cohen J (1988) Statistical power analysis for the behavioral sciences, 2nd edn. Lawrence Erlbaum Associates, Hillsdale, NJ
28. Gunasekera AM, Myrick A, Le Roch K et al (2007) *Plasmodium falciparum*: genome wide perturbations in transcript profiles among mixed stage cultures after chloroquine treatment. Exp Parasitol 117:87–92
29. Trager W, Jensen JB (1976) Human malaria parasites in continuous culture. Science 193: 673–675
30. Le Roch KG, Zhou Y, Blair PL et al (2003) Discovery of gene function by expression profiling of the malaria parasite life cycle. Science 301:1503–1508
31. Zhou Y, Abagyan R (2002) Match-only integral distribution (MOID) algorithm for high-density oligonucleotide array analysis. BMC Bioinformatics 3:3
32. Huber W, von Heydebreck A, Sultmann H et al (2002) Variance stabilization applied to microarray data calibration and to the quantification of differential expression. Bioinformatics 18(Suppl 1):S96–S104
33. Waller RF, Reed MB, Cowman AF, McFadden GI (2000) Protein trafficking to the plastid of *Plasmodium falciparum* is via the secretory pathway. EMBO J 19:1794–1802
34. van Dooren GG, Marti M, Tonkin CJ et al (2005) Development of the endoplasmic reticulum, mitochondrion and apicoplast during the asexual life cycle of *Plasmodium falciparum*. Mol Microbiol 57:405–419
35. Bozdech Z, Zhu J, Joachimiak MP et al (2003) Expression of the schizont and trophozoite stages of *Plasmodium falciparum* with a long-oligonucleotide microarray. Genome Biol 4:R9
36. Bozdech Z, Llinas M, Pulliam BL et al (2003) The transcriptome of the intraerythrocytic developmental cycle of *Plasmodium falciparum*. PLoS Biol 1:E5
37. Brown DE, Huntley CL (1992) A practical application of simulated annealing to clustering. Pattern Recog 25:401–412
38. Kernighan BW, Lin S (1970) An efficient heuristic procedure for partitioning graphs. Bell Syst Tech J 49:291–307
39. Dutt S (1993) New faster Kernighan-Lin type graph-partitioning algorithms. In: Proceedings of the 1993 IEEE/ACM international conference on computer-aided design, Santa Clara, CA, pp 370–377
40. Aarts EHL, van Laarhoven PJM (1985) A new polynomial time cooling schedule. In: Proceedings of the IEEE international conference on computer-aided design, Santa Clara, CA, pp 206–208
41. Whites SR (1984) Concepts of scale in simulated annealing. In: Proceedings of the IEEE international conference on computer-aided design, Port Chester, NY, pp 646–651
42. Wellems T, Plowe CW (2001) Chloroquine-resistant malaria. J Infect Dis 184:770–776
43. Krogstad DJ, Schlesinger PH, Herwaldt BL (1988) Antimalarial agents: mechanism of chloroquine resistance. Antimicrob Agents Chemother 32:799–801
44. Krogstad DJ, Gluzman IY, Kyle DE et al (1987) Efflux of chloroquine from *Plasmodium falciparum*: mechanism of chloroquine resistance. Science 238:1283–1285
45. Ralph SA, D'Ombrain MC, McFadden GI (2001) The apicoplast as an antimalarial drug target. Drug Resist Updat 4:145–151
46. Fichera ME, Roos DS (1997) A plastic organelle as a drug target in apicomplexan parasites. Nature 390:407–409
47. He CY, Shaw MK, Pletcher CH et al (2001) A plastic segregation defect in the protozoan parasite *Toxomplasma gondii*. EMBO J 20: 330–339
48. Fatumo S, Plaimas K, Mallm JP et al (2009) Estimating novel potential drug targets of *Plasmodium falciparum* by analysing the metabolic network of knock-out strains in silico. Infect Genet Evol 9(3):351–358
49. Fatumo S, Plaimas K, Adebiyi E, König R (2011) Comparing metabolic network models based on genomic and automatically inferred enzyme information from Plasmodium and its human host to define drug targets in silico. Infect Genet Evol 11(4):708–715
50. Fatumo S, Adebiyi E, Schramm G et al (2009) An *in-silico* approach to design efficient malaria drug targets to combat the malaria resistance

problem. Presented at the Computer Science and Information Technology Spring Conference, Singapore, 17–20 Apr 2009. http://ieeexplore.ieee.org/xpls/abs_all.jsp?arnumber=5169419&tag=1

51. Raymond J, Segrè D (2006) The effect of oxygen on biochemical network and their evolution of complex life. Science 311:1764–1767

52. Sturn A, Quackenbush J, Trajanoski Z (2003) Client-server environment for high-performance gene expression data analysis. Bioinformatics 19:772–773

53. Bozdech Z, Llinas M, Pulliam BL et al (2003) The transcriptome of the intraerythrocytic developmental cycle of *Plasmodium falciparum*. PLoS Biol 1:85–100

Chapter 5

An In Silico Model for Interpreting Polypharmacology in Drug–Target Networks

Ichigaku Takigawa, Koji Tsuda, and Hiroshi Mamitsuka

Abstract

Recent analysis on polypharmacology leads to the idea that only small fragments of drugs and targets are a key to understanding their interactions forming polypharmacology. This idea motivates us to build an in silico approach of finding significant substructure patterns from drug–target (molecular graph–amino acid sequence) pairs. This article introduces an efficient in silico method for enumerating, from given drug–target pairs, all frequent subgraph–subsequence pairs, which can then be further examined by hypothesis testing for statistical significance. Unique features of the method are its scalability, computational efficiency, and technical soundness in terms of computer science and statistics. The presented method was applied to 11,219 drug–target pairs in DrugBank to obtain significant substructure pairs, which can divide most of the original 11,219 pairs into eight highly exclusive clusters, implying that the obtained substructure pairs are indispensable components for interpreting polypharmacology.

Key words Frequent pattern mining, Graphs, Strings, Likelihood-ratio test, Polypharmacology, Drug–target networks

1 Introduction

Polypharmacology (or drug promiscuity) is a recently emerging concept in drug–target interactions, due to mainly the following three reasons: (1) multi-targeted drugs have been clinically successful, particularly as dual or multiplex kinase inhibitors (1), (2) a lot of approved drugs are not necessarily so selective (2), where a typical example is cancer drugs such as Gleevec (imatinib) and Sutent (sunitinib) which can bind to multiple kinases (3), (3) network science, particularly scale-freeness of drug–target networks imply the robustness of biological systems (4, 5), by which dysfunction of only a single protein can be in most cases compensated, indicating that inhibiting a single target would be therapeutically insufficient (6).

Recent analysis suggests that targets of promiscuous drugs cannot necessarily be similar to each other (2, 7), meaning that

Sandhya Kortagere (ed.), *In Silico Models for Drug Discovery*, Methods in Molecular Biology, vol. 993,
DOI 10.1007/978-1-62703-342-8_5,

only a small part of each target might be connected to the principle behind polypharmacology. Furthermore recent research shows that smaller drugs in molecular weight are likely to be more promiscuous (5), suggesting that only small fragments in each ligand would be related to drug promiscuity. They have brought us a hypothesis that fragments in drug–target pairs, or paired fragments, must be important factors behind polypharmacology. Thus, naturally an in silico approach for analyzing polypharmacology based on this hypothesis is to use molecular graphs for drugs (or chemical compounds) and amino acid sequences for targets (or proteins) and examine paired fragments (or substructures) in molecular graphs and amino acid sequences of drug–target pairs (8). We introduce are a data-driven approach for mining substructure pairs which are significantly shared in currently available drug–target (graph–sequence) pairs. A unique feature of this approach is scalability and efficiency for covering all possible substructure (subgraph–subsequence) pairs which significantly co-occur in given drug–target pairs. Furthermore, in (8), obtained significant substructure pairs were used for clustering current drug–target interactions into eight classes, which are highly exclusive each other, implying that each cluster corresponds to one unique type of promiscuous drugs (or targets) forming polypharmacology.

2 Materials

The “small molecules” dataset of DrugBank (9) (version 2.5 as of January 29, 2009), a standard database on drug information, contains 11,219 drug–target interactions, which are the input as interacting pairs, including 4,191 compounds which were linked to 4,362 targets. On the other hand, noninteracting pairs are all possible combinations from 4,191 compounds and 4,362 targets except the 11,219 drug–target pairs (see Note 1). In drug–target pairs, 1,447 (34.5%) out of 4,191 drugs were promiscuous drugs, i.e., each with at least two targets, and this ratio was consistent with 35% in (7). These promiscuous drugs were involved with 8,475 interactions (75.5% of all 11,219 drug–target pairs) and 171,029 interaction pairs. Drugs are treated as molecular graphs (see Note 2) and targets are represented by amino acid sequences.

3 Methods

The main input of the method is drug–target (or compound–protein) pairs, which are turned into graph–sequence pairs in the method. The method tries to find subgraph–subsequence pairs (see Note 3), which significantly occur in given drug–target pairs, comparing with noninteracting pairs. This method has two steps: (1) all subgraph–subsequence pairs frequently occurring in drug–target pairs are enumerated, and (2) the significance of a frequent subgraph–subsequence pair is evaluated. We describe the above Steps 1 and 2 in Subheadings 3.1 and 3.2, respectively. Note that step 1 corresponds to the entire procedure of the method, and in steps 1 and 2 is performed every time a frequent subgraph–subsequence pair is obtained. Note that step 2 uses both drug–target pairs and noninteracting pairs, while step 1 uses drug–target pairs only.

3.1 Mining Frequent Subgraph–Subsequence Pairs

3.1.1 Preliminaries

Given a dataset of graph–sequence pairs, we can count the number of graph–sequence pairs which contain a certain subgraph–subsequence pair. We call this number *support* of the corresponding subgraph–subsequence pair, following the literature of *frequent pattern mining* (10). When the support of a subgraph–subsequence pair is larger than or equal to a given threshold value, which is called *minimum support*, this pair is called a *frequent* subgraph–subsequence pair. That is, the support of a frequent subgraph–subsequence pair must be larger than or equal to the minimum support. We can further define *marginal support* of a subgraph as the number of given graph–sequence pairs which have this subgraph. The marginal support can be defined for subsequences as well.

The first, key idea for enumerating all frequent subgraph–subsequence pairs efficiently is the following property, which is called *downward closure*:

Proposition 1 (Downward closure)

A subgraph–subsequence pair is infrequent if this pair contains any smaller infrequent subgraph–subsequence pairs.

This property is powerful, because if you find an infrequent subgraph–subsequence pair, you do not have to search pairs with larger subgraphs and subsequences which include this pair. Then this idea naturally leads to a so-called *pattern-growth* approach, in which we can start with smallest subgraph–subsequence pairs and extend them to larger pairs, and if we come across an infrequent pair, then we can stop going on to larger pairs due to the downward closure property.

Without loss of generality, we explain this approach more, focusing on sequences only (rather than graph–sequence pairs). The pattern growth procedure naturally generates a hierarchy, which can be represented in a rooted ordered tree, called an *enumeration tree*. The enumeration tree is a rooted ordered spanning tree over all frequent subsequences, roughly with the following two features: (1) the null sequence is on the root, and sequences with only one letter are the children of the root, (2) each node

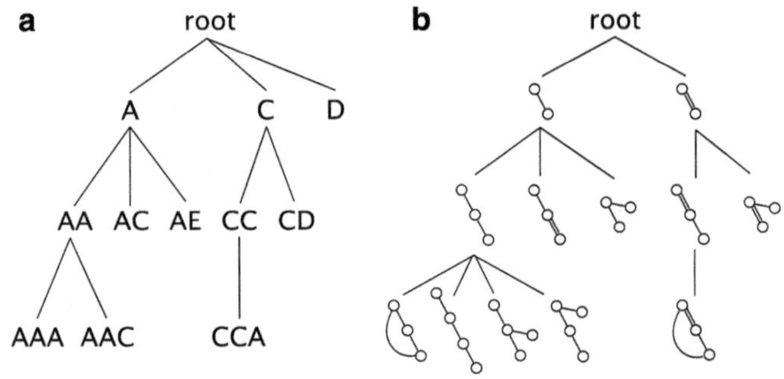

Fig. 1 Samples of enumeration trees for (**a**) frequent subsequences and for (**b**) frequent subgraphs

corresponds to a frequent subsequence in a one-to-one manner, where a subsequence with a larger number of letters is attached to nodes in a deeper level (see Note 4).

An important point of the enumeration tree is that we can enumerate all subsequences completely without any duplication by traversing an enumeration tree as a search space. This tree-shaped search space thus ensures the uniqueness of each subsequence attached to a node and the completeness on searching all frequent subsequences.

In fact, an enumeration tree can be generated by considering the following three points: (1) one node has only one parent node but can have more than one child node, (2) a subsequence of a child must be a larger but the minimum. For example, AC in Fig. 1a is a longer sequence but the minimum of longer sequences, (3) an order of sibling nodes is defined by using some criterion, by which for example, AC can be a child of A but cannot be a child of C in Fig. 1a. That is, A is prior to C, by which AC is generated from A, being faster than that AC is generated from C.

The enumeration tree can be generated for subgraphs in a similar manner, as shown in Fig. 1b, and these subgraphs are used in the next subsection (see Note 5).

We here define some notations which will be used in the next subsection. Let $\mathcal{Q}$ be given subgraph–subsequence pairs. Let $\mathcal{Q}_g$ and $\mathcal{Q}_s$ be subgraph–subsequence pairs containing subgraph g and subsequence s, respectively. Similarly let $\mathcal{Q}_{(g,s)}$ be subgraph–subsequence pairs containing both subgraph g and subsequence s. Let T_g and T_s be enumeration trees for subgraphs and subsequences, respectively. Here $\varnothing_g$ and $\varnothing_s$ indicate the root nodes of T_g and T_s, respectively. Similarly, parent (g) represents a subgraph of the parent of a node to which g is assigned in T_g, and parent(s) represents a subsequence of the parent of a node to which s is assigned in T_s. The support of a pair of subgraph g and subsequence s is denoted by support $((g,s)|\mathcal{Q}) = |\mathcal{Q}_{(g,s)}|$. Similarly the marginal support of subsequence s is denoted by support $((*,s)\,|\,\mathcal{Q})$. Let σ be the minimum support.

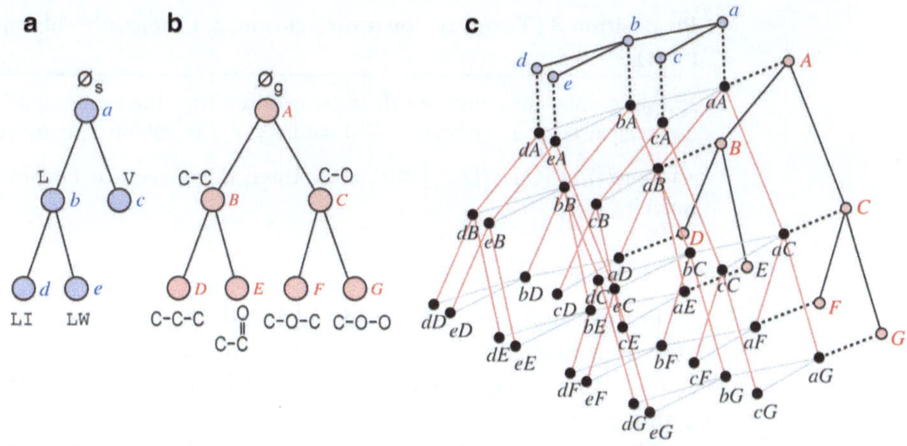

Fig. 2 An example of the search space. The search space (**c**) is defined as the graph product of two enumeration trees for subsequences (**a**) and for subgraphs (**b**). (**c**) Covers all possible frequent subgraph–subsequence pairs

3.1.2 Mining Algorithm

For mining frequent subgraph–subsequence pairs, we combine two enumeration trees, one for frequent subgraphs and the other for frequent subsequences. That is, all combinations of frequent subgraphs and subsequences can cover all frequent subgraph–subsequence pairs, and the search space for all these combinations can be defined by a product graph of the two enumeration trees. Fig. 2a, b show examples of enumeration trees for subsequences and subgraphs, respectively, which can be combined into Fig. 2c, where each subgraph–subsequence pair has two parent nodes, and hence this is no longer a tree. In this case, theoretically, we can compute the support of each subgraph–subsequence pair in a dynamic-programming manner.

Proposition 2 (Dynamic programming for subgraph–subsequence pairs)

$Q_{(g,s)}$ *can be iteratively computed as follows:*

1. $Q_{(g,s)} = Q_{(\mathrm{parent}(g),s)} \cap Q_{(g,\mathrm{parent}(s))}$.

2. $Q_{(g,\varnothing_s)} = \{(G,S) \in Q \mid g \in G\}$.

3. $Q_{(\varnothing_g,s)} = \{(G,S) \in Q \mid s \in S\}$.

In practice, all combinations of frequent subgraphs and frequent subsequences may have a lot of infrequent subgraph–subsequence pairs, and so we can use the downward closure property on the product graph of the two enumeration trees, which can be clearly stated as follows:

Proposition 3 (Two-way downward closure for subgraph–subsequence pairs)

If a subgraph–subsequence pair (g, s) is infrequent, then subgraph–subsequence pairs (g', s') (where $g \subseteq g'$ and $s \subseteq s'$) are all infrequent. Thus, if $\text{support}((g,s)|Q) = |Q_{(g,s)}| < \sigma$, then there is no need to extend (g, s) further.

For example, if (C–C,L) (at node bB) in Fig. 2c is infrequent, patterns at nodes (C–C–C,L), (C–C=O,L), (C–C,LI), (C–C,LW), (C–C–C,LI), (C–C–C,LW), (C–C=O,LI) and (C–C=O,LW) must be all infrequent.

The recursion rules in Proposition 2 make us keep all instances explicitly in the graph product $T_g \times T_s$. That is, $Q_{(g,s)}$ must be kept and be passed to subsequent nodes. This is a space-consuming procedure, because two enumeration trees are practically very huge. Thus, we can consider a depth-first traversal of the graph product $T_g \times T_s$ by simplifying recursion rules in Proposition 2 into those in the following Proposition 4 (see Note 6).

Proposition 4 (A simplified recursion rule)

The recursion rules in Proposition 2 can be simplified to

$$Q_{(g,s)} = Q_{(\text{parent}(g),s)} \cap \{(G,S) \in Q \mid g \in G\} \text{ and}$$
$$Q_{(\varnothing_g,s)} = \{(G,S) \in Q \mid s \in S\}.$$

We can obtain $Q_{(g,s)}$ efficiently by using Proposition 4 as follows: we can first traverse T_s until subsequence s is found, with computing marginal support, $\text{support}((*,s) \mid Q)$ (see Note 7). We can then have $Q_{(\varnothing_g,s)}$. We can further traverse T_g from node $(\varnothing_g, s)$, keeping $Q_{(\cdot,s)}$. Note here that in this traversal, we can reduce the size of $Q_{(\cdot,s)}$ by using the first rule: $Q_{(x,s)} = Q_{(\text{parent}(x),s)} \cap Q_{(x,\varnothing_s)}$, each time when parent($x$) is extended to x. In this way, we can trace $Q_{(g,s)}$' along with the path from $Q_{(g,\varnothing_s)}$ to $Q_{(g,s)}$. This procedure can be applied to subgraph g, since g and s are symmetric. In addition a larger enumeration tree should be examined first for this procedure, and in reality, we can examine T_s first, since practically T_s is expected to be larger than T_g in drug–target pairs. Finally we can present a pseudocode of the in silico mining method as follows:

Proposition 5 (Pseudocode for enumerating all frequent subgraph–subsequence pairs)

1. Compute $Q_{(g,\varnothing_s)}$ for all possible g using a frequent subgraph mining algorithm in terms of $\text{support}((g,*)\,|\,Q)$.
2. Start a frequent subsequence mining algorithm from $(\varnothing_g,\varnothing_s)$:

   For each $s \in T_s$ in a depth-first traversal order:

   For each $g \in T_g$ in a depth-first traversal order:

   (a) Continue if $g = \varnothing_g$ or $s = \varnothing_s$.

   (b) Reduce the size of $Q_{(g,s)}$ by $Q_{(g,s)} = Q_{(\text{parent}(g),s)} \cap Q_{(g,\varnothing_s)}$.

   (c) Compute $\text{support}((g,s)\,|\,Q) = |Q_{(g,s)}|$.

   (d) Break if $\text{support}((g,s)\,|\,Q) < \sigma$.

		A	B	C	D	E	F	G
		$\varnothing_g$	C-C	C-O	C-C-C	C(=O)-C	C-O-C	C-O-O
a	$\varnothing_s$	1 2 3 4 5 6 7 8 9 10	1 2 3 4 5	4 5 6 7 8 9	1 2 3	3 4 5	4 5 6	6 7 8 9
b	L	4 5 6 7 8 10	4 5	4 5 6 7 8	none	4 5	4 5 6	6 7 8
c	V	7 8 9 10	none	7 8 9	none	none	none	7 8 9
d	LI	4 5 10	4 5	4 5	none	4 5	4 5	none
e	LW	6 7 10	none	6 7	none	none	6	6 7

Fig. 3 A sample for enumerating all subgraph–subsequence pairs with the support of 3 or larger of 10 graph–sequence pairs (1, 2,…, 10). This table corresponds to two enumeration trees of Fig. 2a, b

We can explain this algorithm more by using a toy sample shown in Fig. 3, which has 10 graph–sequence pairs numbered as 1, 2,…,10. Each cell of Fig. 3 shows graph–sequence pairs having the corresponding subgraph–subsequence pair, such as that only graph–sequence pairs 4 and 5 have (C–C, L). Edges of the two enumeration trees in Fig. 2 are also shown by curves at the outside of both rows and columns. The objective here is to find all frequent pairs colored in white: (C–O,L), (C–O,V), (C–O–C,L), (C–O–O,L), and (C–O–O,V).

We first build enumeration tree T_g, which corresponds to generating all subgraphs in the top row of Fig. 3, and then starts traversing enumeration tree T_s from the root. By traversing T_s in a depth-first manner, the first pattern to be found is (C–C,L). Since (C–C,L) is infrequent (i.e., $|Q_{(\text{C-C,L})}| < 3$), we do not have to proceed to subsequent (C–C–C,L) and (C–C=O,L). Then, the next subgraph–subsequence pair is (C–O,L), which turns out to be frequent. We then move on to (C–O–C,L). $Q_{(\text{C-O-C,L})}$ is obtained by $Q_{(\text{C-O-C,L})} = Q_{(\text{C-O,L})} \cap Q_{(\text{C-O-C},\varnothing_s)} = \{4,5,6\}$. Similarly, we can move to (C–O–O,L) where $Q_{(\text{C-O-O,L})} = Q_{(\text{C-O,L})} \cap Q_{(\text{C-O-C},\varnothing_s)} = \{6,7,8\}$. We now finished traversing all nodes of T_g for $L \in T_s$, and then we proceed to the next subgraph–subsequence pair (C–C,LI) by traversing T_s from L to LI. However, subsequent nodes, (C–C,LI), (C–O,LI), (C–C,LW), (C–O,LW), and (C–C,V) are all infrequent, and then the next frequent subgraph–subsequence pair becomes (C–O,V). Then subsequent (C–O–C,V) and (C–O–O,V) are examined in this order, and we can find that (C–O–C,V) is infrequent but (C–O–O,V) is frequent. Then we have no nodes to proceed in $T_g \times T_s$, and the procedure is terminated. Finally we obtain all five frequent patterns (C–O,L), (C–O–C,L), (C–O–O,L), (C–O,V), and (C–O–O,V) in this order.

3.2 Evaluating Significance of Subgraph–Subsequence Pairs

This statistical test is the same as that for detecting "epistasis" in genetics (11), called *likelihood ratio test with logistic regression*. We first explain this test, focusing on drug–target pairs, being followed by the method for maximizing the likelihood of logistic regression from given drug–target pairs.

3.2.1 Likelihood Ratio Test with Logistic Regression

Logistic regression can be defined as the probability p that an event occurs given d explanatory variables $x_1, x_2, \ldots, x_d$, as follows:

$$p = \text{Prob}\{\text{the event occurs} \mid x_1, x_2, \ldots, x_d\} = \frac{\exp(\eta)}{1+\exp(\eta)} = \frac{1}{1+\exp(-\eta)},$$

where $\eta = \theta_0 + \theta_1 x_1 + \cdots + \theta_d x_d$ is a (linear) composite variable. Note that p takes a value between zero and one due to the logistic function, even though η ranges from $-\infty$ to ∞. Note further that this equation can be transformed into $\log\left\{\frac{p}{1-p}\right\} = \eta$ where

$$\frac{p}{1-p} = \frac{\text{Prob}\{\text{the event occurs} \mid x_1, x_2, \ldots, x_d\}}{\text{Prob}\{\text{the event does not occur} \mid x_1, x_2, \ldots, x_d\}}.$$

Let $Y \in \{0,1\}$ be a binary response variable, where the probability of $Y=0$ (and $Y=1$) is modeled by

$$\text{Prob}\{Y = 0\} = 1 - p_\theta(\mathbf{X}) \text{ and } \text{Prob}\{Y = 1\} = p_\theta(\mathbf{X}),$$

$$p_\theta(\mathbf{X}) = \frac{\exp(\theta'\mathbf{Z})}{1+\exp(\theta'\mathbf{Z})},$$

where $\theta = (\theta_0, \theta_1, \ldots, \theta_d)'$ and $\mathbf{Z} = (1, \mathbf{X}')' = (1, X_1, X_2, \ldots, X_d)'$. To fit this model to n given drug–target pairs for Y and $\mathbf{X}$, $\{(y^{(1)}, \mathbf{x}^{(1)}), (y^{(2)}, \mathbf{x}^{(2)}), \ldots, (y^{(n)}, \mathbf{x}^{(n)})\}$ suffices to maximize the likelihood $\ell(\theta)$ in terms of parameters θ. The likelihood of n given pairs is defined by

$$\ell(\theta) := \prod_{i=1}^{n} p_\theta(\mathbf{x}^{(i)})^{y^{(i)}} (1 - p_\theta(\mathbf{x}^{(i)}))^{1-y^{(i)}}. \quad (1)$$

For subgraph–subsequence pair (g, s), we can consider two explanatory variables X_1 and X_2 for subgraph g and subsequence s, respectively, each taking one if a graph–target pair has the corresponding substructure; otherwise zero. We use two logistic regression models for Y where $Y = 1$ for drug–target pairs (and $Y = 0$ for noninteracting pairs), i.e., the probability that $Y = 1$:

$$p_{\theta:0-2}(X_1, X_2) = \frac{\exp(\eta)}{1 + \exp(\eta)} \text{ and}$$

$$p_{\theta:0-3}(X_1, X_2) = \frac{\exp(\eta + \theta_3 X_1 X_2)}{1 + \exp(\eta + \theta_3 X_1 X_2)},$$

where $\eta = \theta_0 + \theta_1 X_1 + \theta_2 X_2$. Note that the second model has interaction term $\theta_3 X_1 X_2$ while the first model has no interaction terms. Parameters of these two models are independently fitted by maximizing the likelihood. Then, the significance of pair (g, s) can be statistically measured by testing whether $\theta_3 = 0$ is kept or not. Note that this can be conducted by using the likelihood ratio test of two maximum likelihoods $\hat{L}_{\theta:0-2}$ for $p_{\theta:0-2}(X_1, X_2)$ and $\hat{L}_{\theta:0-3}$ for $p_{\theta:0-3}(X_1, X_2)$. The test statistic $-2\log(\hat{L}_{\theta:0-2} / \hat{L}_{\theta:0-3})$ follows the chi-squared distribution with one degree of freedom under the hypothesis that $\theta_3 = 0$. Thus, we can compute the p-value of the observed statistic from the chi-squared distribution.

3.2.2 Computing Likelihood Ratio Test Numerically

Given drug–target pairs, to maximize the likelihood (or fit the logistic regression model to given pairs), we can use the Newton–Raphson method, which is a typical and standard manner for parameter estimation of logistic regression. Explanatory variables X_1 and X_2, as well as response variable Y, are all binary ($Y \in \{0,1\}, X_1 \in \{0,1\}$ and $X_2 \in \{0,1\}$), and thus, drug–target pairs for (Y, X_1, X_2) have eight possible combinations only, as shown in Table 1a. Thus, only what we have to do is to count how many times each of the eight combinations occurs in given drug–target pairs.

In fact, we can compute the likelihood ratio test using logistic regression from the counts of eight possible combinations: P_{00}, P_{01}, P_{10}, P_{11}, N_{00}, N_{01}, N_{10}, and N_{11} in Table 1. Denoting $(X_1 = 0, X_2 = 0)$ by $\mathbf{x}_{00}$, the probability that $\mathbf{x}_{00}$ is observed P_{00} times (from observations

Table 1
Tables for counting eight values

Explanatory				Response	
	X_1	X_2	X_1X_2	$\#\{Y=1\}$	$\#\{Y=0\}$
(a) Model without the interaction term					
$\mathbf{x}_{00}$	0	0		P_{00}	N_{00}
$\mathbf{x}_{01}$	0	1		P_{01}	N_{01}
$\mathbf{x}_{10}$	1	0		P_{10}	N_{10}
$\mathbf{x}_{11}$	1	1		P_{11}	N_{11}
(b) Model with the interaction term					
$\mathbf{x}_{000}$	0	0	0	P_{00}	N_{00}
$\mathbf{x}_{010}$	0	1	0	P_{01}	N_{01}
$\mathbf{x}_{100}$	1	0	0	P_{10}	N_{10}
$\mathbf{x}_{111}$	1	1	1	P_{11}	N_{11}

with $Y=1$) and N_{00} times (from observations with $Y=0$) can be written as follows:

$$\overbrace{p(\mathbf{x}_{00})\times\cdots\times p(\mathbf{x}_{00})}^{P_{00}}\times\overbrace{(1-p(\mathbf{x}_{00}))\times\cdots\times(1-p(\mathbf{x}_{00}))}^{N_{00}}$$
$$= p(\mathbf{x}_{00})^{P_{00}}(1-p(\mathbf{x}_{00}))^{N_{00}},$$

and the entire likelihood $\ell(\theta)$ of Eq. 1 is implicitly given for binary variables Y, X_1, and X_2 by

$$p(\mathbf{x}_{00})^{P_{00}}(1-p(\mathbf{x}_{00}))^{N_{00}}\,p(\mathbf{x}_{01})^{P_{01}}(1-p(\mathbf{x}_{01}))^{N_{01}}$$
$$\times\, p(\mathbf{x}_{00})^{P_{10}}(1-p(\mathbf{x}_{00}))^{N_{10}}\,p(\mathbf{x}_{11})^{P_{11}}(1-p(\mathbf{x}_{00}))^{N_{11}}.$$

Thus, letting an index set be $\Lambda := \{00,01,10,11\}$, the log-likelihood can be written as follows:

$$L(\theta) := \log\ell(\theta) = \log\left\{\prod_{\lambda\in\Lambda} p(\mathbf{x}_\lambda)^{P_\lambda}(1-p(\mathbf{x}_\lambda))^{N_\lambda}\right\}$$
$$= \sum_{\lambda\in\Lambda}\{P_\lambda \log p(\mathbf{x}_\lambda) + N_\lambda \log(1-p(\mathbf{x}_\lambda))\}$$
$$= \sum_{\lambda\in\Lambda}\left\{P_\lambda \log\frac{p(\mathbf{x}_\lambda)}{1-p(\mathbf{x}_\lambda)} + (P_\lambda + N_\lambda)\log(1-p(\mathbf{x}_\lambda))\right\},$$

which means that we can compute the p-value of likelihood ratio test using logistic regression by using only P_{00}, P_{01}, P_{10}, P_{11}, N_{00},

N_{01}, N_{01}, N_{11}. To maximize $L(\theta)$ by changing θ, the Newton–Raphson method repeats the following update:

$$\theta^{[k+1]} \leftarrow \theta^{[k]} + (\nabla^2 L(\theta))^{-1} \nabla L(\theta) \text{ until } \left| \frac{L(\theta^{[k+1]}) - L(\theta^{[k]})}{L(\theta^{[k]})} \right| < \varepsilon,$$

where score $\nabla L(\theta) = \nabla \log \ell(\theta)$ and the Hessian matrix (asymptotic Fisher information matrix) are given as

$$\nabla L(\theta) = \begin{bmatrix} \partial L(\theta)/\partial\theta_0 \\ \partial L(\theta)/\partial\theta_1 \\ \partial L(\theta)/\partial\theta_2 \end{bmatrix} = \begin{bmatrix} \sum_{\lambda\in\Lambda}(P_\lambda - (P_\lambda + N_\lambda)p(\mathbf{x}_\lambda)) \\ \sum_{\lambda\in\Lambda}(x_1)_\lambda(P_\lambda - (P_\lambda + N_\lambda)p(\mathbf{x}_\lambda)) \\ \sum_{\lambda\in\Lambda}(x_2)_\lambda(P_\lambda - (P_\lambda + N_\lambda)p(\mathbf{x}_\lambda)) \end{bmatrix}$$

$$\nabla^2 L(\theta) = \begin{bmatrix} \frac{\partial^2 L(\theta)}{\partial\theta_0\theta_0} & \frac{\partial^2 L(\theta)}{\partial\theta_0\theta_1} & \frac{\partial^2 L(\theta)}{\partial\theta_0\theta_2} \\ \frac{\partial^2 L(\theta)}{\partial\theta_1\theta_0} & \frac{\partial^2 L(\theta)}{\partial\theta_1\theta_1} & \frac{\partial^2 L(\theta)}{\partial\theta_1\theta_2} \\ \frac{\partial^2 L(\theta)}{\partial\theta_2\theta_0} & \frac{\partial^2 L(\theta)}{\partial\theta_2\theta_1} & \frac{\partial^2 L(\theta)}{\partial\theta_2\theta_2} \end{bmatrix},$$

$$\frac{\partial^2 L(\theta)}{\partial\theta_i\theta_j} = \sum_{\lambda\in\Lambda}(P_\lambda + N_\lambda)(x_i)_\lambda(x_j)_\lambda\, p(\mathbf{x}_\lambda)(1 - p(\mathbf{x}_\lambda)).$$

Hence, the Newton–Raphson update can be written in a matrix form:

$$\theta^{[k+1]} \leftarrow \theta^{[k]} + (\mathbf{X}'W X)^{-1}\mathbf{X}'(\mathbf{y} - \mathbf{p}), \tag{2}$$

where when we use a logistic model $p_{\theta:0-2}(\cdot)$ for $p(\cdot)$,

$$\mathbf{X} := \begin{bmatrix} 1 & 0 & 0 \\ 1 & 0 & 1 \\ 1 & 1 & 0 \\ 1 & 1 & 1 \end{bmatrix}, \quad \mathbf{W} := \begin{bmatrix} d_{00} & 0 & 0 & 0 \\ 0 & d_{01} & 0 & 0 \\ 0 & 0 & d_{10} & 0 \\ 0 & 0 & 0 & d_{11} \end{bmatrix},$$

$$\mathbf{y} := \begin{bmatrix} P_{00} \\ P_{01} \\ P_{10} \\ P_{11} \end{bmatrix}, \quad \mathbf{p} := \begin{bmatrix} (P_{00} + N_{00})\, p_{\theta:0-2}(\mathbf{x}_{00}) \\ (P_{01} + N_{01})\, p_{\theta:0-2}(\mathbf{x}_{01}) \\ (P_{10} + N_{10})\, p_{\theta:0-2}(\mathbf{x}_{10}) \\ (P_{11} + N_{11})\, p_{\theta:0-2}(\mathbf{x}_{11}) \end{bmatrix},$$

$$d_{00} = (P_{00} + N_{00})\, p_{\theta:0-2}(\mathbf{x}_{00})(1 - p_{\theta:0-2}(\mathbf{x}_{00})),$$
$$d_{01} = (P_{01} + N_{01})\, p_{\theta:0-2}(\mathbf{x}_{01})(1 - p_{\theta:0-2}(\mathbf{x}_{01})),$$
$$d_{10} = (P_{10} + N_{10})\, p_{\theta:0-2}(\mathbf{x}_{10})(1 - p_{\theta:0-2}(\mathbf{x}_{10})),$$
$$d_{11} = (P_{11} + N_{11})\, p_{\theta:0-2}(\mathbf{x}_{11})(1 - p_{\theta:0-2}(\mathbf{x}_{11})),$$

and when we use a logistic model $p_{\theta:0-3}(\cdot)$ for $p(\cdot)$,

$$\mathbf{X} := \begin{bmatrix} 1 & 0 & 0 & 0 \\ 1 & 0 & 1 & 0 \\ 1 & 1 & 0 & 0 \\ 1 & 1 & 1 & 1 \end{bmatrix}, \quad \mathbf{W} := \begin{bmatrix} d_{000} & 0 & 0 & 0 \\ 0 & d_{010} & 0 & 0 \\ 0 & 0 & d_{100} & 0 \\ 0 & 0 & 0 & d_{111} \end{bmatrix},$$

$$\mathbf{y} := \begin{bmatrix} P_{00} \\ P_{01} \\ P_{10} \\ P_{11} \end{bmatrix}, \quad \mathbf{p} := \begin{bmatrix} (P_{00} + N_{00})\, p_{\theta:0-3}(\mathbf{x}_{000}) \\ (P_{01} + N_{01})\, p_{\theta:0-3}(\mathbf{x}_{010}) \\ (P_{10} + N_{10})\, p_{\theta:0-3}(\mathbf{x}_{100}) \\ (P_{11} + N_{11})\, p_{\theta:0-3}(\mathbf{x}_{111}) \end{bmatrix},$$

$$\begin{aligned} d_{000} &= (P_{00} + N_{00})\, p_{\theta:0-3}(\mathbf{x}_{000})(1 - p_{\theta:0-3}(\mathbf{x}_{000})), \\ d_{010} &= (P_{01} + N_{01})\, p_{\theta:0-3}(\mathbf{x}_{010})(1 - p_{\theta:0-3}(\mathbf{x}_{010})), \\ d_{100} &= (P_{10} + N_{10})\, p_{\theta:0-3}(\mathbf{x}_{100})(1 - p_{\theta:0-3}(\mathbf{x}_{100})), \\ d_{111} &= (P_{11} + N_{11})\, p_{\theta:0-3}(\mathbf{x}_{111})(1 - p_{\theta:0-3}(\mathbf{x}_{111})). \end{aligned}$$

The deviance of maximum likelihood $\hat{\theta}$ can be defined by

$$D := -2(L(\hat{\theta}) - L^*)$$

where L^* is the log-likelihood by the so-called full model (or saturated model) where probabilities can be given in the following:

$$p(\mathbf{x}_{00}) = \frac{P_{00}}{P_{00} + N_{00}}, \quad p(\mathbf{x}_{01}) = \frac{P_{01}}{P_{01} + N_{01}},$$

$$p(\mathbf{x}_{10}) = \frac{P_{10}}{P_{10} + N_{10}}, \quad p(\mathbf{x}_{11}) = \frac{P_{11}}{P_{11} + N_{11}},$$

resulting in that the log-likelihood L^* can be obtained as

$$L^* = \sum_{\lambda \in \Lambda} \left\{ P_\lambda \log \frac{P_\lambda}{P_\lambda + N_\lambda} + N_\lambda \log \frac{N_\lambda}{P_\lambda + N_\lambda} \right\}.$$

Finally the deviance thus can be written by

$$\begin{aligned} D(\hat{\theta}) &= 2(L^* - L(\hat{\theta})) \\ &= 2 \cdot \sum_{\lambda \in \Lambda} \left\{ \frac{P_\lambda \log P_\lambda}{(P_\lambda + N_\lambda) p_{\hat{\theta}}(\mathbf{x}_\lambda)} + \frac{N_\lambda \log N_\lambda}{(P_\lambda + N_\lambda)(1 - p_{\hat{\theta}}(\mathbf{x}_\lambda))} \right\}. \end{aligned}$$

4 Notes

1. Unknown drug–target pairs may be in noninteracting pairs. However, we think that they are statistically negligible, since the number of noninteracting pairs is huge.
2. 2D structures of drugs were converted into hydrogen-suppressed molecular graphs, where nodes are labeled with atom types except hydrogens and edges are labeled with bond types.
3. Drug substructures and target substructures mean connected subgraphs and consecutive subsequences, respectively.
4. The support of a subgraph–subsequence pair is monotonically decreasing with increasing the size of the subgraph or the subsequence, meaning that a subgraph on a deeper level in an enumeration tree has a smaller support.
5. In the literature of mining frequent subsequences (or subgraphs), there already exist established algorithms, such as the PrefixSpan algorithm (12) for frequent subsequences and the gSpan algorithm (13) for frequent subgraphs. Here the original PrefixSpan algorithm allows any size of gaps in subsequences, but we restrict to only consecutive subsequences. This is because input sequences are amino acid sequences, which are usually long and consist of only 20 amino acids, meaning that if we allow any size of gaps, small subsequences are likely to be frequent, by which mining subsequences in protein sequences will be infeasible.
6. This depth-first traversal, which is similar to the gSpan and PrefixSpan algorithms, gives a practically efficient algorithm.
7. We can use any algorithm for mining frequent subgraphs (subsequences) to compute the marginal support in Q with the traversal over T_g(and T_s). As mentioned above, for this purpose, we use gSpan and PrefixSpan for graphs and sequences, respectively.

References

1. Apsel B, Blair J, Gonzalez B, Nazif T, Feldman M, Aizenstein B, Hoffman R, Williams R, Shokat K, Knight Z (2008) Targeted polypharmacology: discovery of dual inhibitors of tyrosine and phosphoinositide kinases. Nat Chem Biol 4:691–699
2. Campillos M, Kuhn M, Gavin A, Jensen L, Bork P (2008) Drug target identification using side-effect similarity. Science 321: 263–266
3. Frantz S (2005) Drug discovery: playing dirty. Nature 437:942–943
4. Yildirim M, Goh K, Cusick M, Barabasi A, Vidal M (2007) Drug-target network. Nat Biotechnol 25:1119–1126
5. Morphy R, Rankovic Z (2007) Fragments, network biology and designing multiple ligands. Drug Discov Today 12:156–160
6. Hopkins A (2008) Network pharmacology: the next paradigm in drug discovery. Nat Chem Biol 4:682–690
7. Paolini G, Shapland R, van Hoorn W, Mason J, Hopkins A (2006) Global mapping of pharmacological space. Nat Biotechnol 24:805–815

8. Takigawa I, Tsuda K, Mamitsuka H (2011) Mining significant substructure pairs for interpreting polypharmacology in drug-target network. PLoS One 6(2):e16999
9. Wishart D, Knox C, Guo A, Cheng D, Shrivastava S, Tzur D, Gautam B, Hassanali M (2008) DrugBank: a comprehensive resource for in silico drug discovery and exploration. Nucleic Acids Res 36:D901–D906
10. Han J, Cheng H, Dong X, Yan X (2007) Frequent pattern mining: current status and future directions. Data Min Knowl Discov 15(1):55–86
11. Cordell H (2002) Epistasis: what it means, what it doesn't mean, and statistical methods to detect it in humans. Hum Mol Genet 11:2463–2468
12. Pei J, Han J, Mortazavi-Asl B, Wan J, Pinto H, Chen Q, Dayal U, Hsu M-C (2004) Mining sequential patterns by pattern-growth: the PrefixSpan approach. IEEE Trans Knowl Data Eng 16(11):1424–1440
13. Yan X, Han J (2002) gSpan: graph-based substructure pattern mining. In: IEEE International Conference on Data Mining (ICDM'02), Washington, DC, USA, 9-12, December pp. 721–724

Chapter 6

On Exploring Structure–Activity Relationships

Rajarshi Guha

Abstract

Understanding structure–activity relationships (SARs) for a given set of molecules allows one to rationally explore chemical space and develop a chemical series optimizing multiple physicochemical and biological properties simultaneously, for instance, improving potency, reducing toxicity, and ensuring sufficient bioavailability. In silico methods allow rapid and efficient characterization of SARs and facilitate building a variety of models to capture and encode one or more SARs, which can then be used to predict activities for new molecules. By coupling these methods with in silico modifications of structures, one can easily prioritize large screening decks or even generate new compounds de novo and ascertain whether they belong to the SAR being studied. Computational methods can provide a guide for the experienced user by integrating and summarizing large amounts of preexisting data to suggest useful structural modifications. This chapter highlights the different types of SAR modeling methods and how they support the task of exploring chemical space to elucidate and optimize SARs in a drug discovery setting. In addition to considering modeling algorithms, I briefly discuss how to use databases as a source of SAR data to inform and enhance the exploration of SAR trends. I also review common modeling techniques that are used to encode SARs, recent work in the area of structure–activity landscapes, the role of SAR databases, and alternative approaches to exploring SAR data that do not involve explicit model development.

Key words Structure–activity relationship, QSAR, Inverse QSAR, Structure–activity landscapes, Activity cliff, Structure–activity similarity maps, Structure–activity landscape index, Structure–activity index

1 Introduction

Structure–activity relationships (SARs) are key to many aspects of drug discovery, from primary screening to lead optimization. Working with SARs starts with identifying whether a SAR exists in a collection of molecules and their associated activities and involves elucidating the details of one or more such relationships and using that information to make structural modifications to optimize some property or activity. An understanding of the SAR for a set of molecules allows one to rationally explore chemical space, which in the absence of "sign posts" is essentially infinite (1). Invariably, the

Sandhya Kortagere (ed.), *In Silico Models for Drug Discovery*, Methods in Molecular Biology, vol. 993, DOI 10.1007/978-1-62703-342-8_6, © Springer Science+Business Media, LLC 2013 development of a chemical series involves optimizing multiple physicochemical and biological properties simultaneously (2–4). For example, most lead optimization projects will try to improve potency, reduce toxicity, and ensure sufficient bioavailability, among other properties. While the intuition and experience of a medicinal chemist are vital to these efforts, the data generated by modern high-throughput experimental techniques can overwhelm the capabilities of a single chemist. For example, in a primary high-throughput screen, one may be faced with hundreds of chemical series. How does one rapidly identify the most promising series among them? In these scenarios, in silico methods allow rapid and efficient characterization of SARs. These methods allow one to build a variety of models to capture and encode one or more SARs, which can then be used to predict activities for new molecules. By coupling these methods with in silico modifications of structures, one can easily prioritize large screening decks or even generate new compounds de novo and ascertain whether they belong to the SAR being studied. Computational methods do not *replace* medicinal chemistry domain knowledge; however, they can provide a guide to the experienced user by integrating and summarizing large amounts of preexisting data to suggest useful structural modifications.

Although computational methods can help in identifying, explaining, and predicting SARs, naive usage (or even misuse) of these techniques can lead to misleading results. Fundamentally, SAR models are just those models, reduced or simplified representations of reality, replete with assumptions and limitations. These methods cover the spectrum in terms of complexity and utility. This chapter highlights the different types of SAR modeling methods and how they support the task of exploring chemical space to elucidate and optimize SARs in a drug discovery setting. In addition to considering modeling algorithms, I briefly discuss how to use databases as a source of SAR data to inform and enhance the exploration of SAR trends. I also review common modeling techniques that are used to encode SARs, recent work in the area of structure–activity landscapes, the role of SAR databases, and some alternative approaches to exploring SAR data that do not involve explicit model development.

2 Capturing SARs

Over the past 60 years a multitude of means have been used to capture SARs. We can broadly divide them into two groups: those based on statistical or data mining methods (e.g., regression models) and those based on physical approaches (e.g., pharmacophore models). For a comprehensive review of quantitative structure–activity relationship (QSAR) methodologies the reader is referred to previous reviews (5–7). The choice of modeling technique can influence the extent and detail in which an SAR can be explored. For example, statistical QSAR approaches based on two-dimensional descriptors that ignore stereochemistry can miss key elements of an SAR that depend on chirality (8). Three-dimensional approaches, on the other hand, are generally more explicitly informative, in the sense that one can directly understand the nature of ligand–receptor interactions that underlie an observed SAR. Some three-dimensional approaches are more explicit than others, e.g., docking vs. comparative molecular field analysis. However, three-dimensional approaches are generally preferable when a crystal structure is available and when a few chemical series are being explored.

Much of traditional QSAR analysis is based on statistical models that link chemical structure (characterized by numerical descriptors) to biological activities. In some cases, the model makes distributional assumptions (linear regression), whereas others are "model-free" (9). In either case, one develops a model based on a training set of molecules. The model can then be used to predict the activity for new molecules. Although much of QSAR analysis has focused on various forms of linear regression (ranging from ordinary least-squares to more robust methods such as partial least-squares or ridge regression), there is no reason to assume, a priori, that the SAR is linear. Indeed, for most biological systems it is unreasonable to expect linear relationships, simply because multiple, complex processes occur in vivo. Thus, modern nonlinear methods such as neural networks and support vector machines have seen extensive use and tend to exhibit high accuracy.

However, building a predictive model is just the first step. For certain scenarios, such as virtual screening, one can apply the model and simply obtain numerical predictions of activity. However, the focus of this paper is the use of such predictive models for *exploring* SAR. Key to such exploration is the ability to interpret the model and understand how exactly it correlates activity to specific structural features (10, 11). For interpretive purposes, a model should be understandable in terms of both the descriptors used and the underlying model itself. The predictive ability of the model is not primary (though of course, for statistical models, the model should be statistically significant). Examples include linear regression and random forests. In this type of usage, we are interested in what the model can tell us about the effects of specific structural features on the observed activity. It is thus vital that this information can be teased out from the model. Obviously, for more physical methods such as pharmacophore modeling and docking, the interpretability is much more explicit. Clearly, SAR exploration benefits from models that can be dissected. Examples of such interpretive usages have been reported, for both simple models (12) and traditional black box models (13). However, purely predictive models can also be useful, especially to identify more or less active molecules from a

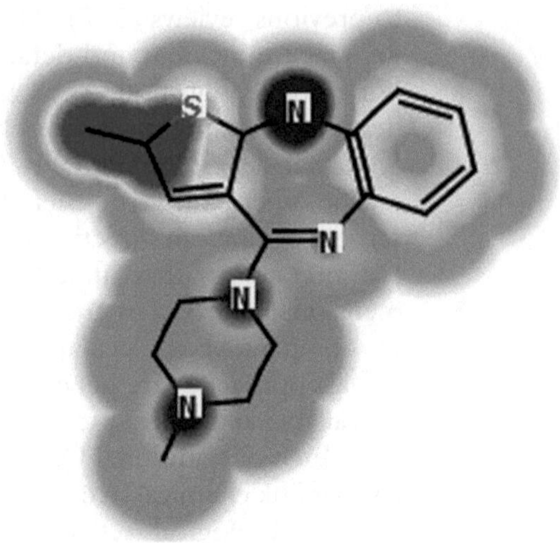

Fig. 1 Representation of a glowing molecule, developed by Optibrium. The *shading* corresponds to the influence of the structural feature on the predicted property (*darker* for a negative influence, *lighter* for a positive influence). Image modified, with permission, from http://www.optibrium.com/community/faq/glowing-molecule

large collection. In such a scenario, users could employ a predictive model to provide an initial ranking, allowing them to focus on a small subset using more interpretive methods.

Closely related to analytical interpretations of QSAR models is the ability to visualize the SAR trends encoded in a model. The "glowing molecule" representation developed by Segall et al. (14) is an example of direct visualization of a predictive model in terms of the actual chemical structure. Figure 1 shows such a representation, where the shading corresponds to the influence of that substructural feature on the predicted property. This type of visualization allows the user to directly understand how structural modifications at specific points will affect the property or activity being optimized.

Although capturing SAR trends in a predictive model and subsequently predicting properties for new molecules is useful, one can also consider the inverse approach, i.e., identifying structures that match a given activity or activity profile. Most formulations of this approach aim to derive a set of descriptor values rather than the structure directly. The challenge is in identifying valid structures from a set of descriptor values. A number of workers have addressed the inverse QSAR problem. Faulon et al. (15) employed the signature molecular descriptors to perform an inverse QSAR analysis of 121 HIV protease inhibitors and Churchwell et al. (16) employed these same descriptors to explore QSARs of peptides inhibiting intercellular adhesion molecule-1 (ICAM-1). Recently, Wong and Burkowski (17) have developed a novel descriptor to address inverse QSAR and coupled this to the kernel method to allow explicit mapping between points in the high-dimensional kernel space (i.e., candidate structures) and the original descriptor space and thence to a set of candidate molecules.

2.1 Is the Model Reliable?

While any statistical method or machine learning algorithm can be used to learn from SAR data and then to predict activities for new molecules, such predictions are not always reliable. These types of approaches to QSAR modeling assume that new molecules to be predicted will have structural features in common with the training set on which they are based. If the new molecule is sufficiently different, one will obtain an unreliable (or even meaningless) prediction. Thus, it is vital to denote the "domain of applicability" of a model, thus letting the user know when the predictions of the model can be relied upon. In scenarios where a model is built at one time point and then used to make predictions over a period of time, defining the domain of applicability can be very useful in determining at what point a model should be rebuilt, because the new molecules diverge sufficiently that they cannot be predicted from the original model (18).

A variety of methods have been developed to define domains of applicability (19–23). The simplest approach is to determine how similar a new molecule is to the training set for the model. Sheridan et al. (24) employed this technique focusing on two approaches: similarity of the molecule to be predicted to the nearest neighbor in the training set and the number of nearest neighbors in the training set (decided by a user-defined similarity cutoff). Their results indicated that either of these approaches leads to a robust measure of reliability of predictions. Xu and Gao (25) also considered similarity to the training set, using a novel distance metric termed the "dimension-related distance," allowing them to measure the similarity of a molecule to the *entire* training set. Other distance metrics have also been employed, including Mahalanobis and Manhattan distances. For models based on linear regression, various diagnostics such as the Cooks distance (26) and leverage (27) have been employed (28–30).

A number of approaches based on descriptor values have also been proposed. The simplest approach is to determine the range of descriptor values in the training set and if the values for the new molecule lie outside the range, the model will have to extrapolate, and hence the prediction will be unreliable. While conceptually simple, this approach easily gives misleading results if nonuniformly distributed descriptor values are used. Alternatives include performing principal-components analysis and using the ranges of the resulting principal components as the space within which reliable predictions can be obtained (31).

3 Exploring SAR Landscapes

QSAR model predictions are a useful guide for lead optimization (32), but alternative views of SAR data can be useful. Over the past few years, the landscape paradigm of SAR data has gained focus, allowing us to explore a number of aspects of SARs. These advances stem from the work of Lajiness (33) viewing chemical structure and bioactivity, simultaneously, in a three-dimensional view, with the structure represented in the *X*–*Y* plane and the activity along the *Z*-axis. The immediate consequence of this is that a SAR dataset can be viewed as a landscape of varying "topography." Smooth regions correspond to molecules that are similar in structure and activity whereas jagged (i.e., discontinuous) regions correspond to structures that are similar but exhibit very different activities (so-called activity cliffs). It has been suggested that the latter regions of the landscape represent the most interesting parts of an SAR, as they provide the possibility of making small structural changes to significantly change activities. At the same time, these discontinuities can be problematic as they can lead to poor performance of many QSAR modeling methods (primarily those based on machine learning or statistical models) (34). As a result, a variety of methods have been developed to characterize and mine SAR landscapes.

Structure–activity similarity (SAS) maps, first described by Shanmugasundaram and Maggiora (35), are pairwise plots of the structure similarity against the activity similarity. The resultant plot can be divided into four quadrants, allowing one to identify molecules characteristic of one of four possible behaviors: smooth regions of the SAR space (rough), activity cliffs, nondescript (i.e., low structural similarity and low activity similarity), and scaffold hops (low structural similarity but high activity similarity). Recently, SAS maps have been extended to take into account multiple descriptor representations (two and three dimensions) (36, 37). In addition to SAS maps, other pairwise metrics to characterize and visualize SAR landscapes have been developed such as the structure–activity landscape index (SALI) (38) and the structure–activity index (SARI) (39).

Visualization of landscapes via network diagrams has also led to novel developments in the exploration of SAR data. Examples include the SALI networks described by Guha and Van Drie (38) and network similarity graphs (NSGs) described by Wawer et al. (40). Both network representations use compounds as nodes and draw edges between them based on a metric that characterizes the pair of nodes in the context of the landscape (SARI for NSGs and SALI for SALI networks). The networks can be then analyzed to identify specific SAR trends. For example, Wawer et al. (41) described an approach to identifying "SAR pathways" (paths in an

NSG that connect regions of low and high SAR discontinuity). Such SAR pathways represent a set of compounds that when ordered appropriately exhibit a continuous series of SAR changes. While network-based analyses of landscapes have seen much activity, an alternative visualization approach described by Seebeck et al. (42) abstracted the idea of the SALI metric and extended it to include the receptor. Using this technique they were able to highlight specific regions within protein-binding sites that are most likely to lead to activity cliffs.

The concept of activity cliffs and the landscape paradigm have also been applied to R-groups, where an "R-cliff" occurs when a pair of compounds differs in a single R-group. This is clearly a specialization of the activity cliff concept, placing this type of analysis in the context of analogue series derived via R-group decompositions (43, 44).

4 Canned SAR

Over the past few years a number of large chemical structure databases have become available. Some of these databases also provide extensive information on compound activities in addition to compound structures. Notable examples include PubChem, ChEMBL, and GVK GOSTAR. The first two are freely available resources, whereas GOSTAR is a commercial offering. While these databases provide structure–activity information, they differ in the nature of the data that are provided. PubChem provides a compound and substance database, where records are individual structures, as well as a bioassay database, which contains assay results for various compound sets, deposited by the Molecular Libraries Initiative (45). The two databases are linked, allowing one to easily identify the assays a compound has been tested in or, conversely, the structures tested in an assay. Assay datasets range in size from two or three compounds to more than 300,000 compounds and assay types range from primary screens to secondary and confirmatory screens, both as single-point and dose–response formats. PubChem data are not curated and thus the assay data can be noisy (which is a given for primary screening data). ChEMBL and GOSTAR are both curated SAR databases, where structures and their activities have been manually extracted from the literature and stored in a standardized form. A variety of annotations have also been added such as assay target and species. These databases have "canned" SARs, making them readily available for analysis.

One obvious application of these databases is to use them as sources of training data when developing predictive models. For example, Novotarskyi et al. (46) employed PubChem Bioassay data to develop models to predict CYP450 1A2 inhibition, and Shen et al. (47) employed the database to develop a support vector machine model to predict hERG (human ether-à-go-go-related gene) liabilities. Though there are many other examples of QSAR modeling studies using PubChem as the source of data, most focus on specific targets. In contrast, Chen and Wild (48) built a series of models using multiple PubChem assays that could be used to predict an "activity profile." They employed random forest models, aiming for pure predictive ability over explanatory power.

A unique database is the GDB-13 database (49), which is an exhaustive enumeration of small-molecule structures containing up to 13 heavy atoms (restricted to C, H, N, O, S, P, and Cl). Although the database does not contain activity information associated with the structures, it can be used as a source of structures for virtual screening purposes (50). It is similar in nature to databases such as ZINC (51). The key difference is that the latter are all commercially available, whereas the former are completely virtual. This class of databases is useful primarily for virtual screening type methods, where the goal is to identify candidates for more in-depth study, rather than to explicitly understand SAR trends.

5 Alternatives to QSAR?

While QSAR approaches (in all their forms) are by far the most common ways to capture and explore SAR trends, a number of other approaches are possible. Although they are not quantitative, they can be useful as "idea generators."

5.1 Characterizing SAR in Series

One approach is to consider fragments as the basis for SAR exploration. This is not without precedent as substructure-based models have been developed that are useful for both prediction and interpretation (52). One approach to using fragments for exploring SAR is to develop "R-group QSAR" models, whose goal is to determine whether a SAR exists and if so how different R-groups affect it. Given a set of molecules, we perform an R-group decomposition, generating a series of scaffolds and associated substituents. Given a scaffold, we can create an R-group matrix, with observations (i.e., molecules containing the scaffold) in the rows and the R-groups, $R_1, R_2, \ldots, R_n$, in the columns. Element (i, j) of the matrix is set to 1 if the i'th molecule contains the j'th substituent. Given this R-group incidence matrix, along with the observed activities for the molecules, one can develop a predictive model. Given that most such R-group matrices will be small, some form of linear regression is likely most suitable. Figure 2 summarizes this technique. However, the limitations of this approach are obvious. First, the number of observations for many scaffolds will be very small (<10) and hence a reliable model cannot be generated. Second, most R-group matrices will be sparse because all substitution points will not be uniformly filled. Also, a binary incidence

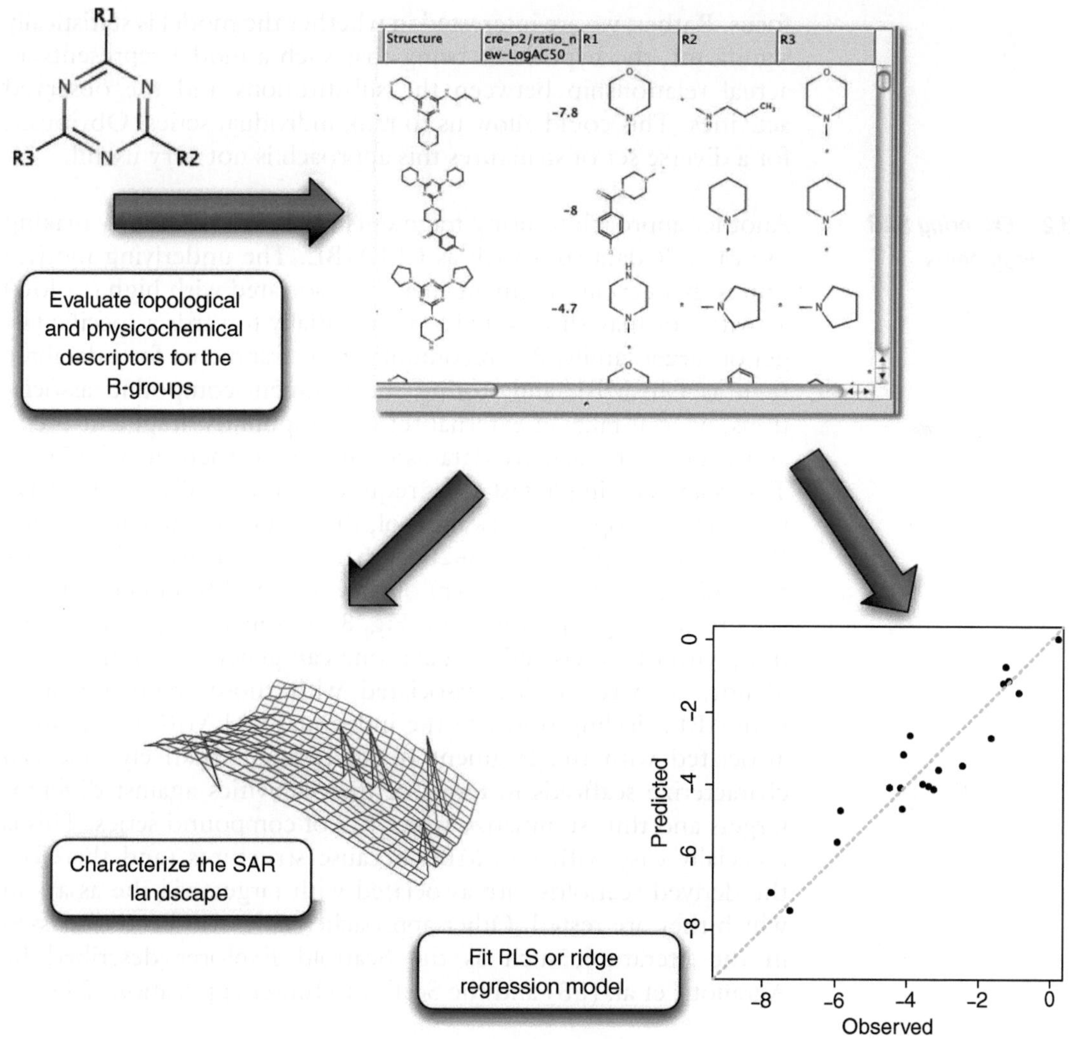

Fig. 2 R-group QSAR workflow that can be used to rank scaffolds (i.e., chemical series) in terms of whether they exhibit an SAR

matrix may not be the most appropriate input for a linear regression model. One way to address this is to replace each of *n* columns with $n \times d$ columns, where *d* represents a *d*-dimensional real-valued descriptor vector derived from the structure of the corresponding R-group. For a scaffold with three or more substitution points, this can lead to an R-group matrix with more columns than rows. Even when this is not the case, such a matrix will usually have a large number of features (depending on how many descriptors are employed). In spite of these problems, for those scaffolds where there is a sufficient number of observations and relatively few substitution points, a linear regression model can be developed. In such cases, the predictive accuracy of the model is not the main focus. Rather, we are interested in whether the model is statistically significant, the hypothesis being that such a model represents an actual relationship between the substitutions and the observed activities. This could allow us to rank individual series. Obviously, for a diverse set of structures this approach is not very useful.

5.2 Exploring SAR via Fragments

Another approach to using fragments to explore SAR is by making use of SAR databases such as ChEMBL. The underlying motivation is that certain fragments may be associated with high (or low) activities or may show activity preferentially towards a specific target or target family. By precomputing the fragments for a database such as ChEMBL and storing the fragment-compound associations, we can take an external set of compounds, fragment them, and explore the activity data associated with them in ChEMBL. This relatively simple task does require a number of steps. We have recently developed a software tool, the Fragment Activity Profiler (http://tripod.nih.gov/?p=206), that allows one to explore structure collections from the point of view of scaffolds, integrated with activity and target information (Fig. 3). Given a compound collection, with observed activity data, one can generate fragments and identify activity profiles associated with those fragments from ChEMBL, drilling down to the individual ChEMBL compounds associated with the fragment of interest. Alternatively, one can characterize scaffolds in terms of their activities against different targets and thus summarize selectivity of compound series. This is especially easy with ChEMBL because structures (and therefore the derived scaffolds) are associated with targets via the assays in which they are tested. Other approaches have also been discussed in the literature, such as the Scaffold Explorer described by Agrafiotis et al. (53) and the Scaffold Hunter application (54).

6 Conclusion

Computational methods play a key role in the process of drug discovery. Although the value added by these models has been discussed, with informed usage of these techniques together with appropriate testing and control methods (55), these methods have much to offer (56). Indeed, with the deluge of data being generated by modern high-throughput experimental techniques and the evolution of large chemical structure databases, it is all the more important to address how computational techniques can enable practicing scientists to explore the wealth of structure–activity data.

Traditional QSAR approaches based on machine learning or statistical methods have been used extensively. However, many are black boxes and can make understanding the actual SAR encoded by the model difficult. This can hinder the use of such models

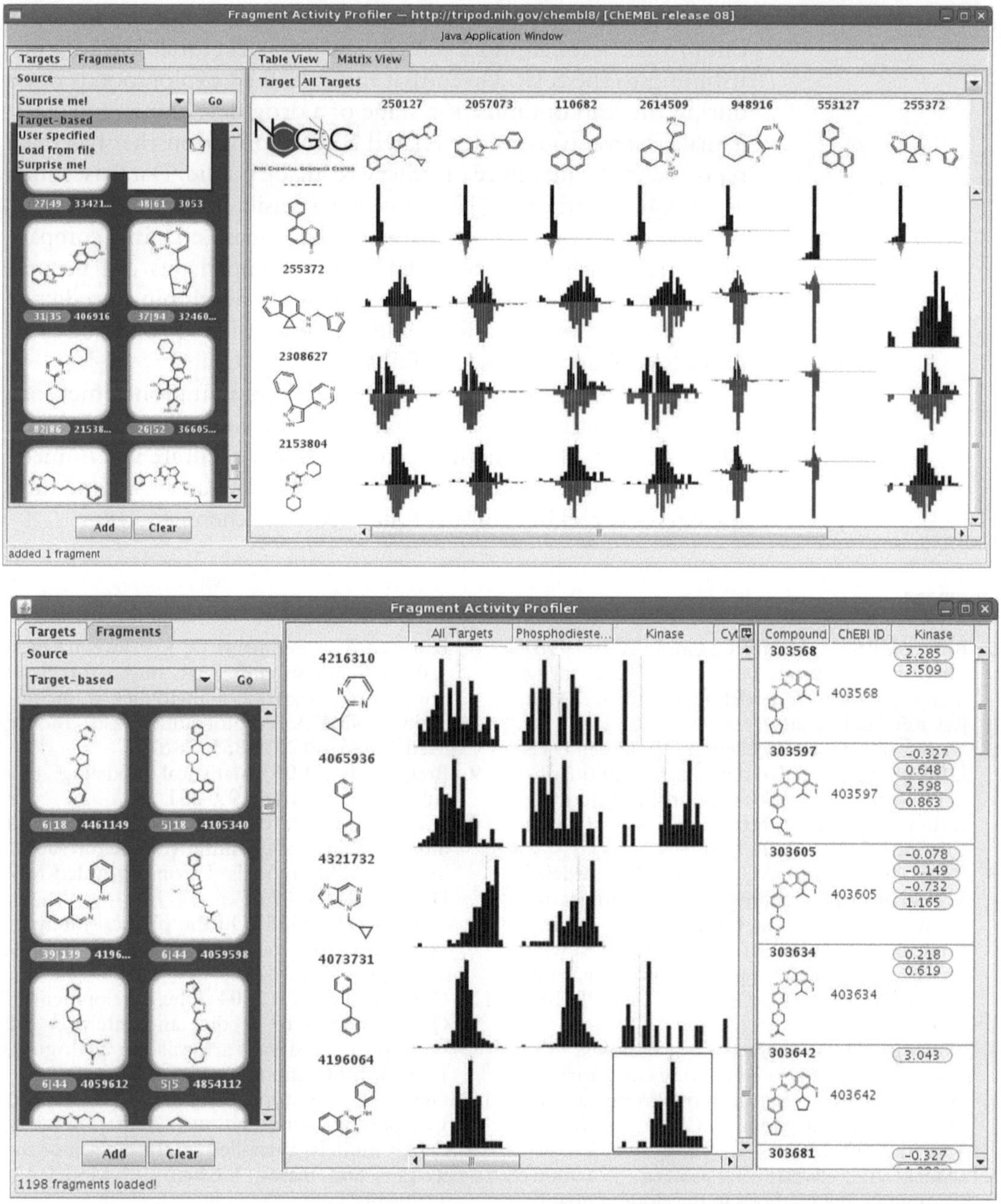

Fig. 3 Screenshots of the Fragment Activity Profiler. *Top*: Single-scaffold view displays activity profiles for a given scaffold versus individual targets (at a specified level of the ChEMBL target hierarchy). *Bottom*: Pairwise view allows comparison of activity profiles of two scaffolds simultaneously

when deciding how to modify a structure to improve an activity or reduce liabilities. Although visualizations in some cases can make this job easier, more physical models such as three-dimensional QSAR methods (comparative molecular field analysis, comparative molecular similarity indices analysis) and pharmacophore methods can provide a much more direct view into structural features contributing to the SAR. Given that much of SAR exploration is done during the lead optimization stage of a drug development project, it makes sense to consider related SAR information that has been reported in the literature or otherwise made public. Databases such as PubChem and ChEMBL allow one to easily access large amounts of SAR data points (both curated and uncurated) and compare those data with one's own data. Such comparisons can be performed in a variety of ways ranging from developing predictive models to browsing scaffolds and viewing activity profiles of compounds containing those scaffolds.

In summary, an extensive array of computational methods allows exploration of SAR trends at varying levels of detail. Although not all the methods are necessarily accurate in a numerical sense, in silico models can serve as idea generators, augmenting the decision-making ability of the medicinal chemist.

References

1. Bohacek RS, McMartin C, Guida WC (1996) The art and practice of structure based drug design: a molecular modeling perspective. Med Res Rev 16(1):3–50
2. Nicolotti O, Gillet VJ, Fleming PJ, Green DVS (2002) Multiobjective optimization in quantitative structure-activity relationships: deriving accurate and interpretable QSARs. J Med Chem 23:5069–5080
3. Cruz-Monteagudo M, Borges F, Cordeiro MN (2008) Desirability-based multiobjective optimization for global QSAR studies: application to the design of novel NSAIDs with improved analgesic, antiinflammatory, and ulcerogenic profiles. J Comput Chem 29(14): 2445–2459
4. Nicolotti O, Giangreco I, Miscioscia TF, Carotti A (2009) Improving quantitative structure-activity relationships through multi-objective optimization. J Chem Inf Model 49(10):2290–2302
5. Dudek AZ, Arodz T, Galvez J (2006) Computational methods in developing quantitative structure-activity relationships (QSAR): a review. Comb Chem High Throughput Screen 9(3):213–228
6. Winkler DA (2002) The role of quantitative structure–activity relationships (QSAR) in biomolecular discovery. Brief Bioinform 3(1):73–86
7. Zvinavashe E, Murk AJ, Rietjens M (2008) Promises and pitfalls of quantitative structure-activity relationship approaches for predicting metabolism and toxicity. Chem Res Toxicol 21(12):2229–2236
8. Banerjee A, Schepmann D, Kurthwein J et al (2010) Synthesis and SAR studies of chiral non-racemic dexoxadrol analogues as uncompetitive NMDA receptor antagonists. Bioorg Med Chem 18(22):7855–7867
9. Breiman L (2001) Statistical modeling: two cultures. Stat Sci 16:199–231
10. Guha R (2008) On the interpretation and interpretability of quantitative structure-activity relationship models. J Comput Aided Mol Des 22(12):857–871
11. Stanton DT (2003) On the physical interpretation of QSAR models. J Chem Inf Comput Sci 43(5):1423–1433
12. Guha R, Jurs PC (2004) The development of QSAR models to predict and interpret the biological activity of artemisinin analogues. J Chem Inf Comput Sci 44:1440–1449
13. Guha R, Stanton DT, Jurs PC (2005) Interpreting computational neural network QSAR models: a detailed interpretation of the weights and biases. J Chem Inf Model 45: 1109–1121
14. Segall M, Champness E, Obrezanova O, Leeding C (2009) Beyond profiling: using ADMET models to guide decisions. Chem Biodivers 6(11):2144–2151
15. Faulon JL, Visco DP Jr, Pophale RS (2003) The signature molecular descriptor. 1. Using extended valence sequences in QSAR and QSPR studies. J Chem Inf Comput Sci 43: 707–720
16. Churchwell CJ, Rintoul MD, Martin S et al (2004) The signature molecular descriptor. 3. Inverse-quantitative structure-activity relation-

ship of ICAM-1 inhibitory peptides. J Mol Graph Model 22(4):263–273
17. Wong WW, Burkowski FJ (2009) A constructive approach for discovering new drug leads: using a kernel methodology for the inverse-QSAR problem. J Cheminform 1:s4
18. Weaver S, Gleeson MP (2008) The importance of the domain of applicability in QSAR modeling. J Mol Graph Model 26(8):1315–1326
19. Schultz TW, Hewitt M, Netzeva TI, Cronin MTD (2007) Assessing applicability domains of toxicological QSARs: definition, confidence in predicted values, and the role of mechanisms of action. QSAR Comb Sci 26:238–254
20. Roberts DW, Patlewicz G, Kern PS et al (2007) Mechanistic applicability domain classification of a local lymph node assay dataset for skin sensitization. Chem Res Toxicol 20(7): 1019–1030
21. Stanforth RW, Kolossov E, Mirkin B (2007) A measure of domain of applicability for QSAR modelling based on intelligent *K*-means clustering. QSAR Comb Sci 26(7):837–844
22. Tetko IV, Bruneau P, Mewes HW et al (2006) Can we estimate the accuracy of ADME-tox predictions? Drug Discov Today 11(15–16): 700–707
23. Jaworska J, Nikolova-Jeliazkova N, Aldenberg T (2005) QSAR applicability domain estimation by projection of the training set in descriptor space: a review. Altern Lab Anim 33:445–459
24. Sheridan RP, Feuston BP, Maiorov VN, Kearsley SK (2004) Similarity to molecules in the training set is a good discriminator for prediction accuracy in QSAR. J Chem Inf Comput Sci 44(6):1912–1928
25. Xu YJ, Gao H (2003) Dimension related distance and its application in QSAR/QSPR model error estimation. QSAR Comb Sci 22: 422–429
26. Cook RD (1977) Detecting influential observations in linear regression. Technometrics 19:15–18
27. Chatterjee S, Hadi AS (1986) Influential observations, high leverage points, and outliers in linear regression. Stat Sci 1(3):379–416
28. Tropsha A, Gramatica P, Gombar VK (2003) The importance of being earnest: validation is the absolute essential for successful application and interpretation of QSPR models. QSAR Comb Sci 22:69–77
29. Eriksson L, Jaworska J, Worth AP et al (2003) Methods for reliability and uncertainty assessment and for applicability evaluations of classification- and regression-based QSARs. Environ Health Perspect 111(10):1361–1375
30. Papa E, Villa F, Gramatica P (2005) Statistically validated QSARs, based on theoretical descriptors, for modeling aquatic toxicity of organic chemicals in Pimephales promelas (fathead minnow). J Chem Inf Model 45(5): 1256–1266
31. Nikolova-Jeliazkova N, Jaworska J (2005) An approach to determining AD for QSAR group contribution models: an analysis of SRC KOWWIN. Altern Lab Anim 33:461–470
32. Brown N, Lewis RA (2006) Exploiting QSAR methods in lead optimization. Curr Opin Drug Discov Devel 9(4):419–424
33. Lajiness MS (1991) Evaluation of the performance of dissimilarity selection methodology. In: Silipo C, Vittoria A (eds) QSAR: rational approaches to the design of bioactive compounds. Proceedings of the VIII European symposium on quantitative structure-activity relationships. Sorrento, Italy, 9–13 Sept 1990. ESCOM, Leiden, pp 201–204
34. Maggiora GM (2006) On outliers and activity cliffs—why QSAR often disappoints. J Chem Inf Model 46(4):1535
35. Shanmugasundaram V, Maggiora GM (2001) Characterizing property and activity landscapes using an information-theoretic approach. In: CINF-032. 222nd National Meeting of the American Chemical Society Chicago, IL, Washington, DC
36. Medina-Franco JL, Martínez-Mayorga K, Bender A et al (2009) Characterization of activity landscapes using 2D and 3D similarity methods: consensus activity cliffs. J Chem Inf Model 49(2):477–491
37. Yongye AB, Byler K, Santos R et al (2011) Consensus models of activity landscapes with multiple chemical, conformer, and property representations. J Chem Inf Model 51(6): 1259–1270
38. Guha R, Van Drie JH (2008) The structure-activity landscape index: identifying and quantifying activity-cliffs. J Chem Inf Model 48(3): 646–658
39. Peltason L, Bajorath J (2007) SAR index: quantifying the nature of structure-activity relationships. J Med Chem 50(23): 5571–5578
40. Wawer M, Peltason L, Weskamp N et al (2008) Structure-activity relationship anatomy by network-like similarity graphs and local structure-activity relationship indices. J Med Chem 51(19):6075–6084
41. Wawer M, Peltason L, Bajorath J (2009) Elucidation of structure-activity relationship pathways in biological screening data. J Chem Inf Model 52(4):1075–1080
42. Seebeck B, Wagener M, Rarey M (2011) From activity cliffs to target-specific scoring models and pharmacophore hypotheses. ChemMed Chem 6(9):1630–1639
43. Agrafiotis DK, Wiener JJM, Skalkin A, Kolpak J (2011) Single r-group polymorphisms (SRPs) and r-cliffs: an intuitive framework for

analyzing and visualizing activity cliffs in a single analog series. J Chem Inf Model 51(5): 1122–1131
44. Sisay MT, Peltason L, Bajorath J (2009) Structural interpretation of activity cliffs revealed by systematic analysis of structure-activity relationships in analog series. J Chem Inf Model 49(10):2179–2189
45. Austin CP, Brady LS, Insel TR, Collins FS (2004) NIH molecular libraries initiative. Science 306:1138–1139
46. Novotarskyi S, Sushko I, Körner R et al (2011) A comparison of different QSAR approaches to modeling CYP450 1A2 inhibition. J Chem Inf Model 51(6):1271–1280
47. Shen M-Y, Su B-H, Esposito EX et al (2011) A comprehensive support vector machine binary hERG classification model based on extensive but biased end point hERG data sets. Chem Res Toxicol 24(6):934–949
48. Chen B, Wild DJ (2010) PubChem BioAssays as a data source for predictive models. J Mol Graph Model 28(5):420–426
49. Blum LC, Reymond J-L (2009) 970 million druglike small molecules for virtual screening in the chemical universe database GDB-13. J Am Chem Soc 131(25):8732–8733
50. Blum LC, van Deursen R, Bertrand S et al (2011) Discovery of α7-nicotinic receptor ligands by virtual screening of the chemical universe database GDB-13. J Chem Inf Model 51(12):3105–3112
51. Irwin JI, Shoichet BK (2005) ZINC—a free database of commercially available compounds for virtual screening. J Chem Inf Model 45(1):177–182
52. von Korff M, Sander T (2006) Toxicity-indicating structural patterns. J Chem Inf Model 46:536–544
53. Agrafiotis DK, Wiener JJM (2010) Scaffold explorer: an interactive tool for organizing and mining structure-activity data spanning multiple chemotypes. J Med Chem 53(13): 5002–5011
54. Wetzel S, Klein K, Renner S et al (2009) Interactive exploration of chemical space with Scaffold Hunter. Nat Chem Biol 5(8): 581–583
55. Jain AN, Cleves AE (2011) Does your model weigh the same as a duck? J Comput Aided Mol Des 26(1):57–67
56. Cramer RD (2011) The inevitable QSAR renaissance. J Comput Aided Mol Des 26(1): 35–38

Chapter 7

Molecular Dynamics Simulations in Drug Design

John E. Kerrigan

Abstract

This minireview focuses on recent developments in the application of molecular dynamics to drug design. Recent applications of endpoint free-energy computational methods such as molecular mechanics Poisson-Boltzmann surface area (MM-PBSA) and generalized Born surface area (MM-GBSA) and linear response methods are described. Recent progress in steered molecular dynamics applied to drug design is reviewed.

Key words Molecular dynamics, Poisson-Boltzmann, Generalized Born, Linear interaction energy, Binding free energy, Periodic boundary conditions, Steered molecular dynamics

1 Introduction

This chapter surveys the recent applications of endpoint free-energy methods, including linear response and molecular mechanics Poisson-Boltzmann surface area (MM-PBSA) and generalized Born surface area (MM-GBSA) used in biomolecular dynamics simulations in drug design. A brief review of recent advances in steered molecular dynamics (a nonequilibrium approach) specific to applications in drug design is included. For an excellent introductory overview on the topic of molecular dynamics simulation in drug discovery, see the recent review by Durrant and McCammon (1).

At present, drug discovery is still an expensive enterprise. According to recent statistics, it takes about 12 years at a cost of ~$1.8 billion on average to bring a drug to market (2). Molecular docking tools used in virtual screening of drug libraries suffer from inadequate scoring functions, which treat the receptor as a rigid or semirigid construct. A recent evaluation of 11 popular docking scoring functions found that the majority of scoring functions failed to correlate better than 0.50 with experimental binding data (3). However, it is difficult to evaluate the docking scoring functions in

Sandhya Kortagere (ed.), *In Silico Models for Drug Discovery*, Methods in Molecular Biology, vol. 993,
DOI 10.1007/978-1-62703-342-8_7, © Springer Science+Business Media, LLC 2013

a fair and statistically relevant manner (4). The need continues for accurate scoring accounting for the "induced fit" of drug/ligand to receptor/enzyme and the unique properties of each small molecule.

2 Endpoint Methods

Free energy is a thermodynamic quantity, a state function, which is dependent only on the energy of the initial and final states of the drug binding process. An example is the formation of the noncovalent enzyme–inhibitor complex ($E-I^*$ in Fig. 1), a final state resulting from the combination of the enzyme (E) and inhibitor (I) in whatever environment (e.g., any acceptable buffer system in an in vitro assay), also known as the initial state or reactants. The path the inhibitor and enzyme take to form the final state of noncovalent complex ($E-I^*$) in the case of the enzyme–inhibitor example is not important for the free energy. The K_i (a dissociation constant) is given by Eq. 1 and the relation of K_i to binding free energy (ΔG_{bind}) is given in Eq. 2, where R is the gas constant ($R=8.314$ J/mol K) and T is the temperature of the system. The gas constant R is derived from the Boltzmann constant (k_B) and the number of particles (N_A) in the system by the relationship $R = N_A k_B$. Additionally defined in Eq. 2 is the relationship between free energy (ΔG), enthalpy (ΔH), temperature (T), and entropy (ΔS). The energies derived from molecular mechanics force fields used in the dynamics simulations are entirely enthalpic (i.e., ΔH). The entropy term (ΔS) remains a challenge to estimate and is often ignored or estimated using parameterization based on experimental data in endpoint method calculations. The more discriminating components of the force field that contribute to enthalpy are the nonbonded forces such as hydrogen bonding, electrostatics, and van der Waals energies.

$$K_i = \frac{[E][I]}{[E-I^*]} \tag{1}$$

$$\Delta G_{\text{bind}} = -RT\ln\left(\frac{1}{K_i}\right) = RT\ln K_i = \Delta H - T\Delta S \tag{2}$$

$$\mathrm{E} + \mathrm{I} \underset{K_i}{\rightleftharpoons} \mathrm{E\text{-}I^*}$$

Fig. 1 Equilibrium between reactants ($E + I$) and noncovalent complex ($E-I^*$) in a typical enzyme–inhibitor binding process

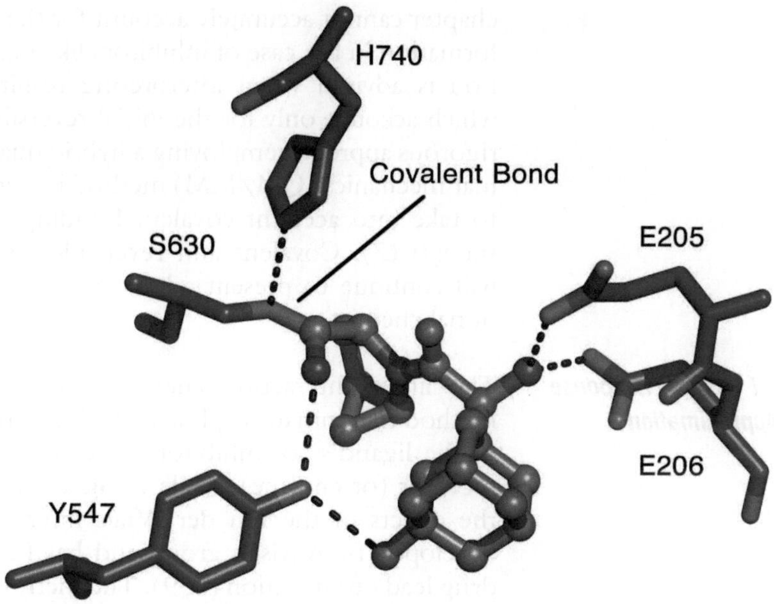

Fig. 2 Structure drawing of saxagliptin. Binding region of DPP-IV showing saxagliptin (ball-stick model) interactions with key amino acid residues (stick model) from X-ray crystal structure (3BJM) (produced with Pymol)

These are simple fundamental relationships, which assume one is working with competitive reversible inhibitors that form noncovalent complexes with the enzyme. The endpoint methods function well enough under these assumptions. Yet, covalent and reversible covalent inhibitors are becoming popular because of their increased potency and efficacy afforded by the covalent bond (5). A recent pharmaceutically relevant reversible covalent binding example is the dipeptidyl peptidase IV (DPP-IV) inhibitor saxagliptin (Onglyza®) developed by Bristol-Myers Squibb. Metzler et al. (6) report that the nitrile functional group of saxagliptin (Fig. 2) forms a "reversible" covalent complex with serine residue S630 of DPP-IV. They also report the importance of histidine residue H740 for covalent bond formation from their mutational analysis. In addition, Metzler et al. report that the BMS-538305 analogue binds tightly to the S630A mutant and equally well to the wild type, whereas saxagliptin binds tightly only to the wild type. The X-ray crystal structure of the complex of saxagliptin with DPP-IV (Fig. 2) verifies the covalent complex formation (6). In addition to the covalent bond with S630, note the electrostatic interaction between the positively charged ammonium ion of saxagliptin with the negatively charged glutamate pairing of E205 and E206 (Fig. 2). The binding free-energy methods discussed in this chapter cannot accurately account for the reversible covalent bond formation in the case of inhibitors like saxagliptin. Therefore, caution is advised when interpreting results using these methods, which account only for the initial reversible binding step. A more rigorous approach employing a hybrid quantum mechanics–molecular mechanics (QM/MM) method is recommended if one desires to take into account covalent bonding and associated transition state(s) (7). Covalent and reversible covalent inhibitors/ligands will continue to present additional challenge(s) to the computational chemist.

2.1 Linear Response Approximation

The linear interaction energy (LIE) approach is an endpoint method that initially exploited the linear response of the energetics of the ligand's (or inhibitor's) electrostatic interactions with the receptor (or enzyme), while using a scaling factor to account for the effects of the van der Waals interactions. This method was developed by Åqvist's group and has found wide applicability in drug lead optimization (8, 9). The method relies purely on the difference in the energetics of the interactions between ligand and receptor in the "bound" versus "unbound" states (i.e., the endpoints).

$$\Delta G_{\text{bind}} = \beta\left(\left\langle V_{\text{bound}}^{\text{elec}}\right\rangle - \left\langle V_{\text{unbound}}^{\text{elec}}\right\rangle\right) + \alpha\left(\left\langle V_{\text{bound}}^{\text{vdw}}\right\rangle - \left\langle V_{\text{unbound}}^{\text{vdw}}\right\rangle\right) + \gamma \quad (3)$$

The β parameter is the scaling factor used for the electrostatic interaction; the α parameter is used to scale the van der Waals component of interaction; and the sometimes-used γ parameter is used to account for binding site hydrophobicity. The γ parameter essentially sets the scale of the absolute free energy. These scaling factors are dependent on the small molecule (10) and the force field used (11). For example, the β parameter is sensitive to polar moieties such as alcohols on the ligand, most likely owing to interaction with water. The α parameter is dependent on the nonpolar solvation characteristic of the binding site. The standard LIE equation parameterization for the α and β parameters is $\alpha = 0.18$ and $\beta = 0.5$ (linear response) (12, 13). In most studies, either previously tested and published parameters are used or new parameters are developed based on regression fit of Eq. 3 to in vitro experimental binding data from models with previously published or known inhibitors.

To use the LIE method, one needs to perform two molecular dynamics simulations per ligand: one simulation of the solvated ligand–receptor complex (the "bound" state) and one simulation of the solvated ligand (the "unbound" state) by itself. These simulations can be simple water sphere or droplet where the long-range electrostatics are accounted for using a local reaction field approach as in Åqvist's molecular dynamics package known as "Q" (14), or they can be fully solvated systems in a periodic box of water using

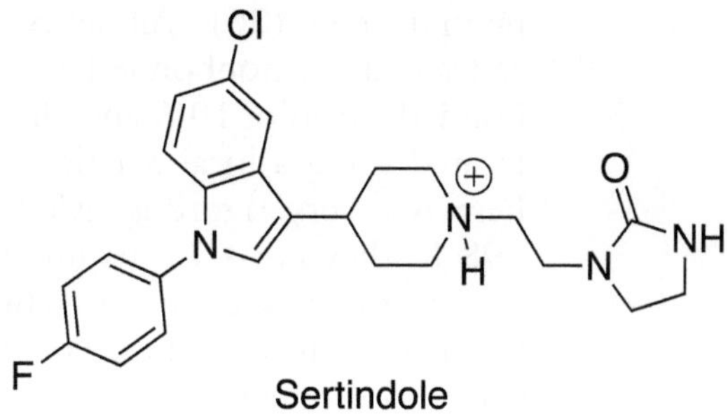

Fig. 3 Structure drawing of sertindole

particle-mesh Ewald (PME) (15, 16) to account for long-range electrostatics.

Timmers et al. (17) published a very nice study of inhibitors of *Mycobacterium tuberculosis* H37Rv cytidine deaminase (CDA). In this study they used the Gromacs biomolecular simulation package (18) and the Gromos 53a6 force field. After docking the ligands to CDA using the Autodock 4 program (19) to create models of the "bound" state, the authors encapsulated their models of the "bound" and "unbound" states in a periodic box of simple point charge extended (SPC/E) water (20) using PME to account for long-range electrostatics in their 10-ns production runs using the NPT ensemble (maintaining constant temperature, pressure, and number of particles). They ran an additional set of simulations for each ligand with the partial atomic charges of the ligand turned off to evaluate the long-range component of the electrostatic energy when using PME. They calibrated their scaling factors using the human CDA structure and five known inhibitors of the enzyme. Although their in vitro screening was incomplete, they identified one hit with ΔG_{bind} (LIE) = −5.9 kcal/mol compared with the experimentally determined ΔG_{obs} = −5.2 kcal/mol.

The LIE method has also been applied to the study of structure–activity relationships for human *ether-à-go-go* related gene (hERG) potassium channel blockers as reported by Boukharta et al. (21) from Åqvist's group. A series of analogues based on sertindole (Fig. 3) were analyzed. Their target structure for hERG potassium channel was derived from a homology model, which they embedded in a phospholipid bilayer composed of phosphatidyl choline (POPC) lipids and refined using molecular dynamics. Jones et al. (22, 23) from Åqvist's group docked the sertindole analogues using the Gold (CCDC) docking package to produce models of the "bound" state followed by molecular dynamics using the "Q" programs (14) and the OPLS-AA (optimized potentials for liquid simulations-all atom) force field (24). Both the "bound" and "unbound" states were solvated in a sphere of TIP3P (transferable intermolecular potential 3 point) water (25) using a 25 Å radius with ionizable residues at the edge of the sphere set to their neutral state (26). All atoms outside the sphere radius were restrained and nonbonded interactions removed. They used a nonbonded cutoff of 10 Å and all interactions beyond this cutoff were treated using a local reaction field. The Åqvist group's approach here was very rigorous. After gradual heating of the system to 298 K, they ran a 1.5-ns unrestrained equilibration run followed by 3 ns of production run. This was repeated five times using different starting velocities. For their LIE calculation, they used the standard parameterization ($\alpha = 0.18$; $\beta = 0.5$), with the γ parameter ($\gamma = -14.1$) determined using regression fit to experimental data derived from in vitro IC_{50} (median inhibitory concentration) data ($\Delta G_{obs} \cong RT\ln[IC_{50}]$). Their results from analysis of sertindole and eight sertindole analogues in comparison with ΔG_{obs} are impressive. The reported LIE estimated ΔG_{bind} values were either on target or within 0.5–1.0 kcal/mol of the ΔG_{obs} with standard deviations based on the five duplicate runs in the same range and an average unsigned error of 0.4 kcal/mol (21).

The LIE method has also been used with continuum electrostatic methods. Nicolotti et al. (27) have published a nice report on this use of the LIE method for screening benzamidine inhibitors of thrombin. Another recent study used the atom-bond electronegativity equalization method σπ force field with the generalized Born (GB) continuum electrostatics approach to predict binding free energies of HIV-1 protease inhibitors (28).

2.2 Molecular Mechanics Poisson-Boltzmann or Generalized Born Surface Area Methods

These two endpoint methods differ only in the mode of calculation used to estimate the polar solvent electrostatic contribution to the free energy of binding. Both solvent electrostatics methods, Poisson-Boltzmann (PB) and generalized Born (GB), are implicit solvation methods that attempt to estimate the effect of solvent dielectric (i.e., the change in going from a polar water encased environment into the less polar environment of the protein binding pocket) on the binding of drug or inhibitor. The PB approach as developed in the program Delphi provides a rigorous treatment of solvent electrostatics using finite-difference methods (29, 30). Another popular PB solver is the Adaptive Poisson-Boltzmann Solver (APBS), which was developed to estimate solvation energies of supramolecular systems (31). Despite the rigor of the approach, the PB method has been used in the scoring of results from high-throughput virtual screening (32). The PB solvers are designed to solve the PB equation (Eq. 4), which relates the electrostatic potential (ϕ) to solvent and solute dielectric (ε) including salt (ions) concentration and its impact on the solute (κ). The $f(x)$ term is the distribution of partial atomic charges. The ∇ is the Laplacian operator for three-dimensional Cartesian coordinates.

$$-\nabla \cdot \varepsilon(x) \nabla \phi(x) + \kappa^2(x) \sin h\phi(x) = f(x) \tag{4}$$

The GB model is a modification of the Coulomb equation to include the Born radius of the particle or atom which estimates the degree of the particle's burial within the molecule. Equation 5 relates ΔG_{elec} to the solvent/solute dielectric (ε), the separation between the partial atomic charges r_{ij}, the effective Born radii $R(i$ and $j)$, and the smoothing function f_{GB}. A Debye-Hückel screening parameter (κ) similar to that used in the PB equation is used to account for the monovalent ions.

$$\Delta G_{elec} = -\frac{1}{2}\sum_{ij}\frac{q_i q_j}{f_{GB}(r_{ij}, R_i R_j)}\left(1 - \frac{\exp[-\kappa f_{GB}]}{\varepsilon}\right) \quad (5)$$

The GB method is regarded as a more computationally efficient approach to obtaining PB-like results at lower computational cost. The GB method has been applied to a variety of biomolecular computational problems including ab initio protein folding (33). The relevant equations for the MM-PBSA and MM-GBSA calculations are as follows (Eqs. 6–9):

$$\Delta G_{bind} = G_{complex} - (G_{receptor} + G_{ligand}) \quad (6)$$

$$E_{MM} = E_{vdw} + E_{elec} \quad (7)$$

$$G = \langle E_{MM} \rangle + \langle G_{psolv} \rangle + \langle G_{npsolv} \rangle - T\langle S \rangle \quad (8)$$

$$G_{npsolv} = \gamma \text{SASA} + b \quad (9)$$

The binding free energy, ΔG_{bind}, is composed of the classical $E_{products}-E_{reactants}$ (the endpoints), where $E_{products} = \Delta G_{complex}$ and $E_{reactants}$ is composed of $G_{receptor}$ and G_{ligand}. The molecular mechanics energy (E_{MM}) is composed of the van der Waals energy (including the internal energy) (E_{vdw}) and the electrostatic energy (E_{elec}). The polar solvation component (G_{psolv}) is calculated using either the PB equation or the GB method. The nonpolar solvation component (G_{npsolv}) is calculated using solvent accessible area with the γ parameter set to 0.00542 kcal/(mol Å^2) for PB and 0.0072 kcal/(mol Å^2) for GB, and the b parameter set to 0.92 kcal/mol for the PB method and 0 kcal/mol for the GB method. The MM-PBSA and MM-GBSA methods are postprocessing methods. The typical procedure is to evaluate evenly spaced snapshots from a molecular dynamics trajectory of a ligand–receptor complex, which was run in either explicit or implicit solvent. This is known as the single trajectory approach and is the most commonly used technique for its speed and simplicity relative to the multiple trajectory approach, which requires additional, separate simulations of the receptor and ligand. For explicit solvent trajectories, only the coordinates of the protein/DNA and the ligand are extracted and used for the implicit solvent calculation run in single-point mode. The Amber 11 suite of biomolecular simulation programs is the most popular set of tools for application of these methods (34, 35). The Amber suite comes with its own Poisson-Boltzmann solver known as *pbsa*; however, the Delphi program can be used as well (29, 36). For the GB method, the two most popular GB implicit solvent models are the HCT (Hawkins, Cramer, Truhlar) (37, 38) and the OBC-I and OBC-II (Onufriev, Bashford, Case) (39) models, which differ in the manner in which the effective Born radii are scaled.

Hou et al. (40) from Wang's group have published an outstanding assessment of the performance of MM-PBSA and MM-GBSA methods accounting for factors such as the length of simulation, the solute dielectric used to calculate G_{psolv}, and entropy calculations. In terms of simulation length, Hou et al. report that longer simulations (4.8 ns) were not better than shorter simulations (400 ps) in predicting ΔG_{bind}. In a previous study of the single-trajectory MM-PBSA and MM-GBSA approach applied to the prediction of ΔG_{bind} of inhibitors of *Plasmodium falciparum* dihydrofolate reductase, Rastelli et al. (41) reported that simple energy minimization of their complexes performed as well as 2-ns simulation in explicit solvent in predicting ΔG_{bind}. Hou et al. report that inclusion of the conformational entropy ($-T\Delta S$) term computed via normal-mode analysis did not improve the results in their study. They cite problems with the normal-mode analysis used, including wide fluctuations between trajectory snapshots and the known issues of anharmonic effects with normal-mode analysis. In their comparison of the PB versus GB methods for estimation of G_{psolv}, Hou et al. report that the PB method provides a more accurate estimate of absolute ΔG_{bind}; however, GB performs better in estimating the relative or $\Delta\Delta G_{bind}$. Of the GB methods evaluated (HCT, OBC-I, and OBC-II), they report that the OBC-I method was most accurate for the systems they studied (except for systems with metals, such as cytochrome C peroxidase, owing to lack of parameters for the metal ion). Not surprisingly, using a polar interior solute dielectric constant ($\varepsilon_{in} = 4$) gave best results for proteins with polar (highly charged) binding sites like thrombin, and lower dielectric ($\varepsilon_{in} = 1$) gave best results for proteins with more hydrophobic binding sites. For systems without metals, Hou et al. (40) recommend using the MM-GBSA OBC-I method with careful selection of solute dielectric.

Yang et al. (42) from Ren's group performed an exhaustive study of the MM-PBSA and MM-GBSA approaches in evaluation of inhibitors of seven families of proteins. They too found that the normal-mode analysis estimation of conformational entropy provided no improvement in correlation with experiment. Yang et al. also studied the impact of solute dielectric of the protein on correlation of their predicted ΔG_{bind} with experimental ΔG_{bind} and found considerable improvement in correlation when using $\varepsilon_{in} = 4$ with their set of enzymes with polar binding sites (42).

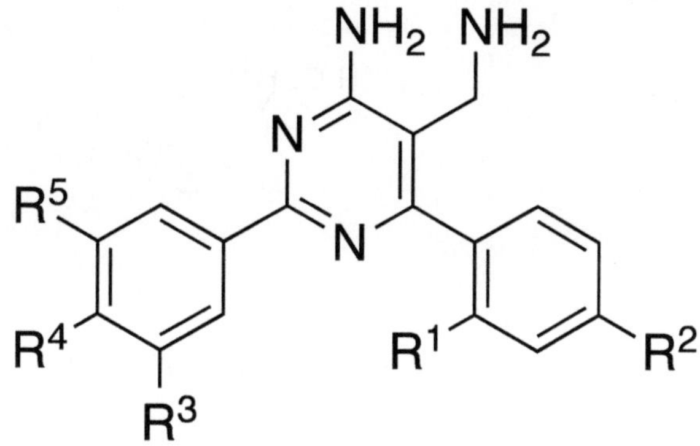

Fig. 4 Structure drawing of aminomethylpyrimidine inhibitor analogues of DPP-IV

Desheng et al. (43) from Mingjuan's group report on the application of the MM-GBSA method in predicting the structure–activity relationship in a series of aminomethylpyrimidine (Fig. 4) inhibitors of DPP-IV. In this work they used the amber *ff03* force field (44) for the protein and the general amber force field (*gaff*) (45) for the small-molecule inhibitors. They performed postprocessing using the MM-GBSA OBC-I model on the last 0.8 ns from 3-ns production dynamics in a periodic box of TIP3P water to predict ΔG_{pred}. Their estimation of ΔG_{pred} successfully predicted the qualitative trend observed in experimental IC_{50} values of the five inhibitor analogues studied.

The MM-PBSA and MM-GBSA methods have been used in combined molecular docking and molecular dynamics studies. A sampling of these studies includes investigations of inhibitors of murine double minute MDM2/MDMX (46), identification of a phosphoinositide-3-kinase inhibitors (47), and the human topoisomerase IIαATPase domain (48). Other studies from direct structures of enzyme–inhibitor complexes include work on inhibitors of Aurora kinase A (49) and epidermal growth factor receptor (50).

2.2.1 Study of Imatinib Binding to c-Abl Kinase

This method is adapted from the MM-PBSA study published by Dubey and Ojha (51). The same starting structure, PDB accession code 2HYY (52), was used. This structure is a tetramer. Chain A was isolated along with the corresponding drug (residue STI) and each structure split into separate PDB files. The missing residue sidechains in the protein were repaired using the DeepView Swiss PDB viewer (53). Hydrogen atoms were added to the imatinib (STI) structure using the Chimera program (University of California, San Francisco) (54). The Amber 11 Antechamber program (55) was used to parameterize (add gaff parameters) and add Austin Model 1 Bond Charge Correction AM1-BCC charges to the imatinib (STI) drug (Fig. 5). The Amber *ff10* force field, a composite

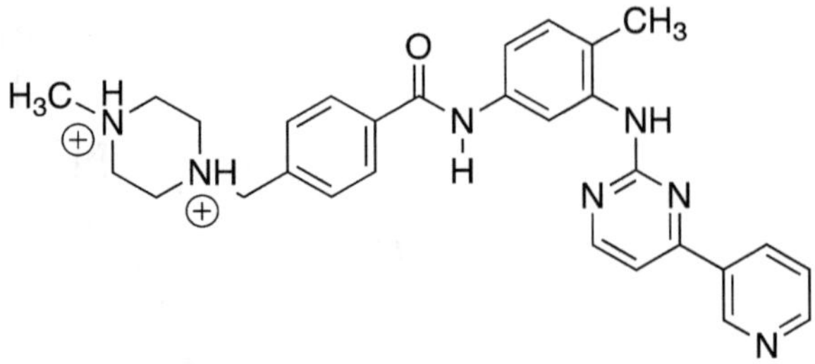

Fig. 5 Structure drawing of imatinib

of *ff99SB* (56), *ff99bsc0* (57), and *chiOL3* (58) parameters, was used to parameterize the protein. The initial structure prepared was an in vacuo structure that was refined using 500 steps of steepest descents followed by 1,000 steps of conjugate gradient minimization using an 18 Å nonbonded cutoff. The refined structure was solvated in an octahedral periodic box of TIP3P water (25) using a 10 Å spacing between protein and box edge with Na^+ ions added to neutralize the overall net charge of the system. Long-range electrostatics in the system were accounted for using the PME method (15, 16) with a 10 Å short-range cutoff. The system was energy minimized keeping the protein and drug atom positions restrained using a harmonic restraint of 200 kcal/mol $Å^2$ followed by energy minimization of the entire system. A 100-ps molecular dynamics run was performed keeping the protein and drug atom positions restrained with a harmonic force of 10 kcal/mol $Å^2$. The NVT ensemble (in which the number of particles "*N*", the volume "*V*," and temperature "*T*" in the ensemble are constant) was used for the restrained dynamics run, with a Langevin thermostat warming the system to 300 K with a collision frequency $\gamma = 5\ ps^{-1}$. A different seed for the random number generator for each dynamics run was used to avoid potential synchronization artifacts with the Langevin thermostat (59). A 2-fs time step was used and the shake constraint (60) was applied to all bonds. The system was subjected to 500 ps of molecular dynamics using the NPT ensemble with a temperature of 300 K and pressure equal to 1 bar. Pressure was maintained using the Berendsen barostat (61). Protein backbone RMSD (root-mean-square deviation) analysis (Fig. 6) reveals that the system equilibrates after 250 ps of simulation. The overall structure model after 500 ps of simulation is shown in Fig. 7. The MMPBSA.py tool in Amber 11 was used to analyze the last 100 snapshots from the 500-ps production dynamics trajectory using both GB (OBC-I method) and PB (Amber pbsa program) methods with an interior (solute) dielectric of 1 and an exterior dielectric of 80. The MMPBSA.py tool uses the energy minimization and normal-mode analysis routines in the nucleic acid builder package (62).

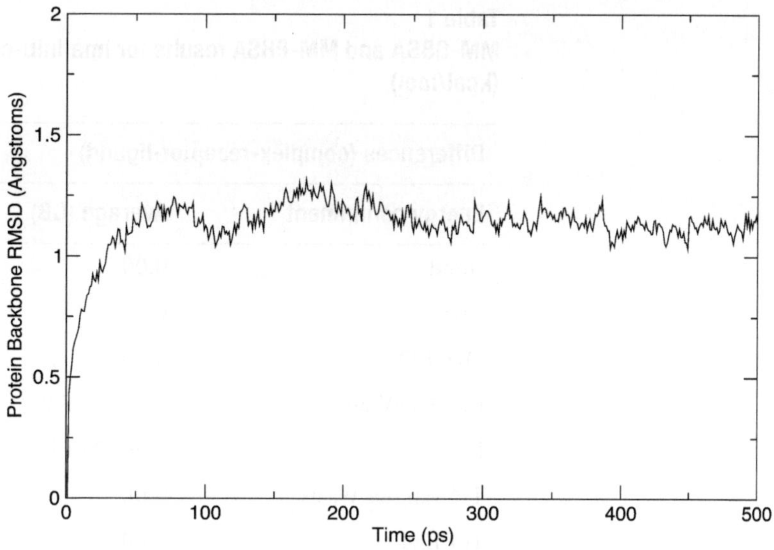

Fig. 6 Protein backbone atom RMSD plot

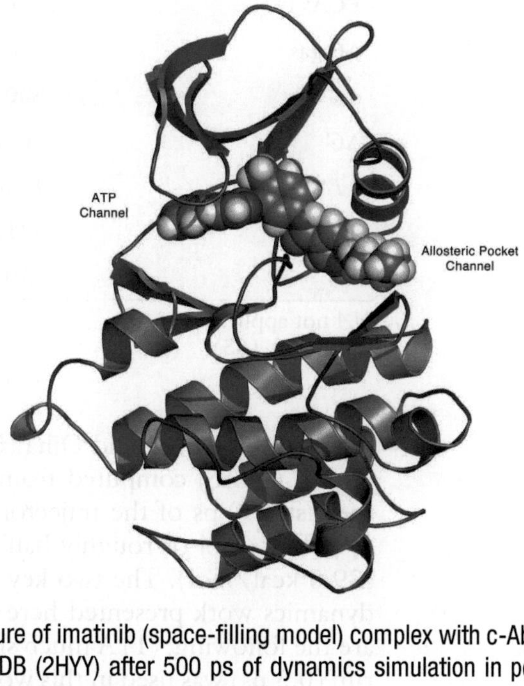

Fig. 7 Structure of imatinib (space-filling model) complex with c-Abl kinase (cartoon) from PDB (2HYY) after 500 ps of dynamics simulation in periodic box of TIP3P water. Figure prepared using Pymol

Table 1 (63) presents the results for the MM-GBSA and MM-PBSA analysis of the trajectory snapshots. The MM-GBSA result overestimates ΔG more than the MM-PBSA result. The MM-PBSA ΔG estimate computed here (−41.7 kcal/mol) is in good agreement with the MM-PBSA ΔG of −45.1 kcal/mol

Table 1
MM-GBSA and MM-PBSA results for imatinib-cAbl kinase complex (kcal/mol)

Differences (complex-receptor-ligand)		
Energy component	**Average (GB)**	**Average (PB)**
Bond	0.00	0.00
Angle	0.00	0.00
DIHED	0.00	0.00
van der Waals	−66.12 ± 3.02	−66.12 ± 3.02
EEL	−293.34 ± 13.1	−293.34 ± 13.1
1-4 van der Waals	0.00	0.00
1-4 EEL	0.00	0.00
EGB	308.72 ± 13.10	323.02 ± 12.82
ESURF	−7.84 ± 0.13	NA
ECAVITY	NA	−5.23 ± 0.20
ΔG gas	−359.46 ± 13.72	−359.46 ± 13.72
ΔG solv	300.88 ± 13.09	317.79 ± 12.78
ΔG	−58.59 ± 2.84	−41.67 ± 4.11
$-T\Delta S$	17.57 ± 2.13	17.57 ± 2.13
ΔG_{bind}	−41.0	−24.1
ΔG_{exp} [a]	−10.37	−10.37

NA not applicable
[a]Pricl et al. (63)

reported by Dubey and Ojha (51). However, the entropic contribution ($-T\Delta S$) computed from 10 evenly spaced snapshots from the last 100 ps of the trajectory using normal-mode analysis was 17.6 kcal/mol or roughly half that reported by Dubey and Ojha (39.0 kcal/mol). The two key differences between the molecular dynamics work presented here and the work of Dubey and Ojha are the following: (1) A much shorter 500-ps production dynamics run (0.5 ns) was used in this work compared with the much longer 4- to 12-ns production run used in the work of Dubey and Ojha. (2) The AM1-BCC charge model was used on the small molecule in this work as compared with the more lengthy ab initio HF/6-31G* calculations followed by RESP (restrained electrostatic potential) fitting in the work of Dubey and Ojha. The present work confirms the observation by Hou et al. (40) that in MM-PBSA calculations, longer simulations do not necessarily produce improved results in terms of the ΔG values excluding the entropy term. From Table 1, the PB value for ΔG_{bind} (−24.1 kcal/mol) including the entropy term is closer to the experimental value (−10.37 kcal/mol) reported by Pricl et al. (63) than is the GB result (−41.0 kcal/mol); however, both values computed from the shorter simulation overestimate the binding free energy. These results indicate that longer simulation time might be needed for a more reliable estimate of the entropic contribution ($-T\Delta S$). The shortcomings of normal-mode analysis discussed earlier and the significant computational cost involved make a case for leaving the entropy term out of the equation where this might be tolerated by the research problem at hand.

3 Steered Molecular Dynamics

Steered molecular dynamics (SMD), a nonequilibrium simulation, is a powerful tool for studying the elasticity of proteins (64), protein unfolding (65, 66), DNA unwrapping in histones (67), ion channels (68, 69), membrane permeability (70), and ligand–protein binding (71–76), among examples too numerous to cover here. The SMD method can be used to study the dissociation of a reversible inhibitor/ligand from the enzyme (77). The technique can provide information about the binding free energy via evaluation of the Potential of Mean Force (PMF) and can also provide information about conformational mechanism and interactions with key amino acid residues during the binding process. The opposite approach to this application of the SMD technique is the pioneering efforts of Shaw's group, who have studied the association of ligands to G protein-coupled receptors (78) and the binding of reversible inhibitors to allosteric and target binding sites of kinases (79) using long equilibrium molecular dynamics simulations without application of unnatural forces.

Allen and Bevan (80) have applied the SMD technique to the study of reversible inhibitors of monoamine oxidase B, and this paper will be used as an example for discussion of the "constant velocity" SMD pulling method. They used the Gromacs suite of biomolecular simulation programs (18) with the united-atom Gromos 43a1 force field to parameterize the lipid bilayer, protein, and small-molecule inhibitors. The protein was inserted into their mixed bilayer composed of phosphatidyl choline (POPC) and phosphatidyl ethanolamine (POPE) lipids in a ratio known to be consistent for a mitochondrial membrane. Each inhibitor-bound system studied was preequilibrated in a periodic box of SPC water (20) with the simulations run using the NPT ensemble at 300 K and 1 atm pressure for 20 ns. Full atomic coordinates and velocities were saved in 200-ps increments giving five replicates for each inhibitor-bound system. A dummy atom was attached to an atom (the SMD atom shown in Fig. 7) of the inhibitor nearest to the

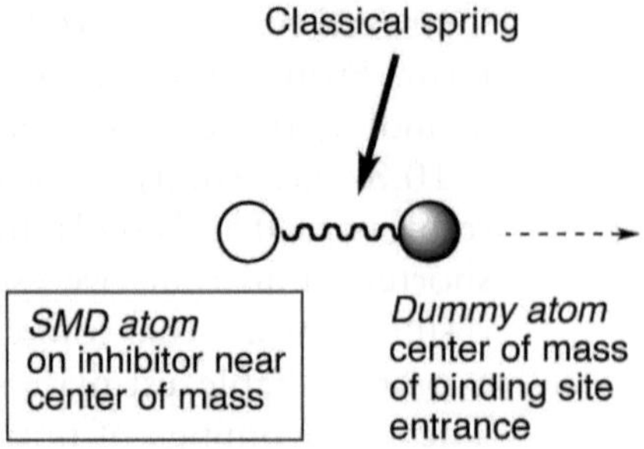

Fig. 8 SMD atom and dummy atom connected by "classical" spring showing direction of pulling (*hashed arrow*)

inhibitors' center-of-mass by means of a classical spring. The dummy atom (Fig. 8) is located at the center of mass of the binding site opening and was used to determine the direction of the pulling vector (hashed arrow in Fig. 8). The pulling force *F* is defined by Eq. 10 where the potential energy *U* is defined in Eq. 11.

$$\bar{F} = -\Delta U \tag{10}$$

$$U = 12k[vt - r - r_0 \bullet n]^2 \tag{11}$$

In Eq. 11, *k* is the spring force constant, *v* is the pulling velocity, *t* is time, *r* is the actual position of the SMD atom, r_o is the initial position of the SMD atom, and *n* is the pulling direction. Allen and Bevan used a force constant of 1,000 kJ/mol nm^2 and a constant pulling velocity of 0.01 nm/ps (80) in their work. In constant velocity SMD simulations, the velocity *v* is held constant and the force *F* changes in response to changes in the potential energy determined by change in distance (δr). The amount of force required to pull the inhibitor out of the binding site will be dependent on the nonbonded forces (i.e., van der Waals and electrostatic forces) from interactions of the inhibitor with amino acid residues of the protein receptor or enzyme and solvent (water). The force *F* and the displacement $\delta r = (r - r_o)$ are plotted and the integration of the curve obtained (Eq. 12) gives the amount of work *W* required to pull the inhibitor from the protein. The Jarzynski equality (81) is used to equate the work *W* to the PMF for a nonequilibrium process as shown in Eq. 13, where k_B is the Boltzmann constant and *T* is temperature.

$$W(r) = v\int_0^r F(r)\mathrm{d}r \tag{12}$$

$$\exp\left(\frac{-\Delta PMF}{k_B T}\right) = \left\langle \exp\left(\frac{-W}{k_B T}\right)\right\rangle \tag{13}$$

Allen and Bevan compared pulling calculations of the inhibitors they studied when the protein was embedded in a lipid bilayer versus alone in a water box (80). They found that the magnitude of the peak of pulling force F required to pull the inhibitor out of the binding site was higher in the case of the bilayer compared with solvent alone (except for one of the seven inhibitors studied). In addition, the peak force occurred before the inhibitor left the binding site, leading Allen and Bevan to conclude that the observed higher force is due to the lipid packing forces on the protein allosterically regulating the residues in the binding site. Allen and Bevan compared experimental ΔG_{dissoc} to peak force and ΔPMF. They found that the ΔPMF value correlated better with ΔG_{dissoc} than did the peak force. Allen and Bevan also reported a strong correlation between ΔG_{dissoc} and the sum of per-residue interaction energies (80). In brief, this SMD study showed that more replicate SMD simulations and reduced velocity pulling and reduced spring force constant would be needed to obtain more accurate values of ΔPMF, with the downside being increased computational cost.

Apart from estimating binding affinity using the Jarzynski equality, the SMD technique can also be used to explore dissociation paths for the inhibitor. One example of this application is presented in a recent study of dissociation paths for imatinib (Fig. 5) from its target kinases c-Kit and c-Abl as reported by Yang et al. (82). A common theme in all SMD simulations is to start with the structure of the enzyme–inhibitor complex and perform an equilibrium molecular dynamics simulation in a water box. Evenly spaced or randomly selected snapshots are taken from the "equilibrium" region of the dynamics trajectory (the area where the protein backbone atom RMSD levels off), and these snapshots are the starting point for the "replicate" nonequilibrium SMD simulations. Yang et al. studied pulling the drug out along the adenosine triphosphate (ATP) channel versus the allosteric-pocket channel (Fig. 7). They used a random sampling of five snapshots from a 4-ns dynamics simulation in a water box for each of their SMD simulations. For their SMD simulations, they used a force constant of 4 kcal/mol Å^2 and a velocity of 0.025 Å/ps with a simulation length of 1 ns. For both c-Kit and c-Abl, Yang et al. reported that the drug preferentially dissociates via the ATP channel rather than the allosteric-pocket channel as it takes less force to move the drug through the ATP channel (82). The predominant force difference contribution arose from the van der Waals interactions.

4 Conclusions

The endpoint methods discussed here continue to be very useful for estimation of binding affinity and exploring interactions that contribute to drug binding in the system studied. These methods are useful tools for more detailed energetic analysis of the results produced from docking studies (17). The LIE method, once fine-tuned by regression fit to known experimental data (training set) for the small-molecule type and enzyme/receptor of interest, is very good for predicting absolute free-energy values (10). The MM-PBSA method is good for predicting absolute free-energy values; however, the MM-GBSA method is better at estimating relative differences between different analogues in the data. Both the MM-PBSA and MM-GBSA methods benefit from careful selection of interior dielectric (solute dielectric) to account for the polarity of the enzyme binding site (40). Estimation of the configurational entropy component of binding free energy remains a challenge. The present normal-mode methods do not improve the accuracy of the results.

The SMD method can be used to estimate binding affinity and is also a nice tool for investigation of dissociation paths for the inhibitor or drug. Like the endpoint methods, the SMD technique is useful for in depth study of complexes produced from molecular docking study (73). The estimation of binding affinity requires a more rigorous approach, whereas more replicate SMD simulations with slower pulling velocities taking a longer time are required to obtain more accurate estimates of the potential of mean force.

Thus the endpoint binding free-energy methods and the SMD method each have strengths and weaknesses. These methods are molecular mechanics based, however, and therefore cannot account for bond formation, transition states, or coordination to metal centers. Higher-level hybrid QM/MM methods are recommended for describing these more complex problems (7, 83). Despite their limitations, the endpoint and SMD methods are versatile tools for the study of ligand–receptor interactions, binding energies, and (specific to the SMD method) ligand dissociation paths.

References

1. Durrant JD, McCammon JA (2011) Molecular dynamics simulations and drug discovery. BMC Biol 9:71
2. Paul SM, Mytelka DS, Dunwiddie CT et al (2010) How to improve R&D productivity: the pharmaceutical industry's grand challenge. Nat Rev Drug Discov 9(3):203–214
3. Wang R, Lu Y, Wang S (2003) Comparative evaluation of 11 scoring functions for molecular docking. J Med Chem 46(12): 2287–2303
4. Cole JC, Murray CW, Nissink JW et al (2005) Comparing protein-ligand docking programs is difficult. Proteins 60(3):325–332
5. Singh J, Petter RC, Baillie TA, Whitty A (2011) The resurgence of covalent drugs. Nat Rev Drug Discov 10(4):307–317
6. Metzler WJ, Yanchunas J, Weigelt C et al (2008) Involvement of DPP-IV catalytic residues in enzyme-saxagliptin complex formation. Protein Sci 17(2):240–250
7. Lodola A, Mor M, Sirirak J, Mulholland AJ (2009) Insights into the mechanism and inhibition of fatty acid amide hydrolase from quantum mechanics/molecular mechanics (QM/MM) modelling. Biochem Soc Trans 37(Pt 2):363–367
8. Åqvist J, Luzhkov VB, Brandsdal BO (2002) Ligand binding affinities from MD simulations. Acc Chem Res 35(6):358–365
9. de Amorim HL, Caceres RA, Netz PA (2008) Linear interaction energy (LIE) method in lead discovery and optimization. Curr Drug Targets 9(12):1100–1105

10. Almlöf M, Carlsson J, Åqvist J (2007) Improving the accuracy of the linear interaction energy method for solvation free energies. J Chem Theory Comput 3(6):2162–2175
11. Almlöf M, Brandsdal BO, Åqvist J (2004) Binding affinity prediction with different force fields: examination of the linear interaction energy method. J Comput Chem 25(10): 1242–1254
12. Åqvist J, Medina C, Samuelsson JE (1994) A new method for predicting binding affinity in computer-aided drug design. Protein Eng 7(3):385–391
13. Hansson T, Marelius J, Åqvist J (1998) Ligand binding affinity prediction by linear interaction energy methods. J Comput Aided Mol Des 12(1):27–35
14. Marelius J, Kolmodin K, Feierberg I, Åqvist J (1998) Q: a molecular dynamics program for free energy calculations and empirical valence bond simulations in biomolecular systems. J Mol Graph Model 16(4–6):213–225, 261
15. Darden T, York D, Pedersen L (1993) Particle mesh Ewald: an N-log(N) method for Ewald sums in large systems. J Chem Phys 98: 10089–10092
16. Essmann U, Perera L, Berkowitz ML et al (1995) A smooth particle mesh Ewald potential. J Chem Phys 103:8577–8592
17. Timmers LF, Ducati RG, Sanchez-Quitian ZA et al (2011) Combining molecular dynamics and docking simulations of the cytidine deaminase from *Mycobacterium tuberculosis* H37Rv. J Mol Model 18:467–479
18. Hess B, Kutzner C, van der Spoel D, Lindahl E (2008) GROMACS 4: algorithms for highly efficient, load-balanced, and scalable molecular simulation. J Chem Theory Comput 4(3): 435–447
19. Morris GM, Huey R, Lindstrom W et al (2009) AutoDock4 and AutoDockTools4: automated docking with selective receptor flexibility. J Comput Chem 30(16):2785–2791
20. Berendsen HJ, Postma JP, van Gunsteren WF, Hermans J (1981) Interaction models for water in relation to protein hydration. In: Pullman B (ed) Intermolecular forces. D. Reidel, Dordrecht, pp 331–342
21. Boukharta L, Keranen H, Stary-Weinzinger A et al (2011) Computer simulations of structure-activity relationships for HERG channel blockers. Biochemistry 50(27):6146–6156
22. Jones G, Willett P, Glen RC (1995) Molecular recognition of receptor sites using a genetic algorithm with a description of desolvation. J Mol Biol 245(1):43–53
23. Jones G, Willett P, Glen RC et al (1997) Development and validation of a genetic algorithm for flexible docking. J Mol Biol 267(3): 727–748
24. Jorgensen WL, Maxwell DS, Tirado-Rives J (1996) Development and testing of the OPLS all-atom force field on conformational energetics and properties of organic liquids. J Am Chem Soc 118(45):11225–11236
25. Jorgensen W, Chandrasekhar J, Madura J, Klein M (1983) Comparison of simple potential functions for simulating liquid water. J Chem Phys 79:926–935
26. Åqvist J (1996) Calculation of absolute binding free energies for charged ligands and effects of long-range electrostatic interactions. J Comput Chem 17(14):1587–1597
27. Nicolotti O, Giangreco I, Miscioscia TF et al (2010) Screening of benzamidine-based thrombin inhibitors via a linear interaction energy in continuum electrostatics model. J Comput Aided Mol Des 24(2):117–129
28. Chen SL, Zhao DX, Yang ZZ (2010) An estimation method of binding free energy in terms of ABEEMsigmapi/MM and continuum electrostatics fused into LIE method. J Comput Chem 32(2):338–348
29. Rocchia W, Alexov E, Honig B (2001) Extending the applicability of the nonlinear Poisson-Boltzmann equation: Multiple dielectric constants and multivalent ions. J Phys Chem B 105:6507–6514
30. Rocchia W, Sridharan S, Nicholls A et al (2002) Rapid grid-based construction of the molecular surface and the use of induced surface charge to calculate reaction field energies: applications to the molecular systems and geometric objects. J Comput Chem 23(1):128–137
31. Baker N, Sept D, Joseph S et al (2001) Electrostatics of nanosystems: application to microtubules and the ribosome. Proc Natl Acad Sci USA 98(18):10037–10041
32. Seifert MH, Kraus J, Kramer B (2007) Virtual high-throughput screening of molecular databases. Curr Opin Drug Discov Devel 10(3):298–307
33. Chen J, Brooks CL III, Khandogin J (2008) Recent advances in implicit solvent-based methods for biomolecular simulations. Curr Opin Struct Biol 18(2):140–148
34. Case, D., Darden, T., Cheatham, T., et al. (2010) AMBER 11. Full featured suite of biomolecular simulation programs. San Francisco., University of California. http://ambermd.org/doc11/Amber11.pdf
35. Case DA, Cheatham TE III, Darden T et al (2005) The Amber biomolecular simulation programs. J Comput Chem 26:1668–1688
36. Honig B, Nicholls A (1995) Classical electrostatics in biology and chemistry. Science 268(5214):1144–1149
37. Hawkins GD, Cramer CJ, Truhlar DG (1995) Pairwise solute descreening of solute charges from a dielectric medium. Chem Phys Lett 246:122–129

38. Hawkins GD, Cramer CJ, Truhlar DG (1996) Parametrized models of aqueous free energies of solvation based on pairwise descreening of solute atomic charges from a dielectric medium. J Phys Chem 100:19824–19839
39. Onufriev A, Bashford D, Case DA (2004) Exploring protein native states and large-scale conformational changes with a modified generalized born model. Proteins 55(2):383–394
40. Hou T, Wang J, Li Y, Wang W (2011) Assessing the performance of the MM/PBSA and MM/GBSA methods. 1. The accuracy of binding free energy calculations based on molecular dynamics simulations. J Chem Inf Model 51(1):69–82
41. Rastelli G, Del Rio A, Degliesposti G, Sgobba M (2010) Fast and accurate predictions of binding free energies using MM-PBSA and MM-GBSA. J Comput Chem 31(4):797–810
42. Yang T, Wu JC, Yan C et al (2011) Virtual screening using molecular simulations. Proteins 79(6):1940–1951
43. Desheng L, Jian G, Yuanhua C et al (2011) Molecular dynamics simulations and MM/GBSA methods to investigate binding mechanisms of aminomethylpyrimidine inhibitors with DPP-IV. Bioorg Med Chem Lett 21(22):6630–6635
44. Duan Y, Wu C, Chowdhury S et al (2003) A point-charge force field for molecular mechanics simulations of proteins based on condensed-phase quantum mechanical calculations. J Comput Chem 24(16):1999–2012
45. Wang J, Wolf R, Caldwell J et al (2004) Development and testing of a general amber force field. J Comput Chem 25(9):1157–1174
46. Lu SY, Jiang YJ, Zou JW, Wu TX (2011) Molecular modeling and molecular dynamics simulation studies on pyrrolopyrimidine-based alpha-helix mimetic as dual inhibitors of MDM2 and MDMX. J Mol Graph Model 30:167–178
47. Li Y, Zhang J, He D et al (2012) Characterization of molecular recognition of phosphoinositide-3-kinase alpha inhibitor through molecular dynamics simulation. J Mol Model 18(5): 1907–1916
48. Baviskar AT, Madaan C, Preet R et al (2011) N-fused imidazoles as novel anticancer agents that inhibit catalytic activity of topoisomerase IIalpha and induce apoptosis in G1/S phase. J Med Chem 54(14):5013–5030
49. Cheng Y, Cui W, Chen Q et al (2011) The molecular mechanism studies of chirality effect of PHA-739358 on Aurora kinase A by molecular dynamics simulation and free energy calculations. J Comput Aided Mol Des 25(2): 171–180
50. Wan S, Coveney PV (2011) Rapid and accurate ranking of binding affinities of epidermal growth factor receptor sequences with selected lung cancer drugs. J R Soc Interface 0(61): 1114–1127
51. Dubey KD, Ojha RP (2012) Conformational flexibility, binding energy, role of salt bridge and alanine-mutagenesis for c-Abl kinase complex. J Mol Model 18(5):1679–1689
52. Cowan-Jacob SW, Fendrich G, Floersheimer A et al (2007) Structural biology contributions to the discovery of drugs to treat chronic myelogenous leukaemia. Acta Crystallogr D Biol Crystallogr 63(Pt 1):80–93
53. Guex N, Peitsch MC (1997) SWISS-MODEL and the Swiss-Pdb viewer: an environment for comparative protein modeling. Electrophoresis 18:2714–2723
54. Pettersen E, Goddard T, Huang C et al (2004) UCSF Chimera-a visualization system for exploratory research and analysis. J Comput Chem 25:1605–1612
55. Wang J, Wang W, Kollman PA, Case DA (2006) Antechamber, an accessory software package for molecular mechanics calculations. J Mol Graph 25:247–260
56. Hornak V, Abel R, Okur A et al (2006) Comparison of multiple Amber force fields and development of improved protein backbone parameters. Proteins 65(3):712–725
57. Perez A, Marchan I, Svozil D et al (2007) Refinement of the AMBER force field for nucleic acids: improving the description of alpha/gamma conformers. Biophys J 92(11):3817–3829
58. Banas P, Hollas D, Zgarbova M et al (2010) Performance of molecular mechanics force fields for RNA simulations: stability of UUCG and GNRA hairpins. J Chem Theory Comput 6(12):3836–3849
59. Sindhikara DJ, Kim S, Voter AF, Roitberg A (2009) Bad seeds sprout perilous dynamics: stochastic thermostat induced trajectory synchronization in biomolecules. J Chem Theory Comput 5(6):1624–1631
60. Miyamoto S, Kollman P (1992) SETTLE: an analytical version of the SHAKE and RATTLE algorithms for rigid water models. J Comput Chem 13:952–962
61. Berendsen HJC, Postma JPM, vanGunsteren WF et al (1984) Molecular dynamics with coupling to an external bath. J Chem Phys 81(8):3584–3590
62. Macke T, Case DA (1998) Modeling unusual nucleic acid structures. In: Leontes N, SantaLucia J Jr (eds) Molecular modeling of nucleic acids. American Chemical Society, Washington, D.C., pp 379–393
63. Pricl S, Fermeglia M, Ferrone M, Tamborini E (2005) T315I-mutated Bcr-Abl in chronic myeloid leukemia and imatinib: insights from a computational study. Mol Cancer Ther 4(8):1167–1174

64. Sotomayor M, Schulten K (2007) Single-molecule experiments in vitro and in silico. Science 316(5828):1144–1148
65. Bung N, Priyakumar UD (2012) Computational investigation of the effect of thermal perturbation on the mechanical unfolding of titin I27. J Mol Model 18(6):2823–2829
66. Zhao GJ, Cheng CL (2012) Molecular dynamics simulation exploration of unfolding and refolding of a ten-amino acid miniprotein. Amino Acids 43(2):557–565
67. Ettig R, Kepper N, Stehr R et al (2011) Dissecting DNA-histone interactions in the nucleosome by molecular dynamics simulations of DNA unwrapping. Biophys J 101(8): 1999–2008
68. Dryga A, Warshel A (2010) Renormalizing SMD: the renormalization approach and its use in long time simulations and accelerated PMF calculations of macromolecules. J Phys Chem B 114(39):12720–12728
69. De Fabritiis G, Coveney PV, Villa-Freixa J (2008) Energetics of K+ permeability through Gramicidin A by forward-reverse steered molecular dynamics. Proteins 73(1):185–194
70. Wells DB, Abramkina V, Aksimentiev A (2007) Exploring transmembrane transport through alpha-hemolysin with grid-steered molecular dynamics. J Chem Phys 127(12):125101
71. Istyastono EP, Nijmeijer S, Lim HD et al (2011) Molecular determinants of ligand binding modes in the histamine H(4) receptor: linking ligand-based three-dimensional quantitative structure-activity relationship (3D-QSAR) models to in silico guided receptor mutagenesis studies. J Med Chem 54(23):8136–8147
72. Patel JS, Branduardi D, Masetti M et al (2011) Insights into ligand-protein binding from local mechanical response. J Chem Theory Comput 7(10):3368–3378
73. Whalen KL, Chang KM, Spies MA (2011) Hybrid steered molecular dynamics-docking: an efficient solution to the problem of ranking inhibitor affinities against a flexible drug target. Mol Inform 30(5):459–471
74. Mai BK, Li MS (2011) Neuraminidase inhibitor R-125489–a promising drug for treating influenza virus: steered molecular dynamics approach. Biochem Biophys Res Commun 410(3):688–691
75. Li D, Ji B, Hwang KC, Huang Y (2011) Strength of hydrogen bond network takes crucial roles in the dissociation process of inhibitors from the HIV-1 protease binding pocket. PloS One 6(4):e19268
76. Li C, Shen J, Li W et al (2011) Possible ligand release pathway of dipeptidyl peptidase IV investigated by molecular dynamics simulations. Proteins 79(6):1800–1809
77. Colizzi F, Perozzo R, Scapozza L et al (2010) Single-molecule pulling simulations can discern active from inactive enzyme inhibitors. J Am Chem Soc 132(21):7361–7371
78. Dror RO, Pan AC, Arlow DH et al (2011) Pathway and mechanism of drug binding to G-protein-coupled receptors. Proc Natl Acad Sci USA 108(32):13118–13123
79. Shan Y, Kim ET, Eastwood MP et al (2011) How does a drug molecule find its target binding site? J Am Chem Soc 133(24): 9181–9183
80. Allen WJ, Bevan DR (2011) Steered molecular dynamics simulations reveal important mechanisms in reversible monoamine oxidase B inhibition. Biochemistry 50(29):6441–6454
81. Jarzynski C (1997) Nonequilibrium equality for free energy differences. Phys Rev Lett 78:2690–2693
82. Yang LJ, Zou J, Xie HZ et al (2009) Steered molecular dynamics simulations reveal the likelier dissociation pathway of imatinib from its targeting kinases c-Kit and Abl. PloS One 4(12):e8470
83. Guallar V, Wallrapp FH (2010) QM/MM methods: looking inside heme proteins biochemistry. Biophys Chem 149(1–2):1–11

Chapter 8

Databases and In Silico Tools for Vaccine Design

Yongqun He and Zuoshuang Xiang

Abstract

In vaccine design, databases and in silico tools play different but complementary roles. Databases collect experimentally verified vaccines and vaccine components, and in silico tools provide computational methods to predict and design new vaccines and vaccine components. Vaccine-related databases include databases of vaccines and vaccine components. In the USA, the Food and Drug Administration (FDA) maintains a database of licensed human vaccines, and the US Department of Agriculture keeps a database of licensed animal vaccines. Databases of vaccine clinical trials and vaccines in research also exist. The important vaccine components include vaccine antigens, vaccine adjuvants, vaccine vectors, and vaccine preservatives. The vaccine antigens can be whole proteins or immune epitopes. Various in silico vaccine design tools are also available. The Vaccine Investigation and Online Information Network (VIOLIN; http://www.violinet.org) is a comprehensive vaccine database and analysis system. The VIOLIN database includes various types of vaccines and vaccine components. VIOLIN also includes Vaxign, a Web-based in silico vaccine design program based on the reverse vaccinology strategy. Vaccine information and resources can be integrated with Vaccine Ontology (VO). This chapter introduces databases and in silico tools that facilitate vaccine design, especially those in the VIOLIN system.

Key words Vaccine, Vaccine design, Vaccine database, VIOLIN vaccine database and analysis system, Vaxign, Reverse vaccinology, Immune epitope, Immunoinformatics, Vaccine adjuvant

1 Introduction

Vaccine research entered a new era with the advent of computer-assisted informatics and high-throughput technologies. Discussing rational vaccine design is more feasible and reasonable now than many years ago. Rational vaccine design has benefited from two areas of progress: vaccine databases and in silico vaccine design tools.

Vaccine-related databases have emerged as a way to digest the ever-increasing volume of data associated with vaccines and vaccination. Advanced DNA sequencing and molecular, cellular, and immunological methods have provided a huge amount of vaccine-related data. For example, high-throughput "-omics" technologies

Sandhya Kortagere (ed.), *In Silico Models for Drug Discovery*, Methods in Molecular Biology, vol. 993,
DOI 10.1007/978-1-62703-342-8_8, © Springer Science+Business Media, LLC 2013

(genomics, transcriptomics, proteomics, and metabolomics) have led the generation of DNA, RNA, protein, and metabolite data in different cells under different experimental conditions. Advanced bioinformatics tools have been developed to permit efficient use of these data. As a result, the publication of vaccine-associated papers has increased exponentially. In the first half of the twentieth century, 1,210 vaccine-related papers were published, according to PubMed. Between 2002 and 2011, however, more than 100,000 vaccine-related papers were published. Databases have been developed to store, reorganize, and categorize these data. Advanced literature-mining approaches are used to retrieve and discover new knowledge, which in turn supports more efficient accumulation of vaccine data in databases.

Various in silico vaccine design tools have been developed and are available online. For example, many immunoinformatics tools have been developed since the 1980s to predict T- and B-cell immune epitopes (1). Many predicted T- and B-cell immune epitopes are possible epitope vaccine targets. The availability of hundreds of pathogen genomes allows us to implement genome-wide scanning for the most promising vaccine targets. The analysis of high-throughput omics data enables the testing and screening of millions of possible vaccine targets in real time. The vaccine development strategy that starts with such scanning has been named *reverse vaccinology* (2). Mathematical modeling also supports vaccine design (3).

Databases and in silico tools play different but complementary roles in vaccine design. Databases collect experimentally verified vaccines and vaccine components. These classified, well-curated data provide the standards and training and testing data for future in silico prediction programs. In contrast, in silico tools provide computational methods to predict and design new vaccines and vaccine components. Predicted results can be experimentally verified and provide results that can be further stored in databases. The Vaccine Investigation and Online Information Network (VIOLIN; http://www.violinet.org) is a comprehensive Web-based vaccine database and analysis system targeted primarily for the use of vaccine researchers (4). VIOLIN contains databases of vaccines and vaccine components. In addition, VIOLIN includes Vaxign, a Web-based, genome-based vaccine design program based on reverse vaccinology and immunoinformatics strategies (5). This chapter summarizes databases and in silico tools that can be used for vaccine design, focusing on VIOLIN components.

2 Databases of Vaccines and Vaccine Components

Vaccine-related databases can be split into two groups: whole vaccine databases and databases of vaccine components. This section introduces these two groups of databases individually.

2.1 Databases of Vaccines

A vaccine database is a compilation of information about a list of vaccines, in the form of a relational database (Table 1) or a collection of HTML and PDF files. Vaccine databases generated by government agencies or international organizations are often targeted for

Table 1
Databases of vaccines and vaccine components

Resource name	Website URL	Comment
Vaccines		
US CDC-listed vaccines	www.cdc.gov/vaccines/vpd-vac/vaccines-list.htm	Vaccines used in USA
US FDA-licensed vaccines	www.fda.gov/cber/vaccines.htm	Licensed vaccines used in USA
Public Health Agency of Canada licensed vaccines	www.phac-aspc.gc.ca/dpg-eng.php#vaccines	Licensed vaccines used in Canada
UK Department of Health	www.dh.gov.uk/en/Publichealth/Immunisation	Official UK vaccination site
PATH Vaccine Resource Library	www.path.org/vaccineresources	Collection of vaccine resources
GAVI Alliance	www.gavialliance.org	Goal: save children's lives
VIOLIN	www.violinet.org	Comprehensive vaccine data
Huvax	www.violinet.org/huvax	Licensed human vaccines
Vevex	www.violinet.org/vevax	Licensed veterinary vaccines
Vaccine antigens		
AntigenDB	www.imtech.res.in/raghava/antigendb	A database of antigens
Protegen	www.violinet.org/protegen	Protective protein antigens
IEDB (Immune Epitope Database and Analysis Resource)	www.immuneepitope.org	Immune epitope database
"Virmugens" for generating live attenuated vaccines		
VirmugenDB	www.violinet.org/virmugendb	*See* text for details[a]
Vaccine adjuvants		
Vaxjo	www.violinet.org/vaxjo	Adjuvants and related vaccines
Vaccine vectors		
Vaxvec	www.violinet.org/vaxvec	Vaccine vectors database
Vaccine-induced animal responses		
Vaxar	www.violinet.org/vaxar	Vaccine-induced animal responses

(continued)

Table 1 (continued)

Resource name	Website URL	Comment
Host–vaccine interactions		
Vaxism	www.violinet.org/vaxism	Protective immunity
Vaccine manufacturers		
Vaxcom	www.violinet.org/vaxcom	Commercial vaccine companies
Vaccine experts		
Vaxperts	www.violinet.org/vaxperts	Vaccine experts with publications

CDC US Centers for Disease Control and Prevention, *FDA* US Food and Drug Administration

[a]Virmugens are pathogen genes that encode for virulence factors; their mutation in virulent pathogens allows for the generation of effective live attenuated vaccines

licensed vaccines and are presented to the public in the form of HTML and PDF files. For example, information on the vaccines licensed in the US market is provided by the US Food and Drug Administration (FDA) (http://www.fda.gov/cber/vaccines.htm). The US Centers for Disease Control and Prevention (CDC) Vaccine Information Statements (VISs) system provides information sheets that explain to vaccine recipients, their relatives, and legal representatives both the benefits and risks of a vaccine (http://www.cdc.gov/vaccines/pubs/vis/). Federal law in the USA requires that VISs be handed out for all vaccines before they are administered. The PATH Vaccine Resource Library (VRL; http://www.path.org/vaccineresources/) offers high-quality, scientifically accurate documents and links to specific diseases and topics in immunization. The Global Alliance for Vaccines and Immunization (GAVI) is an effort to support global collaboration for vaccines and immunization with a goal of saving children's lives (http://www.gavialliance.org).

With the major focus on vaccine research, VIOLIN stores and analyzes research data concerning commercial vaccines and vaccines that are being assessed in clinical trials or are in early stages of development (4). Constructed using the MySQL database management system, the VIOLIN vaccine database currently contains more than 3,000 vaccines or vaccine candidates for more than 170 pathogens manually curated from more than 1,600 peer-reviewed papers and government Web sites and databases. For each vaccine collected in the VIOLIN database, data include vaccine preparation, pathogen genes used and gene engineering, vaccine adjuvants and vectors, vaccine-induced host immune responses (including vaccine-related adverse events), and vaccine efficacy as a result of controlled experiments or clinical trials.

Because licensed vaccines are the focus of much attention in the public as well as in the vaccine research community, VIOLIN has two subdatabases that focus on licensed human and animal vaccines. The Huvax subdatabase (http://www.violinet.org/huvax) stores all licensed human vaccines in the USA and Canada. The data for these licensed human vaccines are annotated through manual curation and a vigorous review process. Huvax includes manually curated data including manufacturer, trade name, storage information, age at which the vaccine should be administered, and other relevant information. These vaccines are also listed by the CDC CVX codes (codes that indicate the product used in a vaccine) for tracking vaccination records. Huvax has provided cross-references based on these CDC CVX codes. Huvax provides ways for users to search, compare, and analyze licensed human vaccines. Various criteria can be used for querying human vaccines. Different vaccines can also be compared side by side. Vevax is a VIOLIN database of licensed veterinary vaccines (http://www.violinet.org/vevax).

2.2 Databases of Vaccine Components

As a manufactured product, a vaccine is composed of several components. A vaccine antigen(s) is a required component for any vaccine. Such a vaccine antigen can be part of a whole organism (e.g., live attenuated vaccine), a protective protein antigen, or an immune epitope. A live attenuated vaccine usually does not require a vaccine adjuvant that functions as an immune stimulant. However, killed whole organism or subunit vaccines usually need vaccine adjuvants. Otherwise, a weak or nonexistent immune response will be induced. Other vaccine components include vaccine vectors and vaccine preservatives. Databases of these vaccine components are briefly introduced here.

2.2.1 Databases of Vaccine Antigens

There are several types of vaccine antigens. The most common antigens used for vaccine development are protein antigens. Immune epitopes may also be manufactured and used for vaccine development. Other molecules can also be used as vaccine antigens (e.g., lipopolysaccharide) (6). The commonly used databases for protein and epitope antigens are introduced here.

AntigenDB (http://www.imtech.res.in/raghava/antigendb/) is an immunoinformatics database of pathogen antigens that stores sequences, structures, origins, and epitopes (7). However, although a protein antigen can induce a strong immune response, the induced immunity may not be protective against virulent pathogen challenge. Protegen (http://www.violinet.org/protegen) is a database of protective antigens in VIOLIN (8). Protective antigens are antigens that can induce protection against virulent pathogen infection. A protective antigen is the core part of a vaccine. Protegen curates more than 600 protective antigens. A customized Protegen BLAST program has been developed for searching similarities between a query sequence and the sequences of all Protegen proteins at the DNA or amino acid level.

At the epitope level, the Immune Epitope Database and Analysis Resource (IEDB) is the largest immune epitope database so far (http://www.immuneepitope.org/). IEDB has stored more than 65,000 antibody and T-cell epitopes since the database was established in 2004 (9). These immune epitopes cover various species (e.g., humans, nonhuman primates, and rodents) that have been well studied in the areas of infectious diseases (9). Because all stored epitopes in IEDB are manually curated from experimental studies, the data in IEDB have often been used as the gold standard for evaluating in silico epitope prediction tools (10).

2.2.2 Databases of Vaccine Adjuvants

Vaccine adjuvants enhance host immune responses to coadministered antigens in vaccines. Vaxjo (http://www.violinet.org/vaxjo/) is a Web-based central database and analysis system that curates, stores, and analyzes vaccine adjuvants and vaccines that use the adjuvants. The adjuvant information stored in Vaxjo includes adjuvant names, components, structure, appearance, storage, preparation, function, safety, and vaccines that use each adjuvant. All data are manually curated and are cited from reliable references including peer-reviewed publications, company Web sites, and government agencies. Bioinformatics scripts are developed and used to link vaccine adjuvants to adjuvanted vaccines stored in the general VIOLIN vaccine database. Presently, more than 100 vaccine adjuvants have been curated in Vaxjo. These adjuvants have been used in more than 380 vaccines stored in VIOLIN against more than 81 pathogens, cancers, or allergies. A user-friendly Web query and visualization interface permits interactive vaccine adjuvant search.

2.2.3 Databases of Other Vaccine Components

The VIOLIN database also includes databases for many other vaccine components. For example, VIOLIN includes a vaccine vector database, Vaxvec (http://www.violinet.org/vaxvec/). A vaccine vector is a weakened or killed version of a virus or bacterium that cannot cause disease in hosts (e.g., human) that is used in genetically engineered vaccines to transport genes coding for antigens into the body to induce an immune response. VirmugenDB (http://www.violinet.org/virmugendb) is a unique VIOLIN database that includes data for pathogen genes that encode for virulence factors, and their mutation in virulent pathogens allows for the generation of effective live attenuated vaccines. We coined the term "virmugen" for these genes. Currently VirmugenDB includes more than 200 virmugens.

Vaxar (http://www.violinet.org/vaxar) is a VIOLIN program that focuses on annotation and analysis of vaccine-induced animal responses. Vaxism (http://www.violinet.org/vaxism/) is another program targeting for vaccine-related host–pathogen interaction mechanisms (e.g., pathogenesis and protective immunity). VIOLIN also includes databases for vaccine experts, vaccine manufacturers, etc. (Table 1).

3 In Silico Tools for Vaccine Design

Two major in silico tools in vaccine design are based on two related but distinct strategies: immunoinformatics-based immune epitope prediction and reverse vaccinology-based protective protein antigen screening.

3.1 Immune Epitope Prediction Tools

Many immunoinformatics algorithms and resources are available for predicting T- and B-cell immune epitopes. T-cell epitopes are bound in a linear form to major histocompatibility complex (MHC) class I or class II molecules or to the human leukocyte antigen (HLA) if the host is human. The interface between ligands and T cells can now be modeled with high accuracy. However, more than 90% of B-cell epitopes are nonlinear and require the knowledge of native three-dimensional protein structure. It remains a challenge to predict B-cell immune epitopes in silico.

3.1.1 T-Cell Immune Epitopes and Their In Silico Prediction

T cells are activated by direct interaction with MHC class I or class II molecules on the surface of antigen-presenting cells (APCs). T-cell receptors can interact with peptides derived from endogenous and exogenous proteins bound in the cleft of MHC class I or class II molecules. MHC class I molecules present peptides that are eight to ten amino acids in length, leading to the activation of $CD8^+$ cytotoxic T lymphocytes (CTLs). MHC class I molecules typically present peptides derived from proteolytic digestion of endogenously synthesized proteins. In contrast, peptides presented by MHC class II molecules are longer, more variable in size, and have more complex anchor motifs than peptides presented by MHC class I molecules (11–13). MHC class II molecules bind peptides consisting of 11–25 amino acids and are recognized by $CD4^+$ T-helper cells. MHC class II molecules generally bind peptides derived from the cell membrane or exogenous proteins internalized by APCs.

As described above, peptide epitopes bound to MHC class I or II molecules are able to stimulate $CD8^+$ and $CD4^+$ T cells, respectively. Both $CD4^+$ T-helper cells and $CD8^+$ CTLs are critical to protective immunity and to vaccine efficacy. $CD4^+$ T-helper cells play an important role in the development of memory B-cell (antibody) and memory CTL (cytotoxic T-cell) responses. $CD4^+$ T-helper cells are also active against pathogens on their own. Therefore, $CD4^+$ T-helper cells have been called the "conductors of the immune system orchestra" (14). $CD8^+$ CTLs are able to directly kill infected target cells and thus are critical in the containment of viral and bacterial infection (15). Therefore, the prediction of peptide epitopes bound to MHC class I or II molecules is critical for effective vaccine design.

A deep understanding of the MHC antigen presentation pathways provides additional methods for T-cell immune epitope predictions. For example, the generation of an MHC class I epitope starts with the degradation of endogenous proteins into oligomeric fragments by cytosolic proteases, mainly the proteasome. These oligomeric fragments may escape from the attack of amino peptidases by entering the endoplasmic reticulum by the transporter associated with antigen presentation (TAP) (16). Therefore, besides the prediction of MHC class I binding affinity, the predictions of the MHC class I pathway can be improved by predictions of proteasomal cleavage and TAP transport efficiency (17–19).

Many tools for MHC class I and II epitope prediction have been developed (3). Examples include the SYFPEITHI website (20), BIMAS for predicting HLA binding (21), a proprietary algorithm EpiMatrix (22), Vaxitope in the VIOLIN Vaxign vaccine design system (5), and many tools available through the IEDB system (23). Each of these tools has been described and validated (10). Although none of these sites yield exactly the same predictions, all predictions can achieve a high degree of accuracy (often in the range of 90–95% positive predictive value for well-studied MHC class I or II) (10, 22).

In general, good performance has been achieved for MHC class I predictions. However, success is still limited for prediction of epitopes for MHC class II (24, 25). The low prediction accuracy of HLA-II binding peptides results from many factors, including insufficient or low-quality training data; limited stringency of binding due to relative permissiveness of the binding groove of HLA-II molecules for peptide binding; difficulty in identifying 9-mer binding cores within longer peptides; and the lack of consideration of the influence of flanking residues (24).

3.1.2 B-Cell Immune Epitopes and Their In Silico Prediction

Antibody response represents the first line of defense against most viral and bacterial pathogens. B-cell epitopes are separated into two groups: linear and discontinuous epitopes. A linear (also called sequential) epitope has a single continuous stretch of amino acids within a protein sequence. A discontinuous (also called conformational) epitope is an epitope whose residues are distantly separated in the sequence but have physical proximity through three-dimensional protein folding. Approximately 90% of B-cell epitopes are nonlinear epitopes, which makes the prediction of B-cell epitopes very difficult.

Owing to the difficulty of evaluating discontinuous epitopes, most experimentally detected epitopes are linear epitopes. Tools for prediction of linear B-cell epitopes include 3DEX, CEP, and Pepito (26–29). IEDB has collected a list of Web tools for B-cell epitope prediction (http://tools.immuneepitope.org/main/

html/bcell_tools.html). In general, these tools are not predictive and need to be refined (30, 31). The huge complexity of predicting B-cell epitopes results partly from the inherent flexibility in the complementarity-determining regions of the antibody and to post-translational modifications such as glycosylation. These factors can result in modification of B-cell epitopes.

To counter the low success rate of B-cell epitope prediction, MHC class II epitopes for $CD4^+$ T cells may be emphasized when antibody response is key in vaccine design. $CD4^+$ T-helper cells are critical to induce the activation of B cells that produce antibodies. B-cell antigens that contain significant MHC class II epitopes may outperform B-cell antigens without cognate help. An identified T-cell epitope may sometimes contain a B-cell epitope. In addition, B-cell epitopes may colocalize near or overlap MHC class II epitopes (32, 33).

3.2 Reverse Vaccinology for Vaccine Prediction

Since the complete genome of *Haemophilus influenzae* (a pathogenic bacterium) was published (34), thousands of pathogen genomes have been sequenced. Reverse vaccinology is a new strategy for vaccine development that starts with predicting putative vaccine candidates by in silico genomics analysis based on various criteria. Reverse vaccinology was first applied to the development of vaccines against serogroup B *Neisseria meningitidis* (35). Host protection against *N. meningitidis* infection mainly requires antibody response. The surface proteins and outer membrane proteins of *N. meningitidis* are most likely antigens that stimulate antibody responses and interact with the antibodies. Therefore, the prediction of bacterial surface or secreted proteins in *N. meningitidis* becomes the main criterion in the first application of reverse vaccinology in *N. meningitidis* vaccine targets. After candidate proteins are predicted, a high-throughput experimental screening approach can be used to evaluate the vaccine protection efficacy (35).

Since the first report of the application of reverse vaccinology, in 2000, many more criteria have been developed. For example, based on the observation that outer membrane proteins containing more than one transmembrane helix are often difficult to clone and purify (35), the number of transmembrane domains of a protein is often chosen by a research designer as another filtering criterion. Bacterial adhesins are proteins responsible for adherence, colonization, and invasion of microbial pathogens to host cells (36). Therefore, whether or not a candidate is an adhesin is now considered as an important criterion. Because many genomes have usually been sequenced for any pathogen, it is possible to select those conserved proteins that can be used to induce protection against multiple pathogenic strains. It is also important to screen for sequence similarities between pathogen proteins (vaccine candidate) and host (e.g., human) proteome. If a pathogen protein is homologous to a host protein, when administered in vivo, this pathogen protein is likely to induce one of two extreme reactions: an autoimmune disease or immune tolerance. Furthermore, immune epitope prediction based on amino acid sequences is also considered as a criterion for the application of reverse vaccinology in rational vaccine design (5).

Vaxign is the first Web-based, genome-wide vaccine target prediction program that uses a reverse vaccinology strategy and immunoinformatics-based prediction of MHC class I and II epitopes (http://www.violinet.org/vaxign) (5). Overall, the predicted features of Vaxign include:

- Protein subcellular location.
- Transmembrane helices.
- Adhesin probability.
- Conservation to human or mouse proteins.
- Sequence exclusion from genome(s) of nonpathogenic strain(s).
- Epitope binding to MHC class I and class II.

Vaxign uses a module-based design, i.e., each feature is implemented using an independent program and a user can choose which features to include in final result generation. To avoid reinventing the wheel, Vaxign uses existing open-source software programs as much as possible.

Vaxign has been demonstrated to successfully predict vaccine targets for many pathogens, including *Brucella* spp. (37, 38) and uropathogenic *Escherichia coli* (5). Currently more than 250 genomes have been precomputed using the Vaxign pipeline and are available for query at the Vaxign Web site. In addition, Vaxign allows a user to provide his or her own input sequences (up to 500 sequences at a time) and performs dynamic vaccine target prediction based on the input sequences.

4 Vaccine Ontology and its Application in Vaccine Design

Vaccine databases provide help with various aspects of vaccine knowledge and research. However, if a database is a collection of HTML or PDF files, such files are difficult to parse and analyze. If a database is a relational database, the database schema is useful for internal data exchange but outside data access and process are usually not feasible. Therefore, how to integrate the large volume of vaccine information is a challenge. A biomedical ontology is a consensus-based controlled vocabulary of terms and relations, with associated definitions that are logically formulated in such a way as to promote automated reasoning (39). Biomedical ontologies structure complex biomedical knowledge in specific domains in such a way as to permit shared understanding of biomedical concepts and data. The Vaccine Ontology (VO; http://www.violinet.org/vaccineontology) is a biomedical ontology in the domain of vaccines and vaccination (40). The collaborative, community-based

VO development aims to promote vaccine data standardization, integration, and computer-assisted reasoning.

The VO database can be used for vaccine data integration. As of January 2013, VO contains more than 5,000 terms, including more than 1,000 vaccines and vaccine candidates that are logically defined and represented. These vaccines or vaccine candidates are targeted for 70 pathogens. VO also stores data for vaccine components, including protective antigens, vaccine adjuvants, and vaccine vectors. The other data stored in VO include vaccine-induced immune responses, vaccine adverse events, and protection efficacy. Each VO term is logically linked with one or more terms in the same ontology. Many terms in VO are imported from other ontologies. The information is provided in a computer-understandable language called Web Ontology Language (OWL). Therefore, computer programs can be developed to parse and use the information in VO for appropriate vaccine design. An example of VO-supported automated reasoning is its application to vaccine literature mining. Compared with a direct PubMed search, the application of VO in searching the PubMed literature database for "live attenuated *Brucella* vaccine" increased the number of hits by more than tenfold with high specificity (41, 42). The application of VO also enhances the discovery of interferon-γ- and vaccine-associated gene networks (43). The role of VO in rational vaccine design consists of its application in computer-assisted automated reasoning.

5 Notes

This chapter introduces the following topics:

1. Commonly used vaccine-related databases.
2. In silico tools for vaccine design.
3. Vaccine Ontology and its application in vaccine design.

Many other method that are not introduced in this chapter can be used to improve vaccine design. For example, beside genomics-based vaccine target prediction in the scope of reverse vaccinology, high-throughput transcriptomics and proteomics technologies (e.g., microarray) can be used to analyze expression patterns of thousands of genes in a genome. The upregulated genes represent genes that encode for possible protein antigens. High-throughput screening of human or animal serum samples for antibodies provides a more direct way to identify antigenic vaccine targets because antibodies are the products of immune responses (44). The availability of three-dimensional structures of proteins may facilitate epitope prediction and antigen discovery (45, 46). Mathematical modeling can also be used to simulate host–pathogen interactions and can contribute dramatically to the understanding of fundamental protective immunity. Mathematical modeling approaches have also supported the optimization of vaccination procedures (3)

Despite extensive efforts, many infectious diseases, including AIDS, tuberculosis, and malaria, still lack effective and safe vaccines. Many noninfectious diseases, including cancer, autoimmune diseases, and allergy, also lack safe and effective vaccines. One reason for this lack is the absence of precise protective immune mechanism and immunological correlates of protection. These challenges can only be addressed with continuous productive research. Better rational vaccine design methods can only be developed after fundamental protective immunity mechanisms are better understood.

Acknowledgments

This work has been supported by grant R01AI081062 from the USA National Institute of Allergy and Infectious Diseases (NIAID).

References

1. De Groot AS, Sbai H, Aubin CS et al (2002) Immuno-informatics: mining genomes for vaccine components. Immunol Cell Biol 80: 255–269
2. Rappuoli R (2000) Reverse vaccinology. Curr Opin Microbiol 3:445–450
3. He Y, Rappuoli R, De Groot AS, Chen RT (2010) Emerging vaccine informatics. J Biomed Biotechnol 2010:218590
4. Xiang Z, Todd T, Ku KP et al (2008) VIOLIN: vaccine investigation and online information network. Nucleic Acids Res 36:D923–D928
5. He Y, Xiang Z, Mobley HL (2010) Vaxign: the first web-based vaccine design program for reverse vaccinology and applications for vaccine development. J Biomed Biotechnol 2010:297505
6. Bhattacharjee AK, Izadjoo MJ, Zollinger WD et al (2006) Comparison of protective efficacy of subcutaneous versus intranasal immunization of mice with a *Brucella melitensis* lipopolysaccharide subunit vaccine. Infect Immun 74:5820–5825
7. Ansari HR, Flower DR, Raghava GP (2010) AntigenDB: an immunoinformatics database of pathogen antigens. Nucleic Acids Res 38: D847–D853
8. Yang B, Sayers S, Xiang Z, He Y (2011) Protegen: a web-based protective antigen database and analysis system. Nucleic Acids Res 39:D1073–D1078 (database issue)
9. Peters B, Sidney J, Bourne P et al (2005) The immune epitope database and analysis resource: from vision to blueprint. PLoS Biol 3:e91
10. Wang P, Sidney J, Dow C et al (2008) A systematic assessment of MHC class II peptide binding predictions and evaluation of a consensus approach. PLoS Comput Biol 4:e1000048
11. Unanue ER (1992) Cellular studies on antigen presentation by class II MHC molecules. Curr Opin Immunol 4:63–69
12. Brown JH, Jardetzky TS, Gorga JC et al (1993) Three-dimensional structure of the human class II histocompatibility antigen HLA-DR1. Nature 364:33–39
13. Chicz RM, Urban RG, Gorga JC et al (1993) Specificity and promiscuity among naturally processed peptides bound to HLA-DR alleles. J Exp Med 178:27–47
14. Ahlers JD, Belyakov IM, Thomas EK, Berzofsky JA (2001) High-affinity T helper epitope induces complementary helper and APC polarization, increased CTL, and protection against viral infection. J Clin Invest 108:1677–1685
15. Plotnicky H, Cyblat-Chanal D, Aubry JP et al (2003) The immunodominant influenza matrix T cell epitope recognized in human induces influenza protection in HLA-A2/K(b) transgenic mice. Virology 309:320–329
16. Bulik S, Peters B, Ebeling C, Holzhutter H (2004) Cytosolic processing of proteasomal cleavage products can enhance the presentation efficiency of MHC-1 epitopes. Genome Inform 15:24–34
17. Peters B, Bulik S, Tampe R et al (2003) Identifying MHC class I epitopes by predicting the TAP transport efficiency of epitope precursors. J Immunol 171:1741–1749
18. Nielsen M, Lundegaard C, Lund O, Kesmir C (2005) The role of the proteasome in generating cytotoxic T-cell epitopes: insights obtained from improved predictions of proteasomal cleavage. Immunogenetics 57:33–41

19. Stranzl T, Larsen MV, Lundegaard C, Nielsen M (2010) NetCTLpan: pan-specific MHC class I pathway epitope predictions. Immunogenetics 62:357–368
20. Rammensee H, Bachmann J, Emmerich NP et al (1999) SYFPEITHI: database for MHC ligands and peptide motifs. Immunogenetics 50:213–219
21. Parker KC, Bednarek MA, Coligan JE (1994) Scheme for ranking potential HLA-A2 binding peptides based on independent binding of individual peptide side-chains. J Immunol 152:163–175
22. De Groot AS, Martin W (2009) Reducing risk, improving outcomes: bioengineering less immunogenic protein therapeutics. Clin Immunol 131:189–201
23. Zhang Q, Wang P, Kim Y et al (2008) Immune epitope database analysis resource (IEDB-AR). Nucleic Acids Res 36:W513–W518
24. Lin HH, Zhang GL, Tongchusak S et al (2008) Evaluation of MHC-II peptide binding prediction servers: applications for vaccine research. BMC Bioinformatics 9(Suppl 12):S22
25. Gowthaman U, Agrewala JN (2008) In silico tools for predicting peptides binding to HLA-class II molecules: more confusion than conclusion. J Proteome Res 7:154–163
26. Enshell-Seijffers D, Denisov D, Groisman B et al (2003) The mapping and reconstitution of a conformational discontinuous B-cell epitope of HIV-1. J Mol Biol 334:87–101
27. Schreiber A, Humbert M, Benz A, Dietrich U (2005) 3D-Epitope-Explorer (3DEX): localization of conformational epitopes within three-dimensional structures of proteins. J Comput Chem 26:879–887
28. Kulkarni-Kale U, Bhosle S, Kolaskar AS (2005) CEP: a conformational epitope prediction server. Nucleic Acids Res 33:W168–W171
29. Sweredoski MJ, Baldi P (2008) PEPITO: improved discontinuous B-cell epitope prediction using multiple distance thresholds and half sphere exposure. Bioinformatics 24:1459–1460
30. Blythe MJ, Flower DR (2005) Benchmarking B cell epitope prediction: underperformance of existing methods. Protein Sci 14:246–248
31. Reimer U (2009) Prediction of linear B-cell epitopes. Methods Mol Biol 524:335–344
32. Rajnavolgyi E, Nagy N, Thuresson B et al (2000) A repetitive sequence of Epstein-Barr virus nuclear antigen 6 comprises overlapping T cell epitopes which induce HLA-DR-restricted CD4(+) T lymphocytes. Int Immunol 12:281–293
33. Graham CM, Barnett BC, Hartlmayr I et al (1989) The structural requirements for class II (I-Ad)-restricted T cell recognition of influenza hemagglutinin: B cell epitopes define T cell epitopes. Eur J Immunol 19:523–528
34. Fleischmann RD, Adams MD, White O et al (1995) Whole-genome random sequencing and assembly of Haemophilus influenzae Rd. Science 269:496–512
35. Pizza M, Scarlato V, Masignani V et al (2000) Identification of vaccine candidates against serogroup B meningococcus by whole-genome sequencing. Science 287:1816–1820
36. Sachdeva G, Kumar K, Jain P, Ramachandran S (2005) SPAAN: a software program for prediction of adhesins and adhesin-like proteins using neural networks. Bioinformatics 21: 483–491
37. He Y, Xiang Z (2010) Bioinformatics analysis of Brucella vaccines and vaccine targets using VIOLIN. Immunome Res 6(Suppl 1):S5
38. Xiang Z, He Y (2009) Vaxign: a web-based vaccine target design program for reverse vaccinology. Procedia Vaccinol 1(1):23–29
39. Xiang Z, Courtot M, Brinkman RR et al (2010) OntoFox: web-based support for ontology reuse. BMC Res Notes 3:175
40. He Y, Cowell L, Diehl AD et al. (2009) VO: Vaccine ontology. The 1st international conference on biomedical ontology (ICBO 2009). 24-26 July 2009, Buffalo, NY. Nature Precedings. http://precedings.nature.com/documents/3553/version/1. doi: 10.1038/npre.2009.3553.1. Accessed 23 Apr 2012.
41. Xiang Z, He Y (2009) Improvement of PubMed literature searching using biomedical ontology. The 1st international conference on biomedical ontology (ICBO 2009). 24-26 July 2009, Buffalo, NY. Nature Precedings. http://precedings.nature.com/documents/3491/version/1. doi:10.1038/npre.2009.3491.1. Accessed 23 Apr 2012
42. Hur J, Xiang Z, Feldman EL, He Y (2011) Ontology-based Brucella vaccine literature indexing and systematic analysis of gene-vaccine association network. BMC Immunol 12:49
43. Özgür A, Xiang Z, Radev DR, He Y (2011) Mining of vaccine-associated IFN-γ gene interaction networks using the vaccine ontology. J Biomed Semantics 2(suppl 2):S8. doi:10.1186/2041-1480-2-S2-S8
44. Banai M, He Y (2012) Systems biology and bioinformatics help decipher Brucella antigens involved in clinical manifestation of the disease. Front Cell Infect Microbiol 2:34. doi:10.3389/fcimb.2012.00034
45. Serruto D, Rappuoli R (2006) Post-genomic vaccine development. FEBS Lett 580: 2985–2992
46. Mora M, Donati C, Medini D et al (2006) Microbial genomes and vaccine design: refinements to the classical reverse vaccinology approach. Curr Opin Microbiol 9: 532–536

Chapter 9

In Silico Models for B-Cell Epitope Recognition and Signaling

Hifzur Rahman Ansari and Gajendra P.S. Raghava

Abstract

Tremendous technological advances in peptide synthesis and modification in recent years have resolved the major limitations of peptide-based vaccines. B-cell epitopes are major components of these vaccines (besides having other biological applications). Researchers have been developing in silico or computational models for the prediction of both linear and conformational B-cell epitopes, enabling immunologists and clinicians to identify the most promising epitopes for characterization in the laboratory. Attempts are also ongoing in systems biology to delineate the signaling networks in immune cells. Here we present all possible in silico models developed thus far in these areas.

Key words B-cell epitopes, Linear, Conformational, Support vector machines, Hidden Markov models, Immunoinformatics

1 Introduction

The human immune system is responsible for the development of immunity, which includes innate and adaptive components. According to the traditional principles of immunology, vertebrates possess both innate and adaptive immune systems, whereas invertebrates have only an innate immune system. The innate immune system is older, acts more rapidly, and is evolutionarily conserved compared with the adaptive immune system. The adaptive immune system can be further divided into humoral (antibody-mediated) and cell-mediated immunity (involving immune cells). Antigen–antibody interaction is the key to the outcome of immune response. An epitope or antigenic determinant is part of an antigen that is recognized by the components of the immune system, such as antibodies, B cells, and T cells. The corresponding part of the antibody that recognizes the epitope is called the paratope. The epitopes of protein antigens can be divided into two major classes, *linear* and *conformational*, also known as *sequential* and *discontinuous*

Sandhya Kortagere (ed.), *In Silico Models for Drug Discovery*, Methods in Molecular Biology, vol. 993,
DOI 10.1007/978-1-62703-342-8_9, © Springer Science+Business Media, LLC 2013

epitopes, respectively. Linear epitopes are made up of amino acid residues that are sequential in the primary structure of the protein, whereas conformational epitopes are formed by residues that are not sequential but come together in the antigen's tertiary structure. About 90% of B-cell epitopes are conformational, meaning they can recognize epitopes in their native state with antigen.

2 Why Epitopes and Their Mapping Are So Important

Both B- and T-cell epitopes have been used extensively in peptide- or epitope-based vaccines. Other types of vaccine design (live, killed, attenuated, and recombinant) are limited owing to safety issues in children and immune-compromised individuals. However, despite their better safety profile, peptide-based vaccines possess poor stability, poor immunogenicity, and lack of memory response. Fortunately, with advances in technologies such as peptide synthesis, peptide modification, and the science of adjuvants, these limitations are no longer a bottleneck. Peptide epitopes can be modified by attaching one or more chains of polyethylene glycol or by incorporating nonnatural amino acids, leading to increased stability and bioavailability (1, 2). Peptide vaccines are also advantageous in bypassing the requirement of antigen processing and delivery of a precise and chemically defined cargo to the antigen-presenting cells (3). To address the problem of antigenic variation, multiple-epitope vaccines can be used, targeting antigens from several strains at a time. In addition to the use of epitopes in peptide-based vaccines, epitope discovery is needed for the selective deimmunization of therapeutic (4) and autoimmunity proteins (5).

Several experimental methods can be used for the identification or mapping of epitopes. X-ray crystallography, nuclear magnetic resonance, and electron microscopy map the "structural" epitopes that are in contact with antibody, whereas methods such as PEPSCAN and enzyme-linked immunosorbent assay are "functional" in approach (6). These experimental approaches, like others, require resources, time, and money.

3 In Silico Models for B-Cell Epitope Prediction

For decades researchers have been developing in silico models to minimize the number of experiments needed to identify or map the potential epitopes on the antigen surface. Because of the basic differences in the recognition of B- and T-cell epitopes, researchers have derived separate algorithms and tools for the two types of epitope. This chapter discusses only B-cell epitope prediction models (linear and conformational). Although they are not very different from basic B-cell epitope algorithms, T-cell epitope models have been reviewed in detail elsewhere (7, 8).

3.1 Linear B-Cell Epitope Prediction Models

Initially researchers sought clues to antigenicity from the protein sequence. They observed several amino acid properties that correlated with antigenicity and derived from these simple in silico models. In 1981 Hopp and Woods derived a method for locating protein antigenic determinants or epitopes by analyzing amino acid sequences for the region of the local hydrophilicity (9). This was done by assigning each amino acid a numerical value and then repetitively analyzing these along the entire sequence for different window lengths. Later researchers identified several properties or indices that correlated with the epitopic nature of the amino acids, including Chou and Fasman (10), Levitt (11), Parker et al. (12), Emini et al. (13), Karplus and Schulz (14), Thornton et al. (15), Jameson and Wolf (16), who devised an antigenic index, and Pellequer et al. (17), whose model involves prediction of turns in proteins. Pellequer and Westhof (18) developed an automated epitope prediction model integrating different properties, called PREDITOP. A later model named PEOPLE combined several of these properties, including antigen index (19). Later Odorico and Pellequer (20) developed an improved version of PREDITOP, now known as BEPITOPE, using more than 30 physicochemical properties including turns. This model permitted searching for possible epitopes in a single protein or in a complete translated genome.

These early models were built using few antigen examples and lacked clean datasets designed specially for B-cell epitopes. Raghava's group at IMTECH in Chandigarh, India, developed a dedicated database of B-cell epitopes called BCIPEP (21). We checked the performance of physicochemical properties in this dataset and achieved ~58% accuracy. The model was then implemented in the form of the BcePred web server (22). A year later Blythe and Flower (23) performed an exhaustive analysis on 50 antigens using 484 amino acid propensity scales and found that even the best set of indices performed only marginally better than random. The study guided researchers to move from use of classical mean propensity scales to more sophisticated machine learning tools.

3.1.1 Machine Learning Methods

The basic principle behind a machine learning model is to first collect clean, experimentally verified data either from the literature or from dedicated databases. Then obtain the rationally selected negative datasets, such as the control, in "wet lab" experimentation. Code the amino acid sequences or data into machine-readable numbers by calculating residue properties or descriptors such as amino acid composition, physicochemical properties, or binary (sparse) matrix. Finally, train the model using machine learning algorithms such as support vector machines (SVMs), artificial neural networks (ANNs), or hidden Markov models (HMMs) with leave-one-out or n-fold cross-validation techniques. The model is then ready to take blind queries as input and can be implemented as a stand-alone tool or as a Web server (http://imtech.res.in/raghava/gpsr/).

In 2006 Saha and Raghava (24) used a feed-forward neural network on 700 nonredundant B-cell epitopes with an equal number of random peptides from UniProt and obtained ~65% accuracy. In the same year Larsen et al. (25) created an HMM model combined with Parker and Levitt scales to get a more accurate prediction of B-cell epitopes and implemented a Web server (BepiPred). Chen et al. (26) later used SVM with an amino acid pair (AAP) antigenicity scale. They proved that unlike single amino acid residues, few AAPs are significantly preferred in epitopes over nonepitopes and achieve a highest accuracy of 73%. El-Manzalawy et al. (27) later implemented Chen's AAP scale with SVM string kernels on homology-reduced datasets and compared this with earlier methods. They achieved an area under the curve (AUC) of 0.76 and the model was implemented in the form of the Web server BCPREDS. Recently, Wang et al. (28) designed a system called LEPS that combines physicochemical propensity and SVM classification and achieved a highest accuracy of 72.5% on earlier and newly created datasets.

3.1.2 In Silico Models for the Variable-Length B-Cell Epitopes

A fixed-length input vector is a prerequisite for machine learning techniques; therefore, most of the methods assume or fix some length (3–22 amino acids) for the epitope sequences, which are in fact variable (3–80) in length. For length fixation, truncation and extension methods originally reported by Chen et al. (26) were used. The first in silico model that could handle variability in epitope length was developed by El-Manzalawy et al. (29) in 2008 and called FBCpred, an extension of their BCPREDS tool (27). Another approach for allowing flexible epitope length was published by Sweredoski et al. (30) in 2009, called COBEpro. Sweredoski et al. pointed to the issue of redundant data, claiming that earlier methods used redundant datasets and there was a big problem while selecting negative datasets. COBEpro is a two-step system for predicting linear B-cell epitopes. It first uses SVM to make predictions on short peptide fragments within the query antigen sequence and then calculates an epitopic propensity score for each residue based on the fragment predictions. COBEpro achieved a cross-validated AUC up to 0.83 on the fragment epitopic propensity scoring task and an AUC up to 0.63 on the residue propensity scoring task. Very recently Wee et al. (31) developed a Bayes feature extraction methodology coupled with SVM for the prediction of B-cell epitopes of diverse length, which they termed BayesB. Table 1 provides an updated list of tools for linear B-cell epitope prediction.

3.2 In Silico Models for Conformational B-Cell Epitopes

Unlike linear B-cell epitopes, conformational epitope prediction models were limited by the need to understand antigen–antibody (Ag–Ab) complex structures before applying these algorithms. As with the linear epitopes, researchers started by seeking structural

Table 1
Linear B-cell epitope tools

Tool	Web site	Reference
LEPS	http://leps.cs.ntou.edu.tw/	Wang et al. (28)
BayesB	http://www.immunopred.org/bayesb/	Wee et al. (31)
COBEpro	http://scratch.proteomics.ics.uci.edu	Sweredoski and Baldi (30)
BCPREDS/FBCPRED	http://ailab.cs.iastate.edu/bcpreds	El-Manzalawy et al. (27, 29)
ABCpred	http://www.imtech.res.in/raghava/abcpred	Saha and Raghava (24)
BepiPred	http://www.cbs.dtu.dk/services/BepiPred	Larsen et al. (25)
Bcepred	http://www.imtech.res.in/raghava/bcepred	Saha and Raghava (22)
BEPITOPE/PREDITOP	Standalone for Windows systems	Odorico and Pellequer (20)

features of the Ag–Ab complex that could be correlated to antigenicity.

Based on the 21 Ag–Ab structures, Kulkarni-Kale et al. (32) in 2005 introduced "accessibility of residues" as a feature that could be exploited for epitope prediction and named the server CEP. Later Anderson et al. (33) used a combination of amino acid statistics, spatial information, and surface exposure and implemented an algorithm called DiscoTope, with a best average AUC of 0.71. A year later Ponomarenko and Bourne (34) created two benchmark datasets and evaluated eight Web servers available for antibody-protein binding site prediction and observed that no method could achieve AUC greater than 0.7. In 2007 Rapberger et al. (35) combined parameters such as solvent accessibility of residues involved in antibody binding, shape complementarity between epitope and paratope, and contact energies of the interacting residues.

In 2008 Ponomarenko et al. (36) developed the ElliPro Web server with approximation of the protein shape as ellipsoid. They implemented Thornton's method (15), which was originally developed for continuous epitopes and, together with protrusion index and neighboring residue clustering, allows the prediction of antibody epitopes in a given protein sequence or structure.

Later Sweredoski et al. (37) incorporated a combination of amino acid propensity scores and half-sphere exposure values at multiple distances to form the BEpro tool (formerly called PEPITO). Using the Epitopia algorithm, Rubinstein et al. (38) for the first time truly exploited an extensive set of physicochemical and structural geometrical features from an antigen's primary or tertiary structures. They trained the Naïve Bayes classifier using a benchmark dataset of 66 and 194 validated nonredundant epitopes derived from antibody–antigen structures and antigen sequences,

Table 2
Conformational B-cell epitope tools

Tool	Description	Web site	Reference
EPSVR	Support Vector Regression and Meta server consensus	http://sysbio.unl.edu/EPSVR/	Liang et al. (40)
SEPPA	Concept of "unit patch of residue triangle" with spatial clustering coefficient	http://lifecenter.sgst.cn/seppa/	Sun et al. (39)
Epitopia	Physicochemical and structural geometrical features with Naïve Bayes	http://epitopia.tau.ac.il/	Rubinstein et al. (38)
EPCES	Use of six different scoring functions	http://sysbio.unl.edu/EPCES/	Liang et al. (52)
BEpro (formerly known as PEPITO)	Combination of amino acid propensity scores and half-sphere exposure	http://pepito.proteomics.ics.uci.edu/	Sweredoski and Baldi (37)
ElliPro	Residue protrusion index and neighbor clustering	http://tools.immuneepitope.org/tools/ElliPro/	Ponomarenko et al. (36)
PEPOP	Clustering of surface accessible segments	http://pepop.sysdiag.cnrs.fr/PEPOP/	Moreau et al. (53)
DiscoTope	Combination of amino acid statistics, spatial information, and surface exposure	http://www.cbs.dtu.dk/services/DiscoTope/	Anderson et al. (33)
CEP	Surface accessibility of residue	http://115.111.37.205/cgi-bin/cep.pl	Kulkarni-Kale et al. (32)

respectively. Moving ahead from single-residue propensity scales or segment clustering, Sun et al. (39) introduced the concept of "unit patch of a residue triangle" using a typical network parameter of the spatial clustering coefficients and developed the SEPPA algorithm. Liang et al. (40) recently developed the EPSVR/EPMeta algorithm using Support Vector Regression with six attributes. Very recently the Raghava group (41) tried to predict conformational B-cell epitopes using only an antigen's primary sequence and named this tool CBTOPE. We used the amino acid composition of patterns and found that CBTOPE is as good as and can complement other structure-based methods. Table 2 summarizes the tools for conformational B-cell epitopes.

Besides the approaches mentioned above, another approach involves mapping a conformational epitope on the antigen surface. Based on the mimotope concept, this method uses random phage display peptide libraries. The term *mimotope* was coined by Geysen et al. (42) for peptides that bind to the corresponding antibody but possess no homology; that is, mimotopes are structural mimics of

Table 3
Epitope mapping using phage display peptides

Tool	Description	Web site	Reference
LocaPep	Selection of seeds and clusters searching	http://atenea.montes.upm.es	Pacios et al. (54)
MimoPro	Dynamic programming, branch and bound and compactness factor	http://informatics.nenu.edu.cn/MimoPro	Chen et al. (55)
Pep3DSearch	Implementation of ant colony optimization algorithm	http://kyc.nenu.edu.cn/Pep3DSearch/	Huang et al. (56)
PEPITOPE	Implementation of PepSurf and Mapitope	http://pepitope.tau.ac.il/	Mayrose et al. (57)
PepSurf	Stochastic-based color-coding method	http://pepitope.tau.ac.il/	Mayrose et al. (58)
MEPS	Surface ensemble and Cβ distances	http://www.caspur.it/meps	Castrignanò et al. (59)
Mapitope	Physicochemical properties of mimotopes	http://pepitope.tau.ac.il/	Bublil et al. (60)
MIMOP	MimAlign and MimCons	Software available upon request	Moreau et al. (61)
MIMOX	Mimotope alignments and residue clustering	http://web.kuicr.kyoto-u.ac.jp/~hjian/mimox	Huang et al. (62)
3DEX	Physicochemical neighborhood of Cα or Cβ atoms	Windows-based software	Schreiber et al. (63)

the antigen epitopes. After binding to the corresponding antibody, these mimotopes are mapped onto the antigen structure. Several algorithms that have been developed to assist in this process are shown in Table 3.

4 Mathematical Models for B-Cell Receptor Signaling

The signals mediated through the B-cell antigen receptor (BCR) are critical to B-cell development and response to antigens. Defective BCR signaling leads to impaired B-cell development, immunodeficiency, and autoimmunity (43, 44). Several experimental techniques including microscopy and live cell imaging are used; limitations include the dynamic and interactive nature of biological systems. Understanding these networks requires the development of mathematical and computational models, which broadly come under the umbrella of systems biology (45, 46). One of the core tasks in system biology is the reconstruction of the regulatory, interacting, and signaling networks in the cell after perturbation.

A major step toward system biology was taken by the Alliance for Cellular Signaling (AfCS, http://www.signaling-gateway.org/), whose objective is to delineate the signaling pathways for B cells and murine macrophages after stimulation with various ligands (47). The AfCS measures cytokine secretion, protein localization, and protein–protein interaction, in addition to cellular calcium, cyclic adenosine monophosphate, and gene expression levels. Once these high-throughput experiments are completed, data are deposited in the University of California, San Diego, Signaling Gateway repository. Additionally AfCS' projects attempt to understand the BCR clustering after antigen cross-linking using Monte Carlo models (48). Network analysis tools include CellNetAnalyzer (49), SQUAD (50), SEBINI, CABIN (51), and others as reviewed by Suresh Babu et al. (45), which will help us better understand the complex signaling networks present in the immune cells.

Acknowledgments

H.R.A. is financially supported by the Council of Scientific and Industrial Research (CSIR), New Delhi, India.

References

1. Croft NP, Purcell AW (2011) Peptidomimetics: modifying peptides in the pursuit of better vaccines. Expert Rev Vaccines 10:211–226
2. Dudek NL, Perlmutter P, Aguilar MI et al (2010) Epitope discovery and their use in peptide based vaccines. Curr Pharm Des 16:3149–3157
3. Purcell AW, McCluskey J, Rossjohn J (2007) More than one reason to rethink the use of peptides in vaccine design. Nat Rev Drug Discov 6:404–414
4. Bryson CJ, Jones TD, Baker MP (2010) Prediction of immunogenicity of therapeutic proteins: validity of computational tools. BioDrugs 24:1–8
5. Steere AC, Drouin EE, Glickstein LJ (2011) Relationship between immunity to Borrelia burgdorferi outer-surface protein A (OspA) and Lyme arthritis. Clin Infect Dis 52(Suppl 3):s259–s265
6. Gershoni JM, Roitburd-Berman A, Siman-Tov DD et al (2007) Epitope mapping: the first step in developing epitope-based vaccines. BioDrugs 21:145–156
7. Tomar N, De RK (2010) Immunoinformatics: an integrated scenario. Immunology 131: 153–168
8. Yang X, Yu X (2009) An introduction to epitope prediction methods and software. Rev Med Virol 19:77–96
9. Hopp TP, Woods KR (1981) Prediction of protein antigenic determinants from amino acid sequences. Proc Natl Acad Sci USA 78: 3824–3828
10. Chou PY, Fasman GD (1974) Conformational parameters for amino acids in helical, beta-sheet, and random coil regions calculated from proteins. Biochemistry 13:211–222
11. Levitt M (1976) A simplified representation of protein conformations for rapid simulation of protein folding. J Mol Biol 104:59–107
12. Parker JM, Guo D, Hodges RS (1986) New hydrophilicity scale derived from high-performance liquid chromatography peptide retention data: correlation of predicted surface residues with antigenicity and X-ray-derived accessible sites. Biochemistry 25: 5425–5432
13. Emini EA, Hughes JV, Perlow DS, Boger J (1985) Induction of hepatitis A virus-neutralizing antibody by a virus-specific synthetic peptide. J Virol 55:836–839
14. Karplus PA, Schulz GE (1985) Prediction of chain flexibility in proteins - a tool for the selection of peptide antigens. Naturwissenschaften 72:212–213
15. Thornton JM, Edwards MS, Taylor WR, Barlow DJ (1986) Location of 'continuous' antigenic determinants in the protruding regions of proteins. EMBO J 5:409–413

16. Jameson BA, Wolf H (1988) The antigenic index: a novel algorithm for predicting antigenic determinants. Comput Appl Biosci 4:181–186
17. Pellequer JL, Westhof E, Van Regenmortel MH (1993) Correlation between the location of antigenic sites and the prediction of turns in proteins. Immunol Lett 36:83–99
18. Pellequer JL, Westhof E (1993) PREDITOP: a program for antigenicity prediction. J Mol Graph 11(204–210):191–202
19. Alix AJ (1999) Predictive estimation of protein linear epitopes by using the program PEOPLE. Vaccine 18:311–314
20. Odorico M, Pellequer J (2003) BEPITOPE: predicting the location of continuous epitopes and patterns in proteins. J Mol Recognit 16:20–22
21. Saha S, Bhasin M, Raghava GP (2005) Bcipep: a database of B-cell epitopes. BMC Genomics 6:79
22. Saha S, Raghava G (2004) BcePred: prediction of continuous B-cell epitopes in antigenic sequences using physico-chemical properties. In: Proceedings of the 3rd international conference on artificial immune systems (ICARUS), 13–16 Sep 2004, Catania, Italy. LNCS 3239: 197–204
23. Blythe M, Flower D (2005) Benchmarking B cell epitope prediction: underperformance of existing methods. Protein Sci 14:246–248
24. Saha S, Raghava G (2006) Prediction of continuous B-cell epitopes in an antigen using recurrent neural network. Proteins 65:40–48
25. Larsen JE, Lund O, Nielsen M (2006) Improved method for predicting linear B-cell epitopes. Immunome Res 2:2
26. Chen J, Liu H, Yang J, Chou KC (2007) Prediction of linear B-cell epitopes using amino acid pair antigenicity scale. Amino Acids 33:423–428
27. El-Manzalawy Y, Dobbs D, Honavar V (2008) Predicting linear B-cell epitopes using string kernels. J Mol Recognit 21:243–255
28. Wang HW, Lin YC, Pai TW, Chang HT (2011) Prediction of B-cell linear epitopes with a combination of support vector machine classification and amino acid propensity identification. J Biomed Biotechnol 2011(43):28–30
29. El-Manzalawy Y, Dobbs D, Honavar V (2008) Predicting flexible length linear B-cell epitopes. Proceedings of the 7th international conference on computational systems bioinformatics, Imperial College Press, London, pp 121–131
30. Sweredoski M, Baldi P (2009) COBEpro: a novel system for predicting continuous B-cell epitopes. Protein Eng Des Sel 22:113–120
31. Wee LJ, Simarmata D, Kam YW et al (2010) SVM-based prediction of linear B-cell epitopes using Bayes Feature Extraction. BMC Genomics 11(Suppl 4):S21
32. Kulkarni-Kale U, Bhosle S, Kolaskar AS (2005) CEP: a conformational epitope prediction server. Nucleic Acids Res 33:W168–W171
33. Anderson PH, Nielsen M, Lund O (2006) Prediction of residues in discontinuous B-cell epitopes using protein 3D structure. Protein Sci 15:2558–2567
34. Ponomarenko JV, Bourne PE (2007) Antibody-protein interactions: benchmark datasets and prediction tools evaluation. BMC Struct Biol 7:64
35. Rapberger R, Lukas A, Mayer B (2007) Identification of discontinuous antigenic determinants on proteins based on shape complementarities. J Mol Recognit 20:113–121
36. Ponomarenko J, Bui HH, Li W et al (2008) ElliPro: a new structure-based tool for the prediction of antibody epitopes. BMC Bioinformatics 9:514
37. Sweredoski M, Baldi P (2008) PEPITO: improved discontinuous B-cell epitope prediction using multiple distance thresholds and half sphere exposure. Bioinformatics 24:1459–1460
38. Rubinstein ND, Mayrose I, Martz E, Pupko T (2009) Epitopia: a web-server for predicting B-cell epitopes. BMC Bioinformatics 10:287
39. Sun J, Wu D, Xu T et al (2009) SEPPA: a computational server for spatial epitope prediction of protein antigens. Nucleic Acids Res 37:W612–W616
40. Liang S, Zheng D, Standley DM et al (2010) EPSVR and EPMeta: prediction of antigenic epitopes using support vector regression and multiple server results. BMC Bioinformatics 11:381
41. Ansari HR, Raghava GP (2010) Identification of conformational B-cell epitopes in an antigen from its primary sequence. Immunome Res 6:6
42. Geysen HM, Rodda SJ, Mason TJ (1986) A priori delineation of a peptide which mimics a discontinuous antigenic determinant. Mol Immunol 23:709–715
43. Gauld SB, Dal Porto JM, Cambier JC (2002) B cell antigen receptor signaling: roles in cell development and disease. Science 296: 1641–1642
44. Cuesta N, Martin-Cofreces NB, Murga C, van Santen HM (2011) Receptors, signaling networks, and disease. Sci Signal 4:mr3
45. Suresh Babu CV, Joo Song E, Yoo YS (2006) Modeling and simulation in signal transduction pathways: a systems biology approach. Biochimie 88:277–283
46. Goldstein B, Faeder JR, Hlavacek WS (2004) Mathematical and computational models of immune-receptor signalling. Nat Rev Immunol 4:445–456
47. Gilman AG, Simon MI, Bourne HR et al (2002) Overview of the alliance for cellular signaling. Nature 420:703–706

48. Srinivas Reddy A, Tsourkas PK, Raychaudhuri S (2011) Monte Carlo study of B-cell receptor clustering mediated by antigen crosslinking and directed transport. Cell Mol Immunol 8:255–264
49. Klamt S, Saez-Rodriguez J, Gilles ED (2007) Structural and functional analysis of cellular networks with Cell NetAnalyzer. BMC Syst Biol 1:2
50. Di Cara A, Garg A, De Micheli G et al (2007) Dynamic simulation of regulatory networks using SQUAD. BMC Bioinformatics 8:462
51. Taylor R, Singhal M (2009) Biological network inference and analysis using SEBINI and CABIN. Methods Mol Biol 541:551–576
52. Liang S, Zheng D, Zhang C, Zacharias M (2009) Prediction of antigenic epitopes on protein surfaces by consensus scoring. BMC Bioinformatics 10:302
53. Moreau V, Fleury C, Piquer D et al (2008) PEPOP: computational design of immunogenic peptides. BMC Bioinformatics 9:71
54. Pacios LF, Tordesillas L, Palacín A et al (2011) LocaPep: localization of epitopes on protein surfaces using peptides from phage display libraries. J Chem Inf Model 51: 1465–1473
55. Chen WH, Sun PP, Lu Y et al (2011) MimoPro: a more efficient Web-based tool for epitope prediction using phage display libraries. BMC Bioinformatics 12:199
56. Huang YX, Bao YL, Guo SY et al (2008) Pep-3D-Search: a method for B-cell epitope prediction based on mimotope analysis. BMC Bioinformatics 9:538
57. Mayrose I, Penn O, Erez E et al (2007) Pepitope: epitope mapping from affinity-selected peptides. Bioinformatics 23:3244–3246
58. Mayrose I, Shlomi T, Rubinstein ND et al (2007) Epitope mapping using combinatorial phage-display libraries: a graph-based algorithm. Nucleic Acids Res 35:69–78
59. Castrignanò T, De Meo PD, Carrabino D et al (2007) The MEPS server for identifying protein conformational epitopes. BMC Bioinformatics 8:S6
60. Bublil EM, Freund NT, Mayrose I et al (2007) Stepwise prediction of conformational discontinuous B-cell epitopes using the Mapitope algorithm. Proteins 68:294–304
61. Moreau V, Granier C, Villard S et al (2006) Discontinuous epitope prediction based on mimotope analysis. Bioinformatics 22: 1088–1095
62. Huang J, Gutteridge A, Honda W, Kanehisa M (2006) MIMOX: a web tool for phage display based epitope mapping. BMC Bioinformatics 7:451
63. Schreiber A, Humbert M, Benz A, Dietrich U (2005) 3D-Epitope-Explorer (3DEX): localization of conformational epitopes within three-dimensional structures of proteins. J Comput Chem 26:879–887

Chapter 10

The Collaborative Drug Discovery (CDD) Database

Sean Ekins and Barry A. Bunin

Abstract

The broad goals of Collaborative Drug Discovery (CDD) are to enable a collaborative "cloud-based" tool to be used to bring together neglected disease researchers and other researchers from usually separate areas, to collaborate and to share compounds and drug discovery data in the research community, which will ultimately result in long-term improvements in the research enterprise and health care delivery. This chapter briefly introduces CDD software and describes applications in antimalarial and tuberculosis research.

Key words Collaborative Drug Discovery, Malaria, Tuberculosis, Neglected diseases

1 Introduction

The rapid and pervasive emergence of antimalarial resistance has led to a reemergence of the disease in areas where it was once controlled. Of particular concern are chloroquine-resistant (CQR) *Plasmodium* strains, which can result in an increase in malaria mortality. Similarly, the reemergence of *Mycobacterium tuberculosis* (Mtb) and drug-resistant strains is causing concern with the latest estimates suggesting that over 2 billion individuals are infected (1). The development of anti-infectious agents is widely recognized as complex and incredibly difficult for a number of reasons, including the relatively low hit rates from high-throughput and secondary screens, as well as the need to obtain very domain specific *biologically relevant chemical diversity* (2). Agents that are active against malaria and Mtb are urgently needed (3).

Unfortunately, research focused on neglected infectious diseases, in general, and on malaria and Mtb, in particular, is largely composed of disjointed efforts of many academic and other nonprofit laboratories distributed across the globe. These independent efforts, while providing significant contributions, often lack the project management, data handling, decision gates, and pipeline integration functions that are critical to efficiently discovering,

Sandhya Kortagere (ed.), *In Silico Models for Drug Discovery*, Methods in Molecular Biology, vol. 993,
DOI 10.1007/978-1-62703-342-8_10, © Springer Science+Business Media, LLC 2013

developing, and bringing new drugs to market. Pharmaceutical company contributions to these efforts, while not negligible, are rarely shared publicly until insurmountable development issues halt project development. Groups including Medicines for Malaria Ventures (MMV), the Worldwide Anti-malarial Resistance Network (WWARN), and the TB Alliance facilitate projects. For example, MMV manages over 50 antimalarial projects in collaboration with over 80 pharmaceutical, academic, and endemic-country partners in 44 countries. The TB Alliance has a similar role for TB drug discovery with over 20 programs ongoing from drug discovery to clinical development. Effectively these groups are virtual drug discovery units, outsourcing all aspects of preclinical and clinical research.

We believe there is a need for a new approach to foster inter-group collaboration to speed up drug discovery. The impact of accelerating the identification and advancement of promising anti-malarial and Mtb lead compounds through the efforts of a collaborative network of top antimalarial and Mtb or other neglected disease researchers, very simply, can be measured in terms of the potential number of lives saved. Malaria infects ca. 200 million people and causes over 1 million deaths *per year*, disproportionately claiming African children under the age of 5 years. Similarly Mtb kills over 1 million people per year.

Neglected disease researchers represent a relatively small number, compared with scientists in cancer and cardiovascular research. However, in all cases they are plagued by a perceived lack of collaboration and data sharing. Grant funding from the NIH and, increasingly, private foundations is paying for high-throughput screening of large compound libraries and other studies in a relatively uncoordinated fashion, with a likely result that there is considerable repetition in experiments and no cross-disease data sharing. Recent research suggests that more Collaborative Drug Discovery (CDD) will be the future paradigm of biomedical research (4–7). To facilitate this there is a growing number of Internet-based, open-access chemistry databases and Internet-based collaborative tools (8, 9) that are likely to enhance scientific research, e.g., structure–activity relationship (SAR) development (10) and rapid lead identification (11). The challenges associated with bringing chemists and biologists together for virtual drug discovery projects for neglected diseases (12) provide an arena for testing a new approach. Traditionally biological data available for sharing is stored in an unstructured format, e.g., in a single document or Excel (Microsoft, Redmond, WA) files. Compilation of data is sporadic with little, if any, standardization of the data formats (e.g., ontology creation) or critical information such as experimental procedures and statistical analysis to quantify data quality to allow reproducibility and comparisons between groups. Before collaborations begin, data security and integrity always should be considered while intellectual property arrangements (materials transfer and intellectual property [IP] rights agreements) are often (at least in academia) seen as necessary, but as a hindrance to progress.

Existing software tools, both commercial and public, do not address these issues. Available traditional chemistry and biology data management software offers no community-oriented data sharing features; it can also be expensive to license and difficult to deploy and support. To promote data exchange, it is essential to respect IP constraints. Existing databases can also be technically complex and challenging to use, so many academic researchers simply do not bother to adhere to the advanced data management practices that pharmaceutical discovery efforts mandate internally.

Exploiting more than 7 years of experience of using cloud computing and Web 2.0 (8), Collaborative Drug Discovery, Inc. (CDD; Burlingame, CA) has developed a unique Web-based software currently helping scientists optimally identify and advance novel drug candidates. The software not only allows scientists to manage and analyze their data more effectively but also provides the option to share their data effortlessly and securely to the degree they want, with whomever they wish, at the time of their choosing (13). It allows them to easily toggle between and simultaneously mine across private, shared, and public datasets. The CDD software and existing user network is uniquely positioned to improve collaborations in the neglected disease space thereby increasing the efficiency of drug discovery and development (12).

1.1 CDD Database

The development of the CDD database has been described previously with applications for collaborative malaria research (13). The CDD database brings the power of cloud computing to drug discovery, enabling collaborators to share research data securely within and across organizations without the need to install and maintain complex software.

The CDD database can archive and mine a broad range of diverse objects that can later be selectively and securely shared with other researchers (or permanently kept private, which is the default behavior). The CDD Database is a hosted collaborative system with an important advantage over traditional personal computer-based database systems since it can enable secure login into the database from any computer, using any common browser (e.g., Firefox, Internet Explorer, Chrome, or Safari). This unique capability for a database system provides flexibility for users. The CDD Web-based database architecture handles a broad array of data types and is arranged as three specific modules. Firstly, the CDD vault, which securely stores and enables mining of data in a private data area, is hosted and managed by CDD. Second, CDD Collaborate enables confidential exchange of data between vaults as selected by users. Third, CDD public hosts public datasets that can be mined. The CDD platform incorporates Marvin, calculated plug-ins for physical and chemical calculations, and the JChem Cartridge for structure searching from ChemAxon (Budapest, Hungary) within the application as the chemistry engine. This allows one to do SAR analysis, including chemical pattern recognition (e.g., similarity and substructure searching), physical–chemical property calculations, Boolean search, and save capabilities for potency, selectivity, toxicity, and other experimentally derived properties. The database can handle heterogeneous data files as well as standardized csv and sdf file convertible formats that represent the chemical and biological data. In particular CDD is tailored for common data formats used by biologists such as Microsoft Excel (.xls) and text (.txt) files. The technology can mine against a variety of values including concentration, time, percent, real, integer, textline, cpm, rlu, Z/Z′ plate statistics, and IC_{50} (log IC_{50}, R^2 values, Hillslope, etc.). The outputs of such mining can be either: saved, exported, shared inside CDD, or plotted with an integrated plug-in.

The researcher can control which data to keep 100% private, share with groups of individual researchers, or share more generally with the public. A further unique capability of CDD is the ability to compare all or subsets of public access data with private data simultaneously together in a single container as well as analyze multiple vaults to which the user has access. The power of this collaborative approach to drug discovery can be seen in different types of community-based research projects. These range from traditional completely private collaborations, to temporally private collaborations that may become more open following a privacy escrow period, to completely open collaborations where researchers can blog about the experiments as they occur (13).

When users log in to CDD they have the option to select the vault which they want to view; after making their selection they will then see a dashboard which summarizes recent protocols and molecules which they have accessed (Fig. 1), as well as a listing of recent activity and messages.

Data import into CDD is currently a simple four-step process from a .csv or .sdf and mapping a dataset to a user-defined protocol if required. Data can be readily mined in CDD and in addition the user can specify which private vaults and public datasets to use (Fig. 2). A full Boolean search is possible by specifying protocol, run, readout, chemical properties, keywords, etc. If molecules are selected, CDD also provides a link to find more information in external databases such as ChemSpider. Data in CDD can also be plotted graphically using an interactive visualization that also provides a snapshot of the molecule and data upon mousing over an *X*, *Y* coordinate. This may allow a simple SAR analysis.

CDD also has "Projects" functionality that enhances the capability to share research data securely using CDD (Fig. 3). This enables users of CDD to organize their data within a vault into projects and invite individual vault members to be able to access specific projects, allowing for more flexible data sharing and

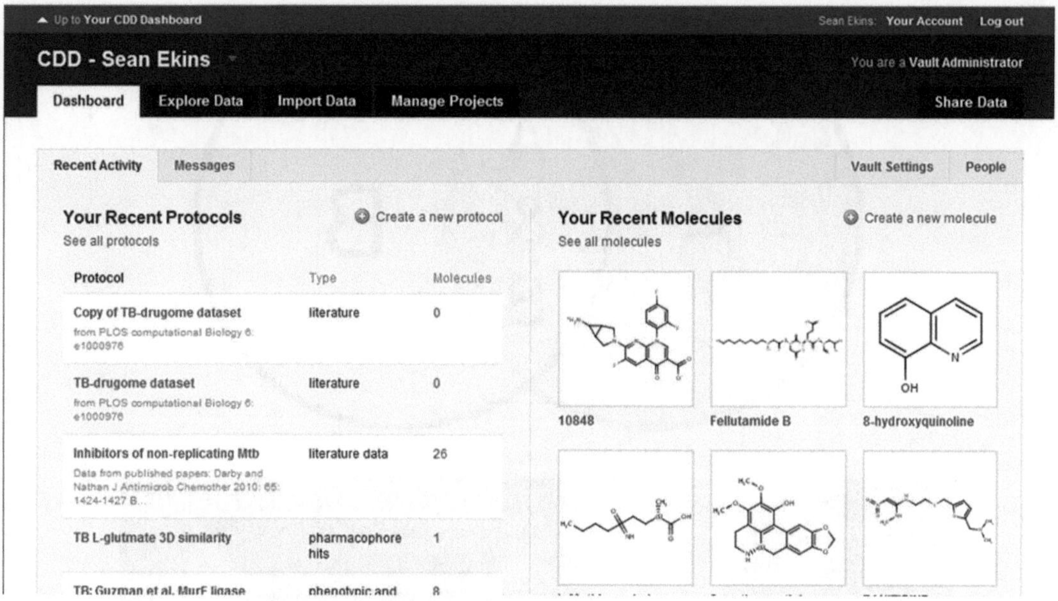

Fig. 1 CDD database screenshot

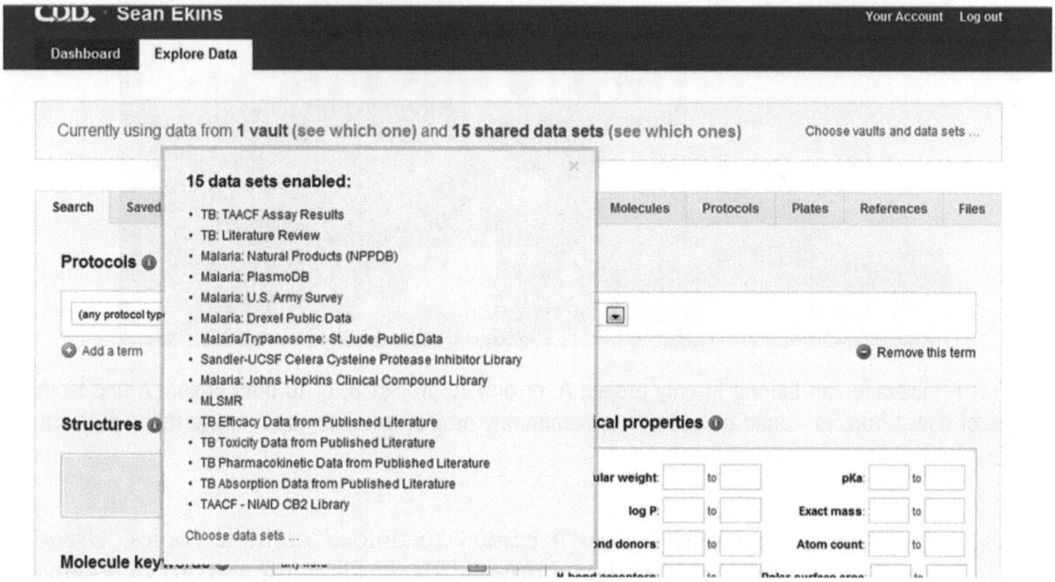

Fig. 2 Multiple neglected disease datasets and private vaults can be queried simultaneously so that scientists see beyond their own data

management both within a group and across groups. Put simply—molecules, batches, plates, and protocols can now belong to multiple projects. They are organized like a Venn diagram (Fig. 3a). For those users who screen compound libraries against multiple targets for different therapeutic areas and want to keep these data in separate projects, only a single version of the library is required.

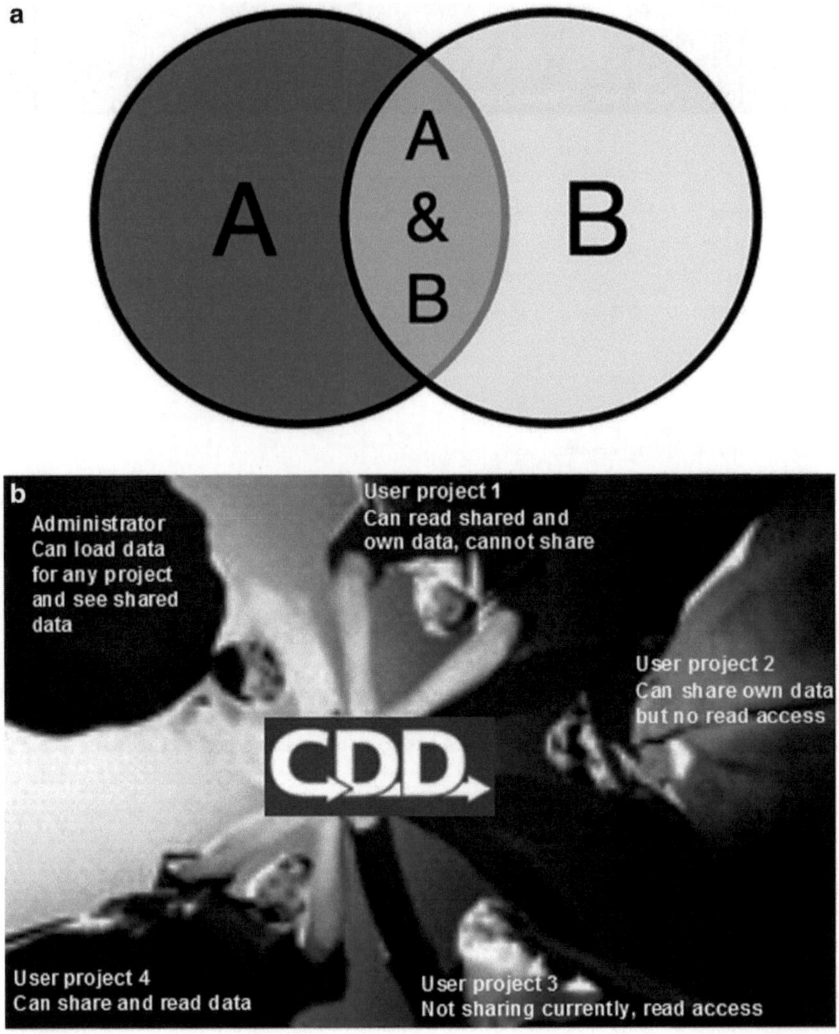

Fig. 3 (**a**) A molecule can belong to only project A, or only to project B, or to both project A and B. (**b**) An example of how "Projects" could be used for a community project to efficiently manage many projects and recreate a "virtual pharma" environment

Previously it was necessary to upload multiple copies. Users are able to share data more selectively, allowing users to view only the data relevant to their projects without compromising the security of data meant to be hidden (Fig. 3b). This results in no longer having to balance several systems for managing data between different groups; no more inviting collaborators into private networks and compromising other data.

1.2 CDD TB DB and CDD Malaria DB as Examples of Community Data Sharing

CDD has been funded by the Bill and Melinda Gates Foundation (BMGF) which has enabled them to build strong relationships with the TB pilot labs and also build the Collaborative Drug Discovery Tuberculosis Database (CDD TB DB, accessed at http://www.collaborativedrug.com/register). We have also curated the literature data on Mtb drug discovery as well as hosting approximately 20 Mtb specific datasets (Table 1) representing well over 300,000 compounds derived from patents, literature, and HTS data. TB researchers use CDD's platform and services because they provide value not available elsewhere. This includes CDD's easy-to-use data mining and unique data sharing capabilities which help multiple groups selectively compare their results, understand their experimental results, and define better future experiments. Leveraging existing literature datasets in CDD allowed researchers to screen already validated hits instead of screening random compounds—a much more effective use of finite screening resources. Similar examples and positive testimonials are available from numerous other TB Pilot laboratories.

Table 1
A subset of the >20 publicly available CDD antimalarial and *Mycobacterium tuberculosis* datasets

Database name/source	Description	Molecules
US Army Survey	An extensive collection of antimalarial-drug animal SAR data, including structures, bioactivity, etc., published originally by the US Army in 1946.	12,318
St. Jude Children's Research Hospital	Supplemental data from Guiguemde et al. (25); structures tested in a primary screen, with additional data in 8 protocols: Bland-Altman analysis, calculated ADME-tox properties, phylochemogenetic screen, sensitivity, synergy, and enzyme assays, as well as a thermal melt analysis.	1,524
Novartis Malaria	Data from Gamo et al. (14), *P. falciparum* strains 3d7 (drug-susceptible), and W2 (chloroquine-, quinine-, pyrimethamine-, cycloguanil-, and sulfadoxine-resistant), obtained from MR4, were tested in an erythrocyte-based infection assay for susceptibility to inhibition of proliferation by selected compounds.	5,695
Johns Hopkins–Sullivan	Percent inhibition of approved drugs at 10 μM.	2,693
MLSMR	A diverse collection tested by the Southern Research Institute against *M. tuberculosis* H37Rv. The most active compounds have dose–response and cytotoxicity data.	214,507
TB efficacy data from the literature	From more than 300 published literature sources, including PubMed citations, targets, cells and organisms tested, MIC, % inhibition, EC_{50}, and IC_{50}.	6,771
TAACF-NIAID-CB2	Results of screening a commercial compound library by the Southern Research Institute to inhibit the growth of *M. tuberculosis* strain H37Rv.	102,634
Novartis *M. tuberculosis*	Aerobic and anaerobic hits	283

ADME-tox absorption, distribution, metabolism, and excretion/toxicologic properties; EC_{50} median effective concentration; IC_{50} median inhibitory concentration; *MIC* minimum inhibitory concentration; *MLSMR* Molecular Libraries Small Molecule Repository; *TAACF-NIAID* Tuberculosis Antimicrobial Acquisition and Coordinating Facility/National Institute of Allergy and Infectious Diseases

CDD is an integral part of over 20 global TB researcher pilot laboratories (including many multigroup collaborations) and also facilitates collaborations with four large global pharmaceutical companies. At the time of writing this is expanding as BMGF foster collaborations between pharma and academics to screen their compounds versus Mtb. In addition, many major individual academic, nonprofit, and commercial groups have used this Web-based database system (13) to facilitate their own research. Functionality developed in CDD not only helps academic scientists, but is increasingly being appreciated by industry as being state-of-the-art for collaborations. Researchers particularly appreciate highly visible new capabilities such as multidimensional scatterplot viewing as well as more subtle improvements such as the streamlined batch data mapper and autoregistration facility. The new, more intuitive CDD design helped increase productivity in the labs. In addition to innovative collaborative capabilities, the intuitive mining tools have become the central "workhorse" that many laboratories use to make decisions about future research directions. The "Projects" functionality was initiated and directed by one of the pilot labs that is managing a portfolio of distributed projects much as a major pharmaceutical company would do internally.

CDD has provided the capabilities and achieved the milestones within budget for our BMGF funded project and updates have been provided on a monthly basis. A recent TB Pilot Laboratory formal survey confirmed CDD's central role defining future TB Research directions. Enthusiastic testimonials independently confirmed that this program provides different insights, which accelerates their project milestones. The CDD database is also now part of the More Medicines for TB project (http://www.mm4tb.org/) in which over 20 groups are collaborating to develop drugs for Mtb.

1.2.1 Dataset Analysis

We have recently seen several large HTS datasets of compounds for TB and malaria become available publically. For example, GlaxoSmithKline (GSK) released over 13,500 in vitro screening hits against malaria using *Plasmodium falciparum* along with their associated cytotoxicity (in HepG2 cells) data from an initial screen of over 2 million compounds (14). Three databases initially hosted the data: European Bioinformatics Institute-European Molecular Biology Laboratory (EBI-EMBL, ChEMBL http://www.ebi.ac.uk/chembl/), PubChem (http://pubchem.ncbi.nlm.nih.gov/), and CDD (13), while other databases also followed suit including ChemSpider from the Royal Society of Chemistry (www.chemspider.com).

We have also undertaken an evaluation of this and other datasets using a simple descriptor analysis as well as readily available substructure alerts or "filters" (15–17). The GSK, St Jude, and Novartis datasets also have very high failure rates with the Abbott Alerts (18, 19) (75–85%) and Pfizer Lint filters (40–57%) (Fig. 4). A set of 14 US Food and Drug Administration (FDA) approved

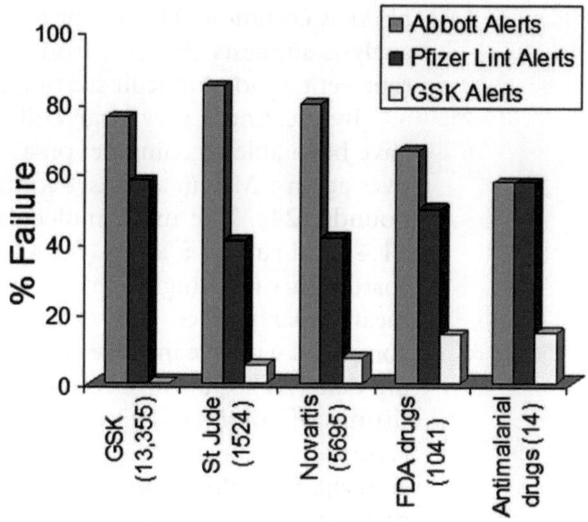

Fig. 4 Percent failure of SMARTS filters (http://pasilla.health.unm.edu/tomcat/biocomp/smartsfilter) for different antimalarial datasets

widely used antimalarial drugs have properties much closer to the St. Jude and Novartis hits. These compounds had fewer failures with the Abbott filters when compared to the GSK, Novartis, and St. Jude antimalarial datasets.

A detailed analysis of our calculated molecular descriptors for the GSK malaria hits (14) shows that most are normally distributed apart from the skewed Lipinski violations data and the bimodal molecular weight. Interestingly, 3,269 (24.3%) of the compounds fail more than one of the Lipinski "rule of 5" characteristics (molecular weight ≤500, log P ≤ 5, hydrogen bond donors ≤5, hydrogen bond acceptors ≤10) (20) using the descriptors calculated in the CDD database. The GSK screening hits are generally large and very hydrophobic as is also suggested in their publication (14), and although they suggested this may be important to reach intracellular targets, there is no discussion of the limitations of such compounds. We have also suggested that these compounds may not be "lead-like" (21, 22) and are closest to "natural product lead-like" (23). These antimalarial hits as a group are also vastly different from the mean molecular properties of compounds that have shown activity against TB, which are generally of lower molecular weight, less hydrophobic, and with lower pK_a and fewer RBN (24).

The GSK antimalarial hits dataset (14) also stood out from the other datasets in terms of physicochemical properties as the mean molecular weight, log P, and number of rotatable bonds were much higher than in the St. Jude (25) and Novartis datasets of antimalarial compounds (26).

Many companies avoid compounds that have reactive groups prior to screening and the availability and use of such computational filters is common. This is not the case in academia, however. Our analysis suggests that hits from some of these HTS datasets may represent a more difficult starting point for lead optimization.

By creating a very large collaborative database CDD TB, we have been able to compare, on a very large scale, actives and inactives against Mtb in a dataset containing more than 200,000 compounds (24). The mean molecular weight (357 ± 85), log P (3.6 ±1.4) and rule of 5 alerts (0.2 ± 0.5) were statistically significantly (based on *t*-test) higher in the most active compounds, while the mean polar surface area (PSA) (83.5 ± 34.3) was slightly lower compared with the inactive compounds for the single point screening data (24). Our most recent analysis for TB used a dataset consisting of another 102,633 molecules screened by the same laboratory against Mtb (17). We were able to analyze the molecular properties, differentiate the actives from the inactives, and show that the actives had statistically significantly (based on *t*-test) higher values for the mean log P (4.0 ± 1.0) and rule of 5 alerts (0.2 ± 0.4), while also having lower hydrogen bond donor count (1.0 ± 0.8), lower atom count (41.9 ± 9.4), and lower PSA (70.3 ± 29.5) than the inactives (17). Overall, when comparing these two datasets the mean values are remarkably similar. Our analysis of this data provided insights into molecular properties and features that are determinants of activity in whole cells (24).

We have also looked at samples of these libraries for analysis with substructure reactivity filters alongside known TB drugs. We showed that they have a pattern similar to the very large antimalarial screening library hits, in the very high failure rates with the Abbott Alerts (Fig. 5) (27, 28), with 81–92% failing the Abbott filters (17), which may be related to mechanism of action.

A more recent analysis of TB screening data (<300 compounds) from Novartis available in CDD suggests that we can also differentiate aerobic and anaerobic compounds based on their statistically different mean molecular properties (29). These analyses may help scientists to focus on compounds with properties that may lead to increased probability of bioactivity against this or other neglected diseases. In addition these large datasets can be used to create computational machine learning models that can identify active molecules against infectious diseases (17, 24, 29) such as TB (Fig. 6) and databases like CDD may have a role for both target-based and phenotypic screening (29).

1.2.2 Collaborations to Find Antimalarials

CDD has developed and deployed a robust preliminary public antimalarial database from five sources which hosts data on approximately 16,000 public compounds. The growth of this database has fostered several key antimalarial discovery collaborations between CDD users. A substructure search for the known chemosensitizer substructure led to the identification of hundreds of compounds for laboratory evaluation by the laboratories of

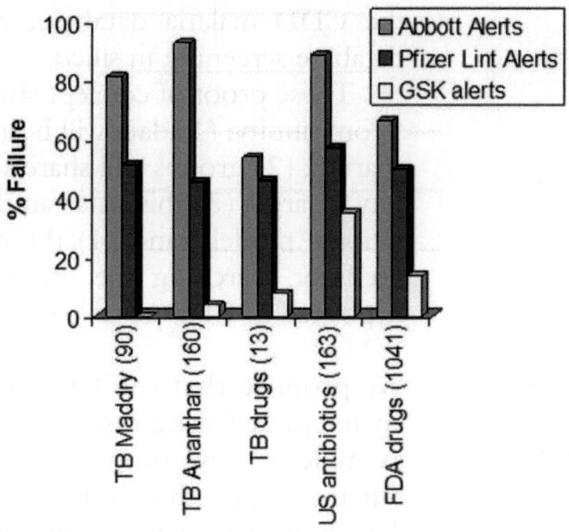

Fig. 5 Percent failure of SMARTS filters (http://pasilla.health.unm.edu/tomcat/biocomp/smartsfilter) for different datasets (27, 28)

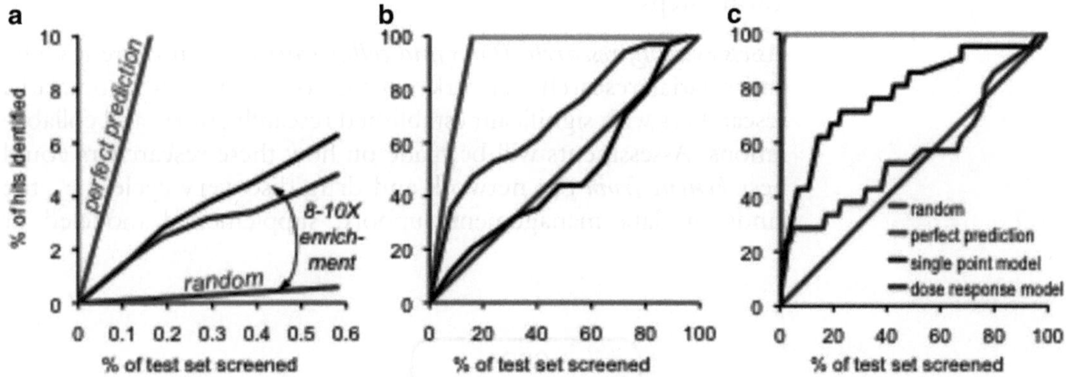

Fig. 6 Test set receiver operator characteristic curves for *M. tuberculosis* Bayesian models. (**a**) Southern Research Institute data for >100,000 molecules (1,702 actives). (**b**) Novartis data for 248 molecules (34 actives). (**c**) FDA-approved drugs, 2,108 molecules (21 actives) (17, 29)

Dr. Peter Smith in Cape Town to overcome the resistance to chloroquine (13). Leading candidates were identified and sent from collaborators for evaluation of efficacy in assays using the resistant African malarial parasite strains in human red blood cells. *This process shaved months off a project timeline relative to synthesizing new compounds from scratch.* Eighteen compounds were identified from a set of FDA approved drugs using substructure searching and half a dozen were purchased and shipped to Africa. When tested in the assay, these known drugs were shown to almost completely reverse (sevenfold reversal) the resistance in human blood cells (13). We have also worked with groups to facilitate computational modeling of malaria data using the public data in the CDD malaria database, which were then used for further database screening in silico.

These proof of concept studies illustrate how CDD can create a community: (1) data will be archived into a database for selective sharing; (2) groups will share some of their data with the community at large; (3) this data can then be used for creation of computational models; and (4) the computational models can then be used for searching the other open datasets or private datasets deposited in CDD to discover new compounds for testing.

1.2.3 Proposed Drug Discovery Cycle Incorporating CDD

We propose that the CDD database could be used as part of an integrated drug discovery process or "cycle" (Fig. 7). In this example we have proposed an antimalarial collaborative network, but this approach could be applicable to any disease of interest. The cycle starts with using data already in the CDD antimalarial database and ends with the data arising from the collaborative efforts being archived in the CDD antimalarial database for enhancement of the next cycle. The cycle would involve the following steps.

Assess existing research efforts and collaborations: Our current CDD antimalarial research network consists of prominent antimalarial researchers with significant established research efforts and collaborations. Assessments will be made on how these researchers could best *benefit from* the network and drug discovery cycle, i.e., the kinds of data management support, supplemental modeled or

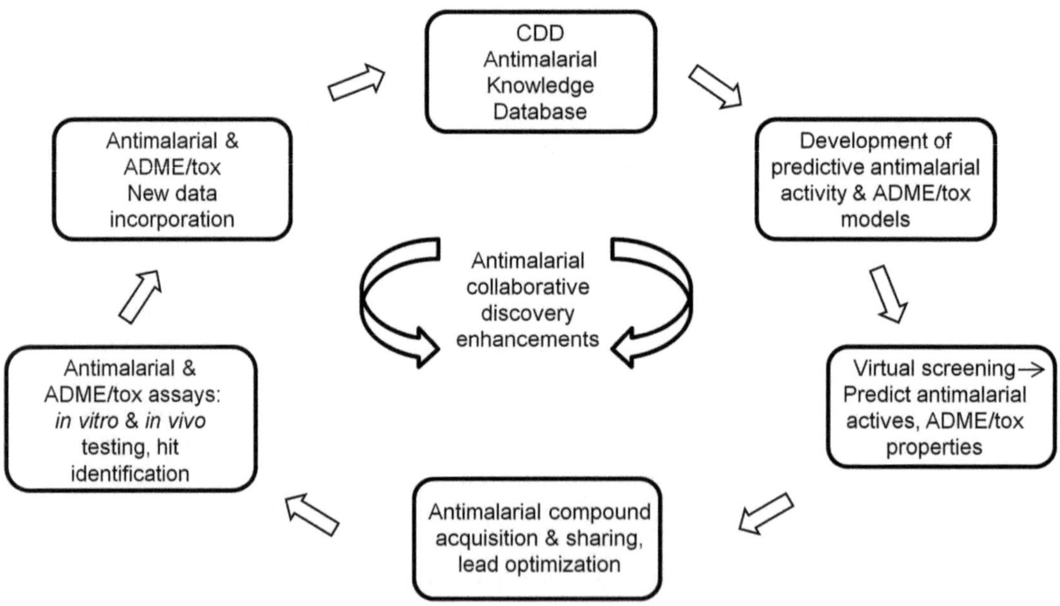

Fig. 7 CDD antimalarial discovery cycle

calculated data, and potential collaborations that could help accelerate their research efforts, as well as what their most valuable *contribution to* the network and drug discovery cycle could be. Existing collaborations between researchers within the CDD antimalarial research network will be assessed to determine how to extend the effort to its next logical discovery component or how to structure the collaboration such that several iterations of data discovery and optimization are achievable. An example is computational chemistry–medicinal chemistry collaborations: Computational chemistry support can span generalized antimalarial data-mining activity models, which will help identify a large number of diverse chemotypes for consideration of purchase or synthesis, to focused activity and/or ADME (absorption, distribution, metabolism, and excretion)/toxicity property models on a chemotype currently under consideration in one of the medicinal chemistry research support groups. Medicinal chemistry groups can provide focused SAR datasets for developing models and synthetic schema for potentially interesting chemotypes for expansion into a virtual library that can be prioritized by predicted activity or filtered by calculated properties.

Identify and facilitate collaborations for each cycle component: Current members of the CDD antimalarial research network span the functional capabilities and expertise of the drug discovery cycle. According to the assessments of existing research efforts and collaborations, a CDD scientific consultant will help identify and facilitate research collaborations between groups for each component of the drug discovery cycle. Specifically, CDD can aid in building collaborations around the following drug discovery cycle components:

1. Model building.
2. Data mining, virtual screening.
3. Compound procurement.
4. Compound profiling.
5. Research data incorporation.

The CDD collaborative research platform and antimalarial database will act as the central hub for information and updates on collaborative research and notification of new data that has been uploaded to the system.

Identify and resolve barriers to rapid discovery turnover: We believe the following may be key issues for facilitation of collaboration between groups:

1. Assimilation of new data-mining and virtual screening tools into CDD for all users.
2. Incorporation of new data structures resulting from antimalarial research collaboration.

3. Subcontracting for compound procurement (purchasing and/ or synthesis) and compound profiling assays (by networked researchers or external contract research organizations).

1.3 Discussion

In the long history of human kind (and animal kind, too) those who have learned to collaborate and improvise most effectively have prevailed.

Charles Darwin

It is also clear that the "new drug discovery" will put a renewed emphasis on collaboration and that research on neglected and rare diseases will require this for success to connect disparate researchers around the globe and create virtual drug discovery teams. Currently available computational database tools for drug discovery, and chemistry in particular, are not collaborative and are of limited application for drug development (13). Therefore at CDD we emphasize *collaboration* as what differentiates us from other companies and technologies currently available.

We recently asked people through an online forum what collaborations meant to them? We had responses such as the following: "Collaboration, to me, means that folks from disparate disciplines or skills work together towards the same end-goal… A collaboration means free and open data sharing, transparent goals and intentions, and a relationship that allows open (frank) and constructive discussion" and "The internet is the perfect place to share (certain) data and many of the new technologies and format available at the Web (REST, SOAP etc.) are perfect to use data collaboratively." In our mind CDD is the perfect marriage of community and technology to address collaboration.

We have described the development of the CDD database as a case study of how such a tool could be used for collaborations. The tool was developed using an agile development process that uses an integrated design-build-test process. In the space of 8 years this database has become a viable technology that has attracted many research foundations, academics, biotech companies, and large pharma customers. In the process we have used it to provide new insights into the vast amounts of screening data being produced (24) as well as to facilitate global collaborations (13) and provide a means for collaboration (29–33). As we see drug discovery become more reliant on networks of collaborators, we think the need for a cloud-based solution will become dominant.

We also think sharing molecules and screening/biological data is just the beginning. Imagine a future in which your computational models and other data types could all be selectively shared in a single database. This is just a glimpse into the future and could be of immense value to the rare and neglected disease communities for collaborating and may have wider implications for more common disease research productivity in the pharmaceutical industry.

Acknowledgments

The authors gratefully acknowledge our colleagues and the many researchers in the CDD community who have collaborated with us and each other.

The CDD TB database is funded by the Bill and Melinda Gates Foundation (Grant#49852 "Collaborative drug discovery for TB through a novel database of SAR data optimized to promote data archiving and sharing").

References

1. Balganesh TS, Alzari PM, Cole ST (2008) Rising standards for tuberculosis drug development. Trends Pharmacol Sci 29:576–581
2. Payne DA et al (2007) Drugs for bad bugs: confronting the challenges of antibacterial discovery. Nat Rev Drug Disc 6:29–40
3. Zhang Y (2005) The magic bullets and tuberculosis drug targets. Annu Rev Pharmacol Toxicol 45:529–564
4. Carpy AJ, Marchand-Geneste N (2006) Structural e-bioinformatics and drug design. SAR QSAR Environ Res 17(1):1–10
5. Ertl P, Jelfs S (2007) Designing drugs on the internet? Free web tools and services supporting medicinal chemistry. Curr Top Med Chem 7(15):1491–1501
6. Munos B (2006) Can open-source R&D reinvigorate drug research? Nat Rev Drug Discov 5(9):723–729
7. Tralau-Stewart CJ et al (2009) Drug discovery: new models for industry-academic partnerships. Drug Discov Today 14(1–2):95–101
8. Williams AJ (2008) Internet-based tools for communication and collaboration in chemistry. Drug Discov Today 13(11–12):502–506
9. Williams AJ (2008) A perspective of publicly accessible/open-access chemistry databases. Drug Discov Today 13(11–12):495–501
10. Ekins S et al (2008) Molecular characterization of CYP2B6 substrates. Curr Drug Metab 9(5):363–373
11. Ekins S et al (2008) Computational discovery of novel low micromolar human pregnane X receptor antagonists. Mol Pharmacol 74:662–672
12. Nwaka S, Ridley RG (2003) Virtual drug discovery and development for neglected diseases through public-private partnerships. Nat Rev Drug Discov 2(11):919–928
13. Hohman M et al (2009) Novel web-based tools combining chemistry informatics, biology and social networks for drug discovery. Drug Disc Today 14:261–270
14. Gamo F-J et al (2010) Thousands of chemical starting points for antimalarial lead identification. Nature 465:305–310
15. Ekins S, Williams AJ (2010) Meta-analysis of molecular property patterns and filtering of public datasets of antimalarial "hits" and drugs. Med Chem Comm 1:325–330
16. Ekins S, Williams AJ (2010) When pharmaceutical companies publish large datasets: an abundance of riches or fool's gold? Drug Disc Today 15:812–815
17. Ekins S et al (2010) Analysis and hit filtering of a very large library of compounds screened against *Mycobacterium tuberculosis*. Mol Biosyst 6:2316–2324
18. Metz JT, Huth JR, Hajduk PJ (2007) Enhancement of chemical rules for predicting compound reactivity towards protein thiol groups. J Comput Aided Mol Des 21(1–3):139–144
19. Huth JR et al (2005) ALARM NMR: a rapid and robust experimental method to detect reactive false positives in biochemical screens. J Am Chem Soc 127(1):217–224
20. Lipinski CA et al (1997) Experimental and computational approaches to estimate solubility and permeability in drug discovery and development settings. Adv Drug Del Rev 23:3–25
21. Oprea TI (2002) Current trends in lead discovery: are we looking for the appropriate properties? J Comput Aided Mol Des 16:325–334
22. Oprea TI et al (2001) Is there a difference between leads and drugs? A historical perspective. J Chem Inf Comput Sci 41:1308–1315
23. Rosen J et al (2009) Novel chemical space exploration via natural products. J Med Chem 52:1953–1962
24. Ekins S et al (2010) A collaborative database and computational models for tuberculosis drug discovery. Mol Biosyst 6:840–851
25. Guiguemde WA et al (2010) Chemical genetics of *Plasmodium falciparum*. Nature 465(7296):311–315

26. Gagaring, K., et al. Novartis-GNF malaria box. [cited]; Available from: ChEMBL-NTD (www.ebi.ac.uk/chemblntd)
27. Maddry JA et al (2009) Antituberculosis activity of the molecular libraries screening center network library. Tuberculosis (Edinb) 89:354–363
28. Ananthan S et al (2009) High-throughput screening for inhibitors of *Mycobacterium tuberculosis* H37Rv. Tuberculosis (Edinb) 89:334–353
29. Ekins S, Freundlich JS (2011) Validating new tuberculosis computational models with public whole cell screening aerobic activity datasets. Pharm Res 28:1859–1869
30. Ekins S, Williams AJ (2010) Reaching out to collaborators: crowdsourcing for pharmaceutical research. Pharm Res 27(3):393–395
31. Williams AJ et al (2009) Free online resources enabling crowdsourced drug discovery. Drug Discov World 10:33–38
32. Louise-May S, Bunin B, Ekins S (2009) Towards integrated web-based tools in drug discovery. Touch Brief Drug Discov 6:17–21
33. Bingham A, Ekins S (2009) Competitive collaboration in the pharmaceutical and biotechnology industry. Drug Disc Today 14: 1079–1081

Chapter 11

Recognition of Nontrivial Remote Homology Relationships Involving Proteins of *Helicobacter pylori*: Implications for Function Recognition

Nidhi Tyagi and Narayanaswamy Srinivasan

Abstract

This chapter explains techniques for recognition of nontrivial remote homology relationships involving proteins of *Helicobacter pylori* and their implications for function recognition. Using the remote homology detection method, employing multiple-profile representations for every protein domain family, remotely related domain family information has been assigned for the 122, 77, and 95 protein sequences of 26695, and J99, and HPAG1 strains of *H. pylori*, respectively. Relationships for some of the *H. pylori* protein sequences with Pfam domain families are reported for the first time. In publicly available domain databases such as Pfam, for some of the *H. pylori* protein sequences functional domain information is associated only with part(s) of the proteins. In the current study other parts of such proteins have been shown to be remotely related to known domain families, raising the possibility of identifying functions for parts of the proteins that do not yet have domains assigned. Further, homologues of enzymes that potentially catalyze step(s) in various metabolic processes in *H. pylori* have been identified for the first time.

Key words *Helicobacter pylori*, Proteins, Remote homology detection, Function assignment

1 Introduction

Helicobacter pylori is a gram-negative, microaerophilic proteobacterium belonging to the gastric *Helicobacter* species that colonize the human gastric mucosa and cause gastroduodenal disease. *H. pylori* infects about 50% of the human population (1–5). It has a characteristic type IV secretion system, as found in the plant pathogen *Agrobacterium tumefaciens*, which enables the immune evasion task important for its survival and hence helps in pathogenesis (6, 7). It tests positive for urease activity, which is crucial for its survival and colonization in the highly acidic, inhospitable gastric environment (8–11). Infection of the host is generally acquired early in life, may persist for a long period, and causes asymptomatic gastritis

Sandhya Kortagere (ed.), *In Silico Models for Drug Discovery*, Methods in Molecular Biology, vol. 993,
DOI 10.1007/978-1-62703-342-8_11, © Springer Science+Business Media, LLC 2013

in some cases. It can also lead to development of mucosa-associated lymphoid tissue lymphoma (12). *H. pylori* has been classified as a class I carcinogen by the World Health Organization (13). About 10% of individuals in the infected population develop gastric or duodenal ulcer and approximately 1% develop gastric cancer (14).

The complete genomic sequences of three widely studied strains of *H. pylori* are available: strains 26695 (15), J99 (16), and HPAG1 (17). These strains differ in their geographic origin; strains 26695 and HPAG1 are closely related to strains isolated in Europe and J99 is similar to strains isolated in West Africa (18).

According to genomic analysis, the numbers of predicted open reading frames in strains 26695, J99, and HPAG1 are 1,590, 1,495, and 1,536, respectively. Of these three strains, only HPAG1 has a plasmid, pHPAG1 (17).

Many databases provide information on functional annotation of the genomes of organisms. One such high-quality standard database is Pfam (19). Pfam is fundamentally a protein domain family database derived based on sequence similarity. Pfam provides details on the functional properties of protein domains of known function. Out of 1,590, 1,495, and 1,536 proteins encoded in the genomes of 26695, J99, and HPAG1 strains of *H. pylori*, respectively, 1,113, 1,130, and 1,143 proteins have at least 1 protein domain associated (defined by Pfam) with the amino acid sequence. Therefore, for these domains of *H. pylori* proteins, preliminary indication of their functions is available.

Many full-length and partial *H. pylori* proteins have no assigned Pfam domain, and hence no functional information is available on these proteins in the Pfam database. A total of 50.1%, 48.6%, and 48.5% of the proteins coded in the genomes of the 26695, J99, and HPAG1 strains, respectively, have no assigned function. For example, an *H. pylori* protein with the National Center for Biotechnology Information (NCBI) code NP_207499.1 is 935 amino acid residues long and has been assigned to the ABC transporter domain family from residues 639 to 852. However, for the regions 1–638 and 853–935, no domain and hence no function have been associated.

This chapter describes a method for identifying such remote homology relationships, using as an example the protein sequences encoded in *H. pylori* as determined using highly sensitive homology detection methods, which will form a crucial step in generating clues to function. The approach employed involves searching in the database of position-specific scoring matrices (PSSMs or profiles) of protein domain families with multiple representations of PSSMs of each family (20–22). PSSMs have been generated using multiple sequence alignment of members of a protein domain family, and multiple representations of PSSMs stem from using different homologues as the reference sequence. Previous assessment of this approach suggests that it is highly sensitive and specific and has a lower error rate compared with single-profile-based searches and hidden Markov model search approaches (20–22). This method has proved to be powerful in identification of function for hypothetical proteins as well as in recognition of metabolic proteins that were not previously recognized in the case of *Plasmodium falciparum* (23).

2 Datasets

2.1 Dataset of H. pylori *Proteome*

The protein sequences of three widely studied strains of *H. pylori*—26695, J99, and HPAG1—were obtained from NCBI (ftp://ftp.ncbi.nlm.nih.gov/genomes). These sequences were predicted from the genome sequence information available for the three strains (15–17).

2.2 Dataset of Protein Domain Families

Pfam Database. Pfam version 23.0 was used for the analysis. Multiple sequence alignments corresponding to each of 10,340 protein domain families were taken from the Pfam database. Each family is represented in terms of PSSMs or profiles so that every *H. pylori* protein sequence can be searched against this profile database. In the case of a single-profile (per family) approach, a reference sequence is chosen arbitrarily for building a PSSM. In the case of a multiple-profiles (per family) approach, however, every sequence from the multiple sequence alignment of the protein domain family concerned is used for building PSSMs, which increases the search space and eliminates bias towards the reference sequence (20–22). A total of 40,587 profiles have been generated from 10,340 protein domain families; thus on average each protein domain family is represented by four profiles.

PALI (Phylogeny and ALIgnment of Homologous Protein Domains) Database. The PALI (v 2.6) database provides three-dimensional structure-based sequence alignments for homologous proteins of known three-dimensional structure (24–26). The protein families have been derived from the SCOP (Structural Classification of Proteins) database (27). There are 2,518 protein families, and using more than one sequence as reference, 37,986 profiles have been generated.

Many domain families are common to Pfam and PALI, although domain boundaries may not be the same for most of the common families in the two databases. In the case of PALI, domain boundaries are largely determined by three-dimensional structure information. In the case of Pfam, sensitivity of homology detection influences identification of domain boundaries.

3 Methods

As described above, a multiple-PSSM search approach was used to search the query (*H. pylori*) sequences using reversed-position-specific (RPS)-BLAST (Basic Local Alignment Search Tool) (28) in the Pfam database of multiple PSSMs and in PALI. Criteria to identify the hits were based on a previous assessment using a profile matching approach (20–22): the *E*-value reported by RPS-BLAST should be less than 10^{-4}, and more than 60% of the profile should be covered by the query in the alignment between the query and the profile. Hits are further characterized by a measure called percentage PSSM factor (PP_{factor}). PP_{factor} is defined as the ratio of the number of profiles of a family obtained as hits for a query protein to the total number of profiles representing that family in the PSSM database (22).

The correctness of hits corresponding to families with known structural information is further verified by employing the protein "fold" recognition method PHYRE (Protein Homology/analogY Recognition Engine) version 0.2, which assesses the compatibility of a sequence to a three-dimensional structure (29).

4 Results

Of 1,576, 1,489, and 1,544 predicted *H. pylori* proteins from the 26695, J99, and HPAG1 strains, respectively, proteins with no domain assignments in Pfam numbered 453, 357, and 400, respectively. Even among the proteins for which at least one domain could be assigned, large numbers of segments have no domains identified. There were 772, 803, and 790 segments in proteins from strains 26695, J99, and HPAG1, respectively, with at least 45 residues in each segment with no domain assignment in the Pfam database. Our focus here is these full-length proteins and segments with no Pfam domain assignment. We used a multiple-PSSM search approach to assign Pfam domains to these proteins and segments. The query set includes 1,210 full-length proteins and 2,365 segments of proteins (totaling 3,575 sequences) with no domain assignment. As no domain is assigned for any of these protein sequences by the Pfam database, which uses highly powerful remote homology detection on the basis of a hidden Markov model, any domain assignments that would be achieved should be considered as nontrivial, extremely difficult cases of remote homology detection. Of these 3,575 query sequences, for 294 sequences it was possible to associate at least one domain family by using a reverse BLAST approach, searching in multiple PSSM databases, an approach that satisfies all the criteria mentioned in Subheading 3.

All such cases with novel domain family assignments are characterized by poor sequence identity and therefore may not be interpreted as bona fide members of the protein domain families concerned. They may be interpreted as remotely related to the family concerned, and hence functional similarity between such *H. pylori* protein sequences with members of the family concerned may or may not exist.

Highly populated protein domain families of *H. pylori* include (1) the cellular component *Helicobacter* outer membrane protein family; (2) the sel1 family, which is associated with β-lactamase activity; (3) members of the CagA and VacA protein families, which are secreted into host cells and are involved in pathogenesis; (4) the ABC_transporter family, which is associated with ATP-dependent transport of molecules across the membrane; (5) the DNA methyltransferase protein domain family; (6) the radical SAM (*S*-adenosylmethionine) family associated with various metabolic functions of pathogens; and (7) the response regulator receiver domain family, which is involved in receiving the signal from the sensor domain in bacterial two-component systems.

Among the protein domain families that have been associated with *H. pylori* proteins for the first time, certain enzymes have been recognized: (1) asparagine synthase; (2) the succinylglutamate desuccinylase/aspartoacylase family; (3) the N-terminal domain of creatinase; (4) the fingers domain of DNA polymerase lambda; (5) the glucosamine-6-phosphate isomerases; (6) peptidase_M14 protein kinase; (7) polynucleotide kinase 3 phosphatase; (8) protein phosphatase 2C; (9) 6-pyruvoyl tetrahydropterin synthase; and (10) the endoribonuclease RegB T4-bacteriophage encoded protein domain family.

4.1 First Recognition of H. pylori Members of Protein Domain Families

Newly discovered evolutionary relationships involving protein domain families for which no bona fide member is currently known to be present in *H. pylori* are discussed in the following section. According to the present analysis, however, specific *H. pylori* proteins are shown to be related to these families.

Peptidase_M14. Members belonging to the domain family peptidase_M14 bind to zinc and have carboxypeptidase activity. A protein sequence from the HPAG1 strain (Uniprot accession code HPAG1_0372) has been shown to be remotely related to the peptidase_M14 family at very low sequence identity of about 13% at an *E*-value of 10^{-4}. The fold recognition program PHYRE also supports the fitness of this sequence on the phosphorylase/hydrolase-like fold with an *E*-value of 10^{-11}. Members of the peptidase_M14 family are classified under "phosphorylase/hydrolase-like fold," according to SCOP. The *Helicobacter* sequences align well with members of the zinc carboxypeptidases (Fig. 1). Two residues in the *Helicobacter* sequence corresponding to Glu-182 and

```
                                                                                179
CBPA1_BOVIN/128-406     IYDFMDLLVAEHPQLVSKLQIGRSYEGRPIYVLKFS--------TGGSNRPAIWIDLGIH
CBPA1_MOUSE/128-406     IYEFMDLLVTEHPQLVSKIQIGSTFEGRPINVLKFS--------TGGTNRPAIWIDTGIH
CBPB1_RAT/123-402       IEAWIQQVATDNPDLVTQSVIGTTFEGRNMYVLKIG--------KTRPNKPAIFIDCGFH
Q7S312_NEUCR/128-414    HLTFLRDLVAAHPNQSEIVTSGTSVNGNAITGIRFWG------SAGKGVKPAIVFHGTVH
YBJ7_SCHPO/189-484      INSWLRLMASLYKDLSELVPVGITAEGRTILGLKLNGRHPSDNGEKIRNKKVIIIQGGSH
Q9VCM8_DROME/185-464    INAWLKKLAETHP-EVLLVELGVSAQGRPILGVQIAFD--------NENRTTVFVESGIH
B4E3Q2_HUMAN/324-632    MRQLMKVVNEMCPNITRIYNIGKSHQGLKLYAVEISDHPG----EHEVGEPEFHYIAGAH
HPAG1_0372/25-150       -------------------KAPTDLEDR--------------------DKAPHLLLAGIQ
                                                 :.                          .      :
                          182                                                     237
CBPA1_BOVIN/128-406     SREWITQATGVWFAKKFTEDYGQDPS-FTAILDSMDIFLEIVTNPDGFAFTHSQNRLWRK
CBPA1_MOUSE/128-406     SREWVTQASGVWFAKKITKDYGQEPT-LTAILDNMDIFLEIVTNPDGFVYTHKTNRMWRK
CBPB1_RAT/123-402       AREWISPAFCQWFVREAVRTYNQEIH-MKQLLDELDFYVLPVVNIDGYVYTWTKDRMWRK
Q7S312_NEUCR/128-414    AREWITTMVAEYQAYYLLTNYGSDAT-VKSFVDKYEFYIFPVVNPDGFIYTQTSNRLWRK
YBJ7_SCHPO/189-484      AREWIGIPSVCYAAWQLLAKYDSDGH-VRKLLDKFEWIFIPVLNVDGYEYTWSNDRLWSK
Q9VCM8_DROME/185-464    AREWIAPATATYIIDQLVN--SKDSA-VQALARSQRWYIFPTVNPDGYQYTFKGDRMWRK
B4E3Q2_HUMAN/324-632    GNEVLGRELLLLLVQFVCQEYLARNARIVHLVEETRIHVLPSLNPDGYEKAYEGGSELG-
HPAG1_0372/25-150       GDEPGGFNATNLFL-------------MHYSVLKGLVEVVPVLNKPSMLRNHRG--LYGD
                        . *                         .     .    .    *  .

CBPA1_BOVIN/128-406     TRSVTSSSLCVGVDANRNWD----AGFGKAG---------ASSSPCSETYHGKYANSEVE
CBPA1_MOUSE/128-406     TRSHTEGSLCVGVDPNRNWD----AAFGMPG---------ASSNPCSETYRGKFPNSEVE
CBPB1_RAT/123-402       TRSTMAGSSCLGVRPNRNFN----AGWCEVG---------ASRSPCSETYCGPAPESEKE
Q7S312_NEUCR/128-414    NRQSNSGSSCVGRDINRNWP----AHWSTSGG--------ASTNPCDEAYKGAKQGDAPE
YBJ7_SCHPO/189-484      NRQPLNNSECFGINLDANWA----FGF--NG----------NIDPCSNEYGGLSPFQANE
Q9VCM8_DROME/185-464    NRALFG--ICRGVDLNRNFP----FHWNVTG---------ASGDPCRYDYSGPSAASEVE
B4E3Q2_HUMAN/324-632    -GWSLGRWTHDGIDINNNFPDLNTLLWEAEDRQNVPRKVPNHYIAIPEWFLSENATVAAE
HPAG1_0372/25-150       MNRKFA-------------------------------------------ALDKNDPEYPT
                                                                           .
                                              306
CBPA1_BOVIN/128-406     VKSIVDFVKD-----HGNFKAFLSIHSYS-QLLLYPYGYTTQSIPDK-TELNQVAKSAVE
CBPA1_MOUSE/128-406     VKSIVDFVTS-----HGNIKAFISIHSYS-QLLLYPYGYTSEPAPDK-EELDQLAKSAVT
CBPB1_RAT/123-402       TKALADFIRNN----LSTIKAYLTIHSYS-QMMLYPYSYDYKLPENY-EELNALVKGAAK
Q7S312_NEUCR/128-414    TTALAAFLNKVKA--AQGLKLFIDWHSYS-QLFMSPYGYTCSSVPAKNSEYQSLARGAVA
YBJ7_SCHPO/189-484      TMALFNLITESLSQEQKKVVGFLDVHSYS-QSVLWPYAYTCDLFPPDTENFEELAIGLVK
Q9VCM8_DROME/185-464    TQRMIEFIRHRVE--SERIRTFISLHSYS-QMIMFPYGHSAERVDNY-HDLTDIGKLAAN
B4E3Q2_HUMAN/324-632    TRAVIAWMEKIPFVLGGNLQGGELVVAYPYDLVRSPWKTQEHTPTPDDHVFRWLAYSYAS
HPAG1_0372/25-150       IQEIKSLIAKP------NIDAVLHLHDGG-------------------GYYRPIYIDATL
                           :   :          .                                  :    .

CBPA1_BOVIN/128-406     ALKSLYGTSYKYGSIIT------TIYQASGGSIDWSYN-QGIKYSFTFELRDTGRYG--F
CBPA1_MOUSE/128-406     ALTSLHGTKFKYGSIID------TIYQASGSTIDWTYS-QGIKYSFTFELRDTGLRG--F
CBPB1_RAT/123-402       ELATLHGTKYTYGPGAT------TIYPAAGGSDDWSYD-QGIKYSFTFELRDTGFFG--F
Q7S312_NEUCR/128-414    AIKSVYGTTFNYGPICT------TVYKATGNSVDYAFDVSGAEYAFTLELRDTGANG--F
YBJ7_SCHPO/189-484      ELHRVNSRYYTYQQACIPY-DGFHKHYLPGTAIDWVYFAADVAWPFNIRLRDMGDYG--Y
Q9VCM8_DROME/185-464    KIKDVSGRIYKSGSIYE------TIYPSSGGSKDWAHGQLKIPITFSYELRGPADSEDLF
B4E3Q2_HUMAN/324-632    THRLMTDARRRVCHTEDFQKEEGTVNGASWHTVAGSLNDFSYLHTNCFELSIYVGCD---
HPAG1_0372/25-150       NPKRWGN-----------------------------------------------------
                              .

CBPA1_BOVIN/128-406     LLPASQIIPTAQETWLG-
CBPA1_MOUSE/128-406     LLPASQIIPTAEETWLA-
CBPB1_RAT/123-402       LLPESQIRQTCEETMLA-
Q7S312_NEUCR/128-414    ILPASQILPSGVEAWAG-
YBJ7_SCHPO/189-484      LLPAKQIVPTAKEFFAM-
Q9VCM8_DROME/185-464    ILSAKEIEPTAAEAFAS-
B4E3Q2_HUMAN/324-632    KYPHESQLPEEWENNRES
HPAG1_0372/25-150       ------------------
```

Fig. 1 Blocks of multiple sequence alignment of protein sequences of carboxypeptidases from *B. taurus*, *Mus musculus*, *Rattus norvegicus*, *Neurospora crassa*, *Schizosaccharomyces pombe*, *Drosophila melanogaster*, and *Homo sapiens* along with protein sequence from *H. pylori* (Uniprot accession code: HPAG1_0372 from strain HPAG1). *Numbers on the top* correspond to amino acid residue number of the carboxypeptidase enzyme from *B. taurus*. *Gray vertical columns* indicate conserved residues. Amino acid residues corresponding to Glu-182 and His-306, which coordinate to zinc, are conserved, whereas another Zn-coordinating amino acid residue corresponding to His-179 is substituted by Gln in the *Helicobacter* sequence. Functionally important residues corresponding to Arg-237 are also conserved

His-306 of carboxypeptidase from *Bos taurus* are conserved. These amino acid residues are implicated in zinc coordination. Yet another functionally important residue that binds to substrate corresponding to Arg-237 of bovine enzyme is also conserved in the *H. pylori* sequence (Fig. 1) (30–32).

KorB. KorB is a DNA-binding protein that is characterized by the presence of the DNA-binding motif helix-turn-helix (HTH) and plays a regulatory role in replication and maintenance of the plasmid.

H. pylori sequences HP1138 (26695), jhp1066 (J99), and HPAG1_1076 (HPAG1) show the possibility of functional similarity with the KorB protein domain family. These sequences share more than 90% sequence identity among themselves. Residues 143 to 223 from these sequences share around 20% sequence identity with the KorB protein domain profile and cover about 98% of the profile of KorB family. The N-terminal region of these *H. pylori* proteins shows functional similarity with the ParB (ParB-like nuclease) domain family, which helps in partitioning or segregating DNA in prokaryotes (33, 34). The present analysis suggests that these proteins are remotely related to the KorB domain family.

Members from the KorB domain family are characterized by the DNA-binding helix-turn-helix motif (α-3 and α-4). Examples of this protein domain family include the well-studied DNA-binding KorB domain protein, the three-dimensional structure of which has been solved with its operator (PDB accession code: 1r71, DNA/RNA-binding 3-helical bundle fold; Fig. 2a). Amino acid residues outside the HTH motif (Thr-211 and Arg-240) determine the sequence-specific DNA binding (35) (Fig. 2a).

Secondary structures have been predicted for these *H. pylori* protein sequences using PSIPRED (36), which suggests the presence of an HTH motif in *H. pylori* sequences (Fig. 2b). The stretch of the sequence that is predicted as an HTH motif aligns well with the DNA-binding motif of KorB, indicating conservative substitutions (Fig. 2c). Multiple sequence alignment of protein sequences from *Helicobacter* along with KorB protein suggests conservation of Arg-240, which is associated with sequence-specific DNA binding (Fig. 2c).

Glucosamine 6-Phosphate Deaminase. Glucosamine 6-phosphate deaminase is an aldose–ketose isomerase that catalyzes conversion of D-glucosamine 6-phosphate into D-fructose 6-phosphate and ammonium ion (37). This enzyme has been reported in several bacterial (38), fungal (39), and animal species (40). This enzyme allows the bacteria to utilize glucosamine (GlcN) or *N*-acetyl-D-glucosamine (GlcNAc) from the medium as a source of carbon. *H. pylori* protein sequences HP1102 (26695), jhp1028 (J99), and HPAG1_1040 (HPAG1) show similarity to this enzyme family at an *E*-value of 10^{-16} and share poor sequence identity of about 15% with the profile of this family.

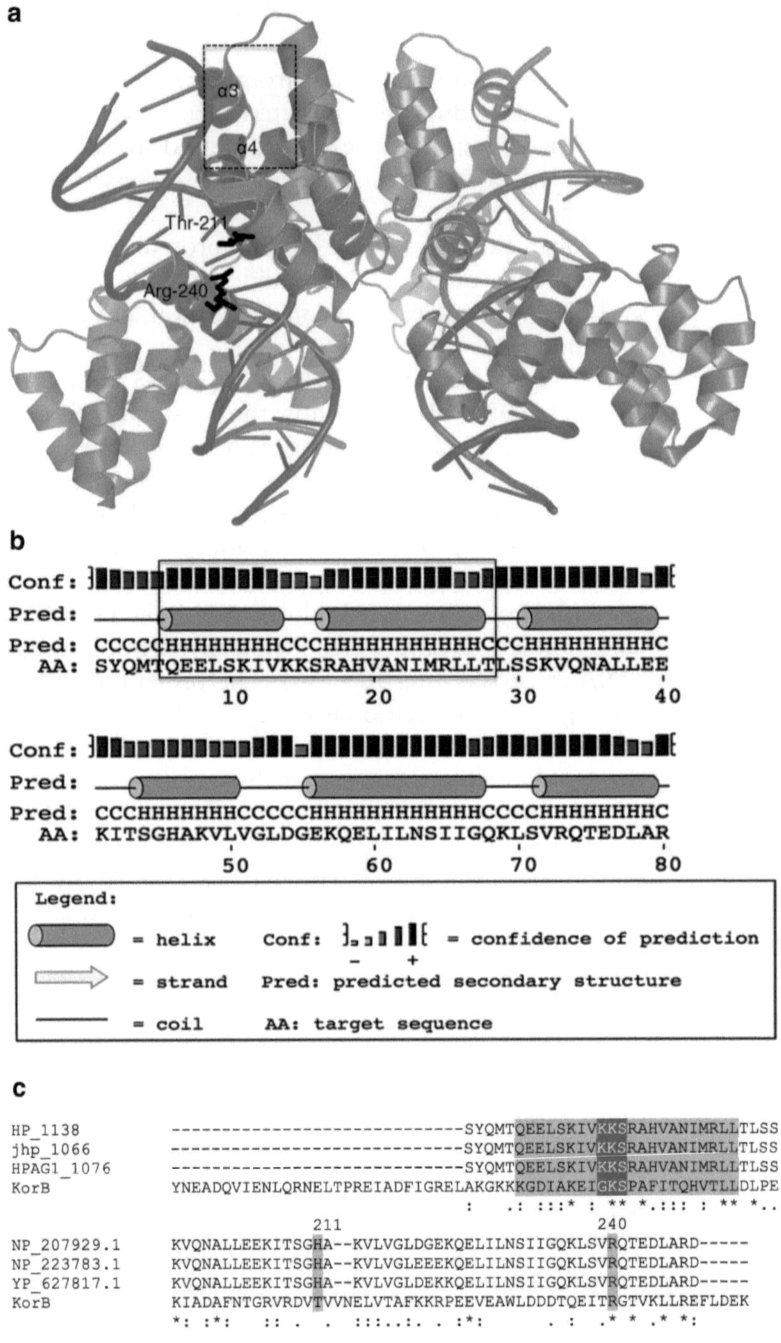

Fig. 2 (**a**) Three-dimensional structure of the KorB domain bound to DNA (PDB accession code: 1r71). The HTH motif is represented within the *dotted box* with the labeled α3 and α4 motifs. Thr-211 and Arg-240 in the three-dimensional structure of KorB protein are depicted. (**b**) Secondary structure prediction of *Helicobacter* sequence (Uniprot accession code: HP1138) is performed using the protein structure prediction server PSIPRED, which predicts an HTH motif (highlighted in the *rectangular box*). (**c**) Multiple sequence alignment of *H. pylori* sequences (Uniprot accession code: HP1138 from strain 26695, jhp1066 from strain J99, and HPAG1_1076 from strain HPAG1) with the well-studied KorB protein. *Light and dark gray* highlights (*white text*) depict helix and turn regions of the HTH motif, respectively, of predicted secondary structure of *H. pylori* sequences and KorB protein (PDB accession code 1r71). Amino acid residues corresponding to Thr-211 and Arg-240 are highlighted in *gray*

Important amino acid residues that participate in catalytic mechanisms of reaction are conserved in *H. pylori* protein sequences (Fig. 3). Residues corresponding to Asp-72 of *Escherichia coli* enzyme (Swiss-Prot code: NAGB_ECOLI) (41), which acts as a proton acceptor, and Asp-141, Glu-148, and His-143, which play an important role in the opening of a pyranose ring of substrate, are found to be conserved (37, 42). This suggests the possibility that these *H. pylori* proteins perform the function of glucosamine 6-phosphate deaminase though they are remotely related to this enzyme family. Fold recognition results confirm the fitness of these sequences on the DNA/RNA-binding 3-helical bundle fold with an *E*-value of 10^{-8} (estimation of precision 100%).

Protein Phosphatase 2C. Protein phosphatase 2C (PP2C) is a divalent cation, Mg^{2+}- or Mn^{2+}-dependent protein serine/threonine phosphatase (43). Members of this family are involved in various functions. PP2C reverses the effect of the protein kinase cascade by dephosphorylating adenosine monophosphate-activated protein kinase, which is responsible for inhibition of fatty acid and cholesterol biosynthesis in mammalian hepatocytes (44). Kinase-associated protein phosphatase from the source *Arabidopsis thaliana* has been shown to interact with serine-threonine receptor-like kinase and may function in the signaling pathway (45). Mutation studies on the protein serine/threonine phosphatase gene *PTC1*, a homologue of mammalian PP2C, show a marked growth defect in *Saccharomyces cerevisiae* (46).

```
                72
NAGB_ECOLI      DEYVGLPKEHPESYYSFMHRNFFDHVDIPAENINLLNGNAPDIDAECRQYEEKIRSYGKI
NAG1_CANAL      DEYLGFAPSDLQSYHYFMYDKFFNHIDIPRENIHILNGLAANIDEECANYEKKIKQYGRI
GNPI1_HUMAN     DEYVGLPRDHPESYHSFMWNNFFKHIDIHPENTHILDGNAVDLQAECDAFEEKIKAAGGI
NAGB_HAEIN      DEYVALPEEHPESYHSFMYNNFFNHIDILPENINILNGNTDDHNAECRRYEEKIKSYGKI
HP_1102         DERI-IDTSHDDSNTKLLHDYLLQNNALKASFIPLLPEKISSDTNALFNFAN--QHFKQP
jhp1028         DERI-IEASHEDSNAKLLHDYLLQNNALKASFTPLLPEKISGDTNELLDFAN--QHFKQP
HPAG1_1040      DERI-IDTSHKDSNTKLLHDYLLQNNALKASFTPLLPKKISSDTNALFHFAN--QHFKQP
                ** : :  .. :*   ::   ::.:  :  .   :*     .       : :  :

                       141 143   148
NAGB_ECOLI      HLFMGGVGNDGHIA--FNEPASSLASRTRIKTLTHDTRVANSRFFDNDVNQVPKYALTVG
NAG1_CANAL      DLFLGGLGPEGHLA--FNEAGSSRNSKTRKVELVESTIKANCRFFGNDESKVPKYALSVG
GNPI1_HUMAN     ELFVGGIGPDGHIA--FNEPGSSLVSRTRVKTLAMDTILANARFFDGELTKVPTMALTVG
NAGB_HAEIN      HLFMGGVGVDGHIA--FNEPASSLSSRTRIKTLTQDTLIANSRFFNNDVTQVPKYALTIG
HP_1102         HLAILGMGTDGHTASLFPETSAFLNEEKENIVLTKP-------------INAPYERLSMS
jhp1028         HLAILGMGTDGHTASLFPETSAFLNEEKENIVLTKP-------------TNAPYERLSMS
HPAG1_1040      HLAILGMGTDGHTASLFPETSAFLNEEKENIVLTKP-------------ANAPYERLSMS
                .* : *:* :** *  * *..:   ....   *.                :.*   *::.
```

Fig. 3 Multiple sequence alignment of bona fide members of the Glucosamine_iso domain family from Pfam and *H. pylori* sequences (Uniprot accession codes: HP1102, jhp1028, and HPAG1_1040 from strain 26695, J99, and HPAG1, respectively). *Numbers on top* correspond to amino acid residue number of *E. coli* enzyme. *Vertical gray columns* represent the conserved residues, which play an important role in catalysis. Uniprot accession numbers of protein sequences used are as follows: NAGB_ECOLI—*E. coli* strain K12; NAG1_CANAL—*Candida albicans*; GNPI1_HUMAN—*H. sapiens*; and NAGB_HAEIN—*Haemophilus influenzae*

The *Helicobacter* protein sequence (Uniprot accession code: HP0431) from strain 26695 shows remote homology with the PP2C family. The *H. pylori* protein sequence shares about 15% sequence identity with the profile of the PP2C family at an *E*-value of 10^{-5} and covers 86% of the profile length. Fold recognition results also confirm the relationship between this *H. pylori* sequence and the PP2C family as the query sequence fits on the PP2C-like fold with an *E*-value of 10^{-16} (estimated precision 100%).

Alignment of *H. pylori* sequences with PP2C protein sequences from *Bacillus subtilis* (yeast and human) suggests that the amino acid residues that bind to metal ions (corresponding to human protein sequence), Asp-38, Asp-60, Gly-61, and Asp-239, are conserved. Glu-37 (corresponding to human protein sequence), which is also implicated in binding to metal ions, is conservatively substituted to Gln. Arg-33 (corresponding to human protein sequence), which binds to phosphate ions, is also conserved (Fig. 4a). Figure 4b represents the three-dimensional structure of human PP2C protein (PDB ID: 1a6q), depicting the position responsible for metal-binding and phosphate-binding residues. Thus homology detection methods establish remote homology relationships between *H. pylori* protein sequences and members of the PP2C family.

4.2 Recognition of Previously Unknown Additional Members of H. pylori Proteins in Protein Domain Families

A number of relationships have been identified involving hypothetical *H. pylori* proteins and protein domain families for which bona fide members are present in *H. pylori*.

Cytochrome C. Members of the cytochrome C family are electron-transfer proteins, which have a characteristic CXXCH heme-binding motif. Two cysteine sulfurs of the CXXCH motif attach covalently with heme vinyl groups through a thioether bond. Along with these, cysteine and histidine of the CXXCH motif are also essential for heme attachment (47–49). *H. pylori* proteins HP0236 (strain 26695), jhp0221 (strain J99), and HPAG1_0239 (HPAG1) show homology to the cytochrome C domain family. These sequences share only 14% sequence identity with this family profile. As shown in Fig. 5a, the CXXCH motif is conserved in all *H. pylori* sequences, indicating that these *H. pylori* sequences might be related to the cytochrome C domain family. Fold recognition results confirm fitness of these sequences on cytochrome fold with an *E*-value of 10^{-9} (estimated precision 100%). Figure 5b represents the three-dimensional crystal structure of cytochrome C6 (PDB ID: 1CYJ) from *Chlamydomonas reinhardtii*. Amino acid residues that are essential for heme binding and that are also conserved in *H. pylori* protein sequences are shown in Fig. 5.

Dynamin_N. Dynamins form a superfamily of large GTPases that includes classical dynamins and dynamin-like proteins (DLPs) (50). They are involved in a variety of functions in eukaryotes such as scission of vesicles and organelles, including clathrin-coated vesicles, caveolae, phagosomes, and mitochondria (50). Previous studies

a

```
                               33 37 38                  60 61
PRPC_BACSU/2-236    LTALKTDTGKIRQHNEDDAGI---FKGKDE-FILAVVADGMGGHLAGDVASKMAVKAMGE
PPM1A_HUMAN/22-284  LRYGLSSMQGWRVEMEDAHTAVIGLPSGLESWSFFAVYDGHAGSQVAKYCCEHLLDHITN
PP2C1_YEAST/21-274  VGVAENKNSKFRRTMEDVHTYVKNFASRLD-WGYFAVFDGHAGIQASKWCGKHLHTIIEQ
HP0431              -----------RLPNQDAFLIRKNQKGLL-----LVLCDGLGSKKDSQIGAKKLCESVEK
                               *   :*         .        .: ** ..   ..   :     : :

PRPC_BACSU/2-236    KWNEAETIPTAPSECEKWLIEQILSVNSKIYDHAQAHEE----------------CQGMG
PPM1A_HUMAN/22-284  N---QDFKGSAGAPSVENVKNGIRTGFLEIDEHMRVMSEKKHGA------------DRSG
PP2C1_YEAST/21-274  N---------ILADETRDVRDVLNDSFLAIDEEINTKLVGNSGCTAAVCVLRWELPDSVS
HP0431              ALNAIDIQG-CNL-------ALLNKVIFSIWECLLYPLEVRNA---------------L
                                          :      * :

PRPC_BACSU/2-236    TTIVCALFTGKTVSVAHIGDSRCYLLQDDDFVQVTEDHSLVNELVRTGEISREDAEHHPR
PPM1A_HUMAN/22-284  STAVGVLISPQHTYFINCGDSRGLLCRNRKVHFFTQDHKPSNPLEKERIQNAGGSVMIQR
PP2C1_YEAST/21-274  DDSMDLAQHQRKLYTANVGDSRIVLFRNGNSIRLTYDHKASDTLEMQRVEQAGGLIMKSR
HP0431              STLQLVFIGKEKAFIGKVGD--GLLAVLGKEH----------------------------
                              .     : **    *    .

                                                                   239
PRPC_BACSU/2-236    KN---VLTKALG------------TDQLVSIDTRS--FDIEPG-DK-LLLCSDGLTNKVE
PPM1A_HUMAN/22-284  VNGSLAVSRALGDFDYKCVHGKGPTEQLVSPEPEVHDIERSEEDDQFIILACDGIWDVMG
PP2C1_YEAST/21-274  VNGMLAVTRSLGDKFF---------DSLVVGSPFTTSVEITSE-DKFLILACDGLWDVID
HP0431              --------RLLREDKQDLVHFTTPFDRNTLLTWEV--LERKKID--ALFLCTDGLDESLI
                            : *              :  .        .:        ::*. **: : :

PRPC_BACSU/2-236    GTELKDILQSDSAPQE----KVNLLVDKANQNGGEDNI
PPM1A_HUMAN/22-284  NEELCDFVRSRLEVTDDLEKVCNEVVDTCLYKGSRDNM
PP2C1_YEAST/21-274  DQDACELIKDITEPNE----AAKVLVRYALENGTTDNV
HP0431              DARRLDFVKDIIS-------------------------
                    .    ::::.
```

b

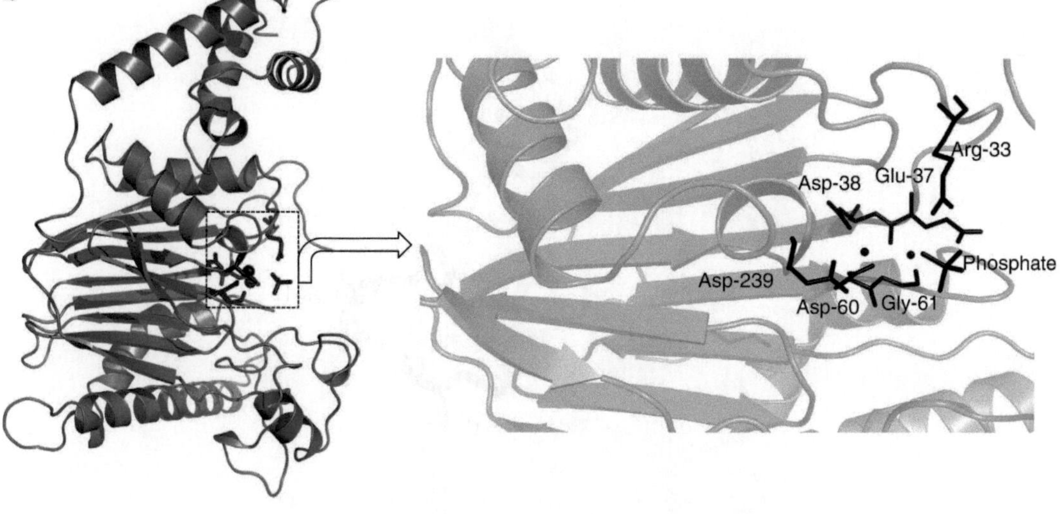

Fig. 4 (**a**) Multiple sequence alignment of bona fide members of the PP2C family and *Helicobacter* sequence (Uniprot accession code: HP0431) from strain 26695. *Vertical gray columns* represent conserved amino acid residues. Amino acid residues that bind to metal ions (corresponding to the human protein sequence) Asp-38, Asp-60, Gly-61, and Asp-239 are conserved. Glu-37 (corresponding to the human protein sequence), which is also implicated in binding to metal ions, is conservatively substituted to Gln. Arg-33 (corresponding to the human protein sequence), which binds to phosphate ions, is also conserved. Uniprot accession numbers of protein sequences used are as follows: PRPC_BACSU—*B. subtilis*; PPM1A_HUMAN—*H. sapiens*; and PP2C1_YEAST—*S. cerevisiae*. (**b**) The three-dimensional structure of human PP2C protein (PDB id: 1a6q) depicting the position of metal-binding and phosphate-binding residues that are conserved/conservatively substituted in an *H. pylori* protein sequence (represented in *black sticks*). Mn^{2+} is represented in *black sphere*

a

```
Q2G890_NOVAD          -APDGRTLF-A---QNCSSCHAAGPGHPG-TARLAIDAGPDF--AVLEQR-
A3Z583_9SYNE          -AADGEQLF-N---NHCAGCHPHGGNIIRR------GRTLKLRALEQRQL-
Q0HNQ4_SHESM          -AQDAEAIF-N---KACTVCHSMGVA-GAPKAHNAAEWEPRLAKGMDTLLH
Q605U5_METCA          -ARHGAKVY-R---SACAGCHDHGKDGAP---RL---DRPESWKSRSFE--
CYC6_CYAME                 -LAHGEQIF-S---ANCAACHAGGNN----------VIMPEKTLKLDALEA
C550A_CYACA           -IKKGKRLF-N---SHCGSCHVGGITKTNPNVGLDLESLNGANPPRNNINA
A3RTX2_RALSO          --NAQARTW-A---ASCASCHHADTAPGAK------TVAPTL--APLAGRP
A6Q8D3_SULNB          GDPAQTLIE-K---NKCMSCHNIMGMKDAPPFAGIAWRNSRINTNTAKATL
Q7MS60_WOLSU          -LEAGELLV-K---MSCSHCHSLEKT-GVYRPLQARLQGMPKEGIKAIIDS
B3EQ05_CHLPB          -AMSGQELY-T---ANCAACHSMRPP-AKTAPPIVGLASRFRTVYGNKSDA
A9AXG3_HERA2          -LEFGQRNY-E---QYCATCHGLAGE-GQGPGN---KAAPKL--SDLQGRL
A2SE78_METPP          -VDRGRALYA----STCVWCHSAGAV-SG-------GYAPDLRASPMLLDL
QCRC_COREF                 -VSRGGDLF-R---LNCASCHNFTGR-GG--ALSSGKYAPVLDPANEQEIY
HP0236                ----GMALYKNPRGVACAKCHGIKGE-QQ-------------EITFYYEKG
jhp0221               ----GLALYKNPRGVACAKCHGIKGE-QQ-------------EITFYYEKG
HPAG1_0239            ----GMALYKNPRGVACAKCHGIKGE-KQ-------------EITFYYEKG
Q7TU05_RHOBA          -SARGSKLF-T---AHCANCHRLEDR-GH-------EIGPNLATVKTRGDD
A4VQ99_PSEU5          -VVRGKAVWEN---NNCIGCHSLLGE-GA-------YFAPEL--GNVFVRR
A3WTY2_9BRAD          ARMGSLAFVKF---RGCVACHQGEKGEG-------GLSGPRL--TDAGARL
A7HE13_ANADF          -LSPGAALLKQ---ESCLQCHALGAE-GE-------RIGPRL--EWIGGRR
Q39TS3_GEOMG          -LIKGRELYKE---MHCAGCHQIDGQ-GG-------NVGPDL--SNFADKD
A6DM55_9BACT          -STIGAELFYD---LGCISCHTRPEE-TQ-------IKGERLLLSNVANKF
A3NRU4_BURP0          -ALRGQRIF-A---DRCAGCHAVRGT-GA-----AGTQAPDL--THVGARR
Q87H21_VIBPA          --IAAGEKQ-A---MVCKACHQFEPN-GVT------VVGPPLWGLAERNIA
Consensus                                 CXXCH
```

b

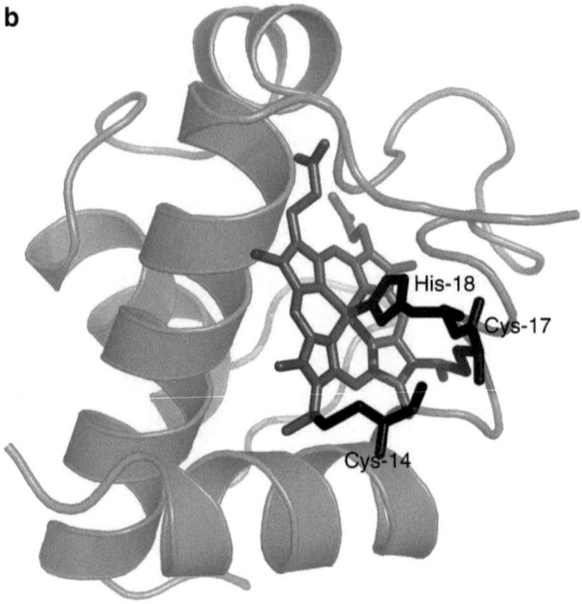

Fig. 5 (**a**) Multiple sequence alignment of members of the cytochrome_C family and putative cytochrome_C *H. pylori* sequences HP0236, jhp0221, and HPAG1_0239 from strains 26695, J99, and HPAG1, respectively. *Vertical gray columns* clearly depict the presence of the CXXCH motif, which is the characteristic feature of this family. Uniprot accession numbers of protein sequences used are as follows: C550A_CYACA—*Cyanidium caldarium*; CYC6_CYAME—*Cyanidioschyzon merolae* (red alga); QCRC_COREF—*Corynebacterium efficiens*;

suggest that dynamin may also be associated with microtubules in vitro (51, 52), and dynamin has also been characterized as phosphoprotein in nerve terminals (53). Dynamin family members have been reported in bacterial species such as *E. coli*, *H. pylori*, and *Mycobacterium tuberculosis*, but the functions of prokaryotic counterparts remain obscure (54). The three-dimensional structure of bacterial dynamin-like protein, from cyanobacteria *Nostoc punctiforme*, is available in a guanosine diphosphate-associated and a nucleotide-free state (55).

Though dynamin-like protein sequences are already reported from *H. pylori*, a paralogous sequence (HP0733 from strain 26695) at very low sequence identity is reported in the present analysis. The entire protein is 521 residues long and the region corresponding to residues 65–188 shows very poor sequence identity of about 16% to the dynamin family profile with an *E*-value of 10^{-4} and covers about 80% of the profile length. The guanosine triphosphate–binding sequence motifs (50, 56) GXXXXGKS and DXXG, corresponding to the G1 (P-loop) and the G3 motif, respectively, are very well conserved; threonine, which is involved in catalysis and is present in the G2 motif, is not conserved, and the N/TKTD consensus sequence pattern in the G4 motif is partially conserved in the *H. pylori* sequence (Fig. 6). Thus, sequence analysis of *H. pylori* protein suggests its remote homology with the protein domain family dynamin_N.

PTPS (6-Pyruvoyl Tetrahydropterin Synthase). 6-Pyruvoyl tetrahydropterin synthase catalyzes formation of tetrahydrobiopterin biosynthesis. Tetrahydrobiopterin is a cofactor for several important enzymes, such as aromatic amino acid hydroxylases and nitric oxide synthase (57). *H. pylori* protein HPAG1_0913 shares homology with members of the protein domain family PTPS. *H. pylori* protein shares poor sequence identity of 14% with the PTPS profile at an *E*-value of 10^{-10} and covers about 95% of the length of the profile. Fold recognition results also confirm the relationship between *H. pylori* protein and the PTPS protein domain family. A fold recognition algorithm ensures fitness of the *H. pylori* protein sequence on the three-dimensional structure of PTPS from

Fig. 5 (continued) A3RTX2_RALSO—*Ralstonia solanacearum*; A3WTY2_9BRAD—*Nitrobacter* spp.; A6DM55_9 BACT—*Lentisphaera araneosa*; A3Z583_9SYNE—*Synechococcus* spp.; A2SE78_METPP—*Methylibium petroleiphilum*; A7HE13_ANADF—*Anaeromyxobacter* spp.; Q87H21_VIBPA—*Vibrio parahaemolyticus*; Q39TS3_GEOMG—*Geobacter metallireducens*; A3NRU4_BURP0—*Burkholderia pseudomallei*; Q7TU05_RHOBA—*Rhodopirellula baltica*; A6Q8D3_SULNB—*Sulfurovum* spp.; Q605U5_METCA—*Methylococcus capsulatus*; B3EQ05_CHLPB—*Chlorobium phaeobacteroides*; Q7MS60_WOLSU—*Wolinella succinogenes*; A9AXG3_HERA2—*Herpetosiphon aurantiacus*; Q2G890_NOVAD—*Novosphingobium aromaticivorans*; Q0HNQ4_SHESM—*Shewanella* spp.; and A4VQ99_ PSEU5—*Pseudomonas stutzeri*. (**b**) Three-dimensional structure of cytochrome C6 (PDB id: 1CYJ) from *C. reinhardtii*. Amino acid residues that are essential for heme binding and that are also conserved in the *H. pylori* protein sequences are marked as *black sticks*. The heme molecule is represented in *gray sticks*

DNM1L_HUMAN	IVVVGTQSSGKSSVLE	TGIVTRRPL	TLVDLPGMTKV	AVITKLDLMD
Q9U4L0_CAEEL	IVVVGSQSAGKSSVLE	TGIVTRRPL	SLVDLPGITKI	AVLTKLDLMD
Q8SSJ7_ENCCU	IVAIGSQSSGKSSVLE	TNLVTRCPV	TLVDLPGLIKV	GVVTKIDLMD
Q9ZP56_ARATH	IVVVGDQSSGKSSVLE	QGICTRVPL	TMVDLPGITRV	AVVTKADMAP
Q7XRQ4_ORYSJ	IVVVGDQSSGKSSVLE	QGICTRVPL	TLVDLPGITRV	AVVTKSDKAP
Q7RHB6_PLAYO	IAVIGAQSVGKTSLLE	EGIATRTPI	TLIDLPGLTKV	GVITKCDVVE
Q84N64_ARATH	VLVVGQQTDGKSALVE	GGTKTRRPI	TIIDTPGLIAP	VVSTKLDTKI
Q7T2M4_CARAU	IAVIGDQSSGKSSVLE	TGIVTRCPL	TLIDLPGIARV	GILTKPDLVD
MX_CHICK	IAVIGDRNSGKSSVLE	KGVITRCPL	TLIDLPGIARE	GVLTKPDLVN
MX_ANAPL	IAVIGDQSSGKSSILE	NGIVTRCPL	TLIDLPGIARV	GILTKPDLVD
MX2_CANFA	IAVIGDQSSGKSSVLE	SGIVTRCPL	TLIDLPGITRV	GVLTKPDLVD
MX1_SHEEP	IAVIGDQSSGKSSVLE	SGIVTRCPL	TLIDLPGITRV	GILTKPDLVD
MX1_RAT	IAVIGDQSSGKSSVLE	SGIVTRCPL	TLIDLPGITRV	GILTKPDLVD
Q91196_ONCMY	IAVIGDQSSGKSSVLE	SGIVTRCPL	TLIDLPGIARV	GILTKPDLVD
HP0733	IAFYGETGAGKSTFIE	QSKVVQQER	TLLDVPGIE--	FYVTKTPNP-
Consensus	GXXXXGKS/T	T	DXXG	N/TKXD

Fig. 6 *H. pylori* sequence HP0733 from strain 26695 aligns well with other bona fide members of the Dynamin family. The *H. pylori* sequence shows the presence of the fully conserved consensus sequence of GXXXXGKS and DXXG and the partially conserved N/TKXD pattern, where aspartate of the N/TKXD sequence pattern is substituted by proline. Uniprot accession numbers of protein sequences used are as follows: ARC5_ARATH—*A. thaliana*; DNM1L_HUMAN—*H. sapiens*; MX1_RAT—*R. norvegicus*; MX1_SHEEP—*Ovis aries*; MX2_CANFA—*Canis familiaris*; MX_ANAPL—*Anas platyrhynchos*; MX_CHICK—*Gallus gallus*; Q8SSJ7_ENCCU—*Encephalitozoon cuniculi*; Q9U4L0_CAEEL—*C. elegans*; Q7RHB6_PLAYO—*Plasmodium yoelii yoelii*; Q9ZP56_ARATH—*A. thaliana*; Q7XRQ4_ORYSJ—*Oryza sativa* subsp. *japonica*; Q7T2M4_CARAU—*Carassius auratus*; and Q91196_ONCMY—*Oncorhynchus mykiss*

the source *Pseudomonas aeruginosa* (PDB ID: 2OBA) with an *E*-value of 10^{-17} (estimated precision 100%). Multiple sequence alignment of the *H. pylori* protein sequence with PTPS protein sequences from humans, *Drosophila melanogaster*, *Caenorhabditis elegans*, and *Shigella flexneri* suggests conservation of metal-binding residues (Cys-23, His-48, and His-50 corresponding to human protein) and active site residues (Cys-42 and His-89 corresponding to human protein) (Fig. 7).

4.3 Recognition of Some of the "Missing" Metabolic Proteins of H. pylori

The present study suggests that remote homologues of certain enzymes are part of metabolic pathways and are not reported so far in any of the *H. pylori* strains. In the following section, one such example is discussed.

AstE_AspA (Succinylglutamate Desuccinylase/Aspartoacylase). The AstE_AspA domain family includes succinylglutamate desuccinylase and aspartoacylase, both of which belong to the Zn-dependent carboxypeptidases family (58). Aspartoacylase (EC 3.5.1.15) catalyzes deacetylation of *N*-acetylaspartic acid (NAA) to produce acetate and l-aspartate (59). Though an aspartate metabolism pathway (Kyoto Encyclopedia of Genes and Genomes [KEGG] pathway identifier: hpy00250) is reported for *H. pylori*, aspartoacylase has not yet been identified in *H. pylori*.

```
                                        23                 42    48 50
PTPS_DROME       -----QQPVAFLTRRETFSACHRLHSPQLSDAENLEVFGKCNNFHGHGHNYTVEITVRGP
PTPS_HUMAN       TEGGGRRCQAQVSRRISFSASHRLYSKFLSDEENLKLFGKCNNPNGHGHNYKVVVTVHGE
PTPS_CAEEL       -----RMPIVTMERVDSFSAAHRLHSEKLSDAENKETFGKCNNSNGHGHNYVWKVKLRGE
PTPS_SHIFL       ------MMSTTLFKDFTFEAAHRLP--HVPEG------HKCG--RLHGHSFMVRLEITGE
HPAG1_0913       --------MMVIRRLYKFCASHVVRN---------CSSVKCAQ-NIHGHNYEVEVFIETN
                            : :  .* *.* :               **   . ***.:   : :
                                            89
PTPS_DROME       IDRRTGMVLNITELKEAIETVIMKRLDHKNLDKDVEYFANTPSTTENLAVYIWDNIRLQL
PTPS_HUMAN       IDPATGMVMNLADLKKYMEEAIMQPLDHKNLDMDVPYFADVVSTTENVAVYIWDNLQKVL
PTPS_CAEEL       VDPTSGMVYDLAKLKKEMS-LVLDTVDHRNLDKDVEFFKTTVSTSENVAIYMFEKLKSVM
PTPS_SHIFL       VDPHTGWIIDFAELKAAFK-PTYERLDHHYLN-DIPGLEN--PTSEVLAKWIWDQVKPVV
HPAG1_0913       RLDSANMALDFGLMQQEMQ-VFIESFDHAHHFWDKESDEFQRFIEDHCVRYVKCSFNLSA
                     :.   ::  ::  :.    . .**     *           :  . ::   ...

PTPS_DROME       KKPELLYEVKIHETPKNIISYRGP-------------------
PTPS_HUMAN       PVG-VLYKVKVYETDNNIVVYKGE-------------------
PTPS_CAEEL       SNPSVLYKVTIEETPKNIFTYKGC-------------------
PTPS_SHIFL       P---LLSAVMVKETCTAGCIYRGE-------------------
HPAG1_0913       EGYALMFLYYLTKILQKSVFSNDEGELKVSSVRVHETKNGYAE
                     ::    : :        ..
```

Fig. 7 Multiple sequence alignment of *H. pylori* protein sequence (HPAG1_0913) with PTPS protein sequences from human (PTPS_HUMAN), *D. melanogaster* (PTPS_DROME), *C. elegans* (PTPS_CAEEL), and *S. flexneri* (PTPS_SHIFL). Conservation of metal-binding residues (Cys-23, His-48, and His-50 corresponding to human protein) and active site residues (Cys-42 and His-89 corresponding to human protein) is highlighted in the *gray background*

A protein (HP1075 from strain 26695) that shows remote homology to the AstE_AspA family is reported in the present study. The *H. pylori* protein sequence shares sequence identity of 17% with the profile of the AstE_AspA domain family with an *E*-value of 10^{-7} and profile coverage of 81%. Also, the *H. pylori* protein sequence fits on the crystal structure of aspartoacylase protein from *Mesorhizobium loti* with an *E*-value of 10^{-15} (estimated precision 100%).

Further analysis was done on the multiple sequence alignment of the *H. pylori* protein with members of the AstE_AspA family. Amino acid residues corresponding to bovine carboxypeptidase A responsible for Zn binding (Glu-72, His-69), carboxylate binding (Arg-145), and catalytic residues (Glu-270) (58) are conserved in the *H. pylori* sequence. However, Zn-binding residues corresponding to His-69 are substituted by Gln, suggesting the possibility that this *H. pylori* protein is related to this enzymatic family (Fig. 8).

The KEGG database provides information on different pathways that are responsible for various cell processes. The aspartate metabolism pathway in *H. pylori* as reported in the KEGG database (60) has been studied. A remote homology relationship between the *H. pylori* protein sequence and the aspartoacylase protein fills the gap in this metabolic pathway.

```
                          69 72                                               145
HP1075              --PHLLLLAGIQGDEPGGFNATNLFLMHYSV--LKGLVEVVPVL-NKPSMLRNHRGLY--
P72108_NATPH        PGPTLFLSAAIHGDELNGIEIVREVAHEWSHDDLAGTLVCLPVL-NVPGFIAQERYLPVY
Q87RV7_VIBPA        EGPVLMVNAAIHGDELNGVEIVRQMINQLNPLKLKGTVIAVPIV-NVFGFIHKSRYLP-D
Q8EGL7_SHEON        PGPTLLVCAAIHGDELNGIEICRRLLGRVNPKNLAGTLIVVPIV-NVFGFIQQSRYLP-D
Q89N67_BRAJA        AGPKLLLMAGNHGDEYEGEITLTRLYRRLDAKAVKGEITILPFA-NLPAVMAARRRSPLD
A9CLI1_AGRT5        DGPSILLMAGNHGDEYEGEINLKRLIRSIEPADITGRLTIIPFA-NMPAVMAARRCSPLD
Q8KGS0_RHILO        EGPRLLVTGGNHGNELEGPLVARRLIEWLPEAQTCGRVIVLPML-NPLAVQAWSRNTPLD
Q98AA0_RHILO        EGPSLLITGGNHGNELQGPILARRLVKWLPEAQRCGRIIIVPEI-NPLAVQAWTRNTPID
Q88EN1_PSEPK        IGPTVLLTGGNHGDEYEGPIALTKLANQLDIETVRGRIIILPFM-NYPAFKNASRTSPID
A9CH33_AGRT5        DGPTALLTGANHGDEYEGPAALFELAYTLDPADVSGRIIIVPAL-NYPAFRAGTRTSPID
Q987A0_RHILO        KGPTALLSGGNHGDEYEGPLALYDLARALDPKQVSGTVIIVPAM-NYPAFRAGTRTSPID
Q92WC2_RHIME        EGPTALLTGGNHGDEYEGPIALFDLARKLDASDVEGRVIIVPAM-NYPAFQAQTRTSPID
ASPA_PROMA          SGIQVLLVAGTHGNEINAPWLFEQWKQKDSLINT-HNINIQTVIGNPVALEQGKRYV---
ACY2_BOVIN          PIKKVAIFGGTHGNELTGVFLVKHWLENSTEIQR-TGLEVKPFITNPRAVKKCTRYI---
ACY3_HUMAN          PLRRVAVTGGTHGNEMSGVYLARHWLHAPAELQR-ASFSAVPVLANPAATSGCRRYV---
ACY3_MOUSE          PLLRVAVTGGTHGNEMCGVYLARYWLQNPGELQR-PSFSAMPVLANPAATAACCRYL---
ASPA_ANASP          KINRVAIVGGTHGNEFTGAFLVKKFQQFPEVIQK-PSFETLTILGNPKAFEAGKRYI---
ASPA_SYNY3          AVRSLAIVGGVHGNERTGVEVVRRWQQKDFARYF-PALKCHYFLANPRAIAANRRYI---
ASTE_VIBVU          ETKNIILSSGIHGDETAPMELIDKIVHDIETGFQDVQARCLFIIAHPEATNAHTRFI---
ASTE_VIBCH          ATKNMVISCGIHGDETAPMELLDKWIDDIVSGFQPVAERCLFIMAHPQATVRHVRFI---
ASTE_SHEON          QGKDVILSCGVHGNETAPIELCNTLIKQLLQQKIIAKQRTLFLIGNPLAINNGTRII---
ASTE_YERPE          YTQSVVISAGIHGNETAPIEILNQLVTDLLAGQLPLSVRLLVLLGNPPAIRKGKRYL---
ASTE_ECOLI          PQGALVISAGIHGNETAPVEMLDALLGAISHGEIPLRWRLLVILGNPPALKQGKRYC---
ASTE_SALTY          VDRALILSAGIHGNETAPVEMLDKLLSALYSGSLTLTWRVLVVLGNPQALAAGIRYC---
ASTE_PSEPK          CGLDLLLSAGIHGNETAPIELLERLLHGVANGKIRPRARVLFLFGNPAAIRKGERFI---
ASTE_PSESM          NGTDLLLSAGIHGNETAPIELLDELIRSIARGDLKPRARILFLFGNPAAMRLGARYV---
ASTE_PSEAE          NGVDLLLSAGIHGNETAPIELLERLIRKVAAGTLKPAARVLFLFGNPEAIRRGERYV---

                       169          270
HP1075              DAVLHLHDGGGYY   HPIE---EY-------
P72108_NATPH        DFGLDFHTSTRG-   ATPTVTVELGEAHR--
Q87RV7_VIBPA        DYILDLHTGAIHR   DIPVLTYEAGEALRF-
Q8EGL7_SHEON        THVIDLHTGAIHR   GIPCILYEAGEALRFS
Q89N67_BRAJA        DVVFDIHSGGTSM   GAIGLSGEFGGGGT--
A9CLI1_AGRT5        DIVFDMHSGGTSM   GAIGISGEFGGGGS--
Q8KGS0_RHILO        DTVFDLHSFGPTW   GKVFVCAEFGGGT---
Q98AA0_RHILO        DTVLDLHSFGPTW   GKTFICTEFGG----G
Q88EN1_PSEPK        DYVLDIHSGGKTL   GKIFVTTELGGGGT--
A9CH33_AGRT5        DIVLDFHSGGRTL   GKVFVTTELGGAGT--
Q987A0_RHILO        DIVFDFHSGGKTL   GKVFVSTELGGGGT--
Q92WC2_RHIME        DIVLDFHSGGKTL   GKIFITTELGGGG---
ASPA_PROMA          QIAIDFHSTTSSM   WPCGFVVEVGPV-PQG
ACY2_BOVIN          DIIFDLHNTTSNM   AKYPVGIEVGP-QPQG
ACY3_HUMAN          DFVLDLHNTTANM   AKNGLGLELGPQ-PQG
ACY3_MOUSE          DFTLDLHNTTANT   SKNGICLEMGPQ-PQG
ASPA_ANASP          DVVIDLHSTTANM   NKLGFVIEVGAVA-QG
ASPA_SYNY3          DFIVDLHTTTAAM   CELGLTIELGPV-PQG
ASTE_VIBVU          RWHLDLHCAIRAS   SAQALTMELGRVARIG
ASTE_VIBCH          RWHLDLHCAIRGS   AAQALTLELGQVARLG
ASTE_SHEON          RIHYDLHTAMRAS   GADAFTIELGKVYPMG
ASTE_YERPE          RLHYDLHTAIRGS   QAASCTLELGKALPFG
ASTE_ECOLI          RWHLDLHTAIRGS   GALACTLELGKALPFG
ASTE_SALTY          RWHLDLHTAIRGS   GALSCTLELGKALPFR
ASTE_PSEPK          RLHYDLHTAIRGS   EAEAFTLELGKARPFG
ASTE_PSESM          RLHYDLHTAIRGS   GAEAFTLELGKARPFG
ASTE_PSEAE          HLHYDLHTAIRGS   GAEAFTLELGKARPFG
```

Fig. 8 Putative member of carboxypeptidase family from *H. pylori* (HP1075) is aligned with homologous members of bovine carboxypeptidase A. Glu-72, His-169 (Zn binding sites), Arg-145 (carboxylate-binding determinant), and Glu-270 (catalytic residue) are conserved in the *H. pylori* homologue. However, Zn binding residues corresponding to His-69 are substituted by Gln in the *Helicobacter* protein sequence

4.4 New Assignments of Domains in H. pylori *Sequences with Prior Assignment of Domains for the Rest of the Sequences*

In some proteins, only a part of the *H. pylori* protein is assigned a functional domain by Pfam and the rest of the protein is not assigned any function. Remote homology detection methods employed in the current analysis have established relationships between these unassigned regions and Pfam domain families.

N6_Mtase. The N6_Mtase family consists of N-6 adenine-specific DNA methylase, which specifically methylates the amino group at the C-6 position of adenines in DNA and is part of the bacterial restriction modification system. These enzymes are characterized by the presence of a conserved motif, Asp/Asn-Pro-Pro-Tyr/Phe (61–63). An *H. pylori* sequence (jhp_0613 from J99) that is 1,167 amino acid residues long has been assigned to the domain family helicase_C (from residues 82 to 162) by Pfam. However, no functional domain has been assigned from residues 1 to 81 and 163 to 1,167. Another sequence, HPAG1_0653 from strain HPAG1, which is 1,389 amino acid residues long, has been assigned as ResIII domain from residues 21 to 171 and as helicase_C domain from residues 332 to 411 by Pfam, and for the rest of the protein sequence no functional information is available.

A functional domain of N6_Mtase has been assigned to the fragment of protein sequence jhp0613 from residues 400 to 771 (sequence identity with N6_Mtase profile of 14%, *E*-value of 10^{-10}, and profile coverage of 92%) and HPAG1_0653 from residues 630 to 1,013 (identity with N6_Mtase profile 17%, *E*-value 10^{-9}, and profile coverage 96%). Both of these sequences show the presence of a conserved motif, Asn-Pro-Pro-Tyr (Fig. 9), which is a characteristic feature of members of the N6_Mtase family. Fold recognition results confirm the fitness of *H. pylori* protein sequences on *S*-adenosyl-l-methionine-dependent methyltransferases fold (*E*-value 10^{-19}, estimated precision 100%). Three-dimensional structures of members from the N6_Mtase family belong to same fold (e.g., PDB ID: 2ar0).

GHMP_kinases_C. The GHMP_kinases superfamily (64, 65) represents galactokinases, homoserine kinases, mevalonate kinases, and phosphomevalonate kinase. Each of these kinases transfers the γ-phosphoryl group of adenosine triphosphate to an acceptor (66). *H. pylori* protein sequences HP1050, HPAG1_0397, and jhp0375 have been assigned as GHMP_kinases_N domain at their N-terminal region by Pfam. The remote homology detection method employed relates the C-terminal region of all these proteins to the GHMP_kinases_C domain family. *H. pylori* protein sequences share about 25% sequence identity with the GHMP_kinases_C profile (*E*-value 10^{-11}, profile coverage 93%). Fold recognition results also ensure fitness of *H. pylori* protein sequences on the three-dimensional structure of members of the GHMP_kinases_C protein domain family (*E*-value 10^{-9}, estimated precision 100%).

Alignment of *H. pylori* protein sequences with homologous sequences from *Methanococcus jannaschii* suggests conservation

```
P95732_SALCH     DVIVTNPPFGGTEEDGIEKN-------------------FPAEMQTRET-ADL-FLQLII
T1M_SALTY        HIVATNPPFGSAAGTNITRT----------------------FVHPTSN-KQLCFMQHII
O25953_HELPY     DYVVANPPFSLKNWTD----GLSIDPKSK-QVINDSFNRFEDGTPPEKN-GDFAFLLHII
O52273_LACLA     DSVVMNPPYSAKWSAADK---FLSDPRFE----------RFGKLAPKSK-ADFAFLLHGF
O68168_9LACT     DAVMANPPYSAHWNN-KD---REDDPRFR-----------EYGIAPKTK-ADYSFLLHCL
T1MD_MYCPN       EVVVSNPPYSISWAGDKKSN-LVSDVRFK----------DAGTLAPNSK-ADLAFVLHAL
O25521_HELPY     DAIVSNPPYSTKWVGDSNP-ILINDERFS----------PAGVLAPKNA-ADLAFTMHML
T1M1_ECOLX       DAIVSNPPYSVKWIGSDDPT-LINDERFA----------PAGVLAPKSK-ADFAFVLHAL
Y1220_METJA      DKVVANPPWNQDGYDENT---LKVNEKYK--------DIYMYGFPNKNS-ADWAWVQLIN
O27025_METTH     DVVIANPPWNQDGYGEDT---LKKGDYWR--------ERFRYGFVNKQS-ADWAWIQHMI
T1MH_METJA       DYVVANPPFNDSEWDANR--IKPDDPRLR-------IGNKKVPVPPNGN-ANYMWILHFI
O33298_MYCTU     DYVMANLPFNIKDWARNE-----EDPRWR------------FGVPPANN-ANYAWIQHIL
P95510_PASHA     DVVMANPPFNQKEWWNES---LANDPRWA------------YGTPPQGN-ANFAWLQHMI
T1MH_HAEIN       DFIMANPHFNDKEWWNES---LADDPRWA------------YGTPPKGN-ANFAWLQHMI
O34139_KLEPN     DYILANPPFNLKDWRNDA--ELTKDPRFA-----------GYRTPPTGN-ANYGWILHML
jhp0613          RVIIGNPPYSSGAKSENDNNQNLSHPKLEKRVY---ETYGKNSTAQNKN-STRDTLIHSI
HPAG1_0653       RVIIGNPPYSAGAKSENDNNQNLTHPKLQKWVY---ETYGKNSTSRNVGQTTRDTLIQSI
```

Fig. 9 A block of multiple sequence alignment of N6_Mtase family members and *H. pylori* sequences (NCBI gene code: jhp0613 and HPAG1_0653 from strains J99 and HPAG1, respectively). *H. pylori* sequences show the presence of a conserved Asn-Pro-Pro-Tyr motif (*gray columns*), which is a characteristic feature of all N-Mtases

```
HP_1050          GYNAAFVEKKKVISLKTKIPSFLKAVMVIPNRAISTKQSRHLLPKRYSVQESVFNLSHAS
HPAG1_0397       GYNAAFVEKKKVISLKTKIPSFLKAVMVIPNRAISTKQSRHLLPKRYSVQESVFNLSHAS
jhp_0375         GYNAAFVEKKKVISLKTKIPSFLKAVMVIPNRVISTKQSRHLLPKRYSVQESVFNLSHAS
Q58504_METJA     -----------------------DILIAIPNISINTKEAREILPKAVGLKDLVNNVGKAC
                                        . ::.***  *.**::*.:***  .::: * *:.:*.

HP_1050          LMTMAIVQGKWDLL-RCCSKDRMHQYKRMQTYPVLFAIQKIALENNALMSTLSGSGSSFF
HPAG1_0397       LMTMAIVQGKWDLL-RCCSKDRMHQYKRMQTYPVLFAIQKIALENNALMSTLSGSGSSFF
jhp_0375         LMTMAIVQGKWDLL-RCCSKDRMHQYKRMQTYPVLFAIQKLALENNALMSTLSGSGSSFF
Q58504_METJA     GMVYALYNKDKSLFGRYMMSDKVIEPVRGKLIPNYFKIKEEVKDK-VYGITISGSGPSII
                  *. *: : . .*: *   .*:: :  * :  *  * *:: . :: .   *:****.*::

HP_1050          NMCYEEDAPKLKQVLSKKFPKFRVAVLDFDNDGVLIEKD
HPAG1_0397       NMCYEEDAPKLKQVLSKKFPKFRVAVLDFDNDGVLIEKD
jhp_0375         NMCYEEDAPKLKQVLSKKFPKFRVAVLDFDNDGVLIEKD
Q58504_METJA     AFPKEEFIDEVENILRDYYENTIRTEVGKGVEVV-----
                  :  **   :::::* . : :   : :. . : *
```

Fig. 10 Alignment of *H. pylori* GHMP kinase sequence with homologous sequences from *M. jannaschii*. Conservation of functionally important amino acid residues is highlighted in the *gray background*

of functionally important amino acid residues. It has been reported that the stretch of amino acid residues Ser/Thr-Gly-Ser-Gly-Pro-Ser, from residues 259 to 264 (corresponding to the homoserine kinase domain from *M. jannaschii*), is conserved. This stretch takes the conformation of a loop and helps in stabilizing the conformation of the phosphate-binding loop in the N-terminal domain. Amino acid residues (Gly-260, Ser-261) in this stretch may also participate in catalysis by providing an additional amide group to interact with the γ-phosphate of adenosine triphosphate (67). *H. pylori* protein sequences show the presence of the Ser-Gly-Ser-Gly-Ser-Ser motif corresponding to the *M. jannaschii* protein (Fig. 10).

Acknowledgments

This research was supported by Microsoft Corporation (Redmond, WA).

References

1. Blaser MJ, Parsonnet J (1994) Parasitism by the "slow" bacterium *Helicobacter pylori* leads to altered gastric homeostasis and neoplasia. J Clin Invest 94:4–8
2. Kusters JG, van Vliet AH, Kuipers EJ (2006) Pathogenesis of *Helicobacter pylori* infection. Clin Microbiol Rev 19:449–490
3. Marshall BJ, Warren JR (1984) Unidentified curved bacilli in the stomach of patients with gastritis and peptic ulceration. Lancet 1:1311–1315
4. Tee W, Lambert JR, Dwyer B (1995) Cytotoxin production by *Helicobacter pylori* from patients with upper gastrointestinal tract diseases. J Clin Microbiol 33:1203–1205
5. Zarrilli R, Ricci V, Romano M (1999) Molecular response of gastric epithelial cells to *Helicobacter pylori*-induced cell damage. Cell Microbiol 1:93–99
6. Bourzac KM, Guillemin K (2005) *Helicobacter pylori*-host cell interactions mediated by type IV secretion. Cell Microbiol 7:911–919
7. Cascales E, Christie PJ (2003) The versatile bacterial type IV secretion systems. Nat Rev Microbiol 1:137–149
8. Eaton KA, Brooks CL, Morgan DR et al (1991) Essential role of urease in pathogenesis of gastritis induced by *Helicobacter pylori* in gnotobiotic piglets. Infect Immun 59:2470–2475
9. Meyer-Rosberg K, Scott DR, Rex D et al (1996) The effect of environmental pH on the proton motive force of *Helicobacter pylori*. Gastroenterology 111:886–900
10. Mobley HL, Island MD, Hausinger RP (1995) Molecular biology of microbial ureases. Microbiol Rev 59:451–480
11. Weeks DL, Eskandari S, Scott DR et al (2000) A H+-gated urea channel: the link between *Helicobacter pylori* urease and gastric colonization. Science 287:482–485
12. Cover TL, Blaser MJ (1996) *Helicobacter pylori* infection, a paradigm for chronic mucosal inflammation: pathogenesis and implications for eradication and prevention. Adv Intern Med 41:85–117
13. Matsumoto Y, Marusawa H, Kinoshita K et al (2007) *Helicobacter pylori* infection triggers aberrant expression of activation-induced cytidine deaminase in gastric epithelium. Nat Med 13:470–476
14. Beswick EJ, Suarez G, Reyes VE (2006) *H. pylori* and host interactions that influence pathogenesis. World J Gastroenterol 12: 5599–5605
15. Tomb JF, White O, Kerlavage AR et al (1997) The complete genome sequence of the gastric pathogen *Helicobacter pylori*. Nature 388:539–547
16. Alm RA, Ling LS, Moir DT et al (1999) Genomic-sequence comparison of two unrelated isolates of the human gastric pathogen *Helicobacter pylori*. Nature 397:176–180
17. Oh JD, Kling-Backhed H, Giannakis M et al (2006) The complete genome sequence of a chronic atrophic gastritis *Helicobacter pylori* strain: evolution during disease progression. Proc Natl Acad Sci USA 103:9999–10004
18. Falush D, Wirth T, Linz B et al (2003) Traces of human migrations in *Helicobacter pylori* populations. Science 299:1582–1585
19. Finn RD, Tate J, Mistry J et al (2008) The Pfam protein families database. Nucleic Acids Res 36:D281–D288
20. Anand B, Gowri VS, Srinivasan N (2005) Use of multiple profiles corresponding to a sequence alignment enables effective detection of remote homologues. Bioinformatics 21: 2821–2826
21. Gowri VS, Krishnadev O, Swamy CS et al (2006) MulPSSM: a database of multiple position-specific scoring matrices of protein domain families. Nucleic Acids Res 34:D243–D246
22. Gowri VS, Tina KG, Krishnadev O et al (2007) Strategies for the effective identification of remotely related sequences in multiple PSSM search approach. Proteins 67:789–794
23. Tyagi N, Swapna LS, Mohanty S et al (2009) Evolutionary divergence of Plasmodium falciparum: sequences, protein-protein interactions, pathways and processes. Infect Disord Drug Targets 9:257–271
24. Balaji S, Sujatha S, Kumarm SS et al (2001) PALI-a database of Phylogeny and ALIgnment of homologous protein structures. Nucleic Acids Res 29:61–65
25. Gowri VS, Pandit SB, Karthik PS et al (2003) Integration of related sequences with protein three-dimensional structural families in an updated version of PALI database. Nucleic Acids Res 31:486–488

26. Sujatha S, Balaji S, Srinivasan N (2001) PALI: a database of alignments and phylogeny of homologous protein structures. Bioinformatics 17:375–376
27. Murzin AG, Brennerm SE, Hubbard T et al (1995) SCOP: a structural classification of proteins database for the investigation of sequences and structures. J Mol Biol 247:536–540
28. Marchler-Bauer A, Panchenko AR, Shoemaker BA et al (2002) CDD: a database of conserved domain alignments with links to domain three-dimensional structure. Nucleic Acids Res 30:281–283
29. Kelley LA, Sternberg MJ (2009) Protein structure prediction on the Web: a case study using the Phyre server. Nat Protoc 4:363–371
30. Bradshaw RA, Ericsson LH, Walsh KA et al (1969) The amino acid sequence of bovine carboxypeptidase A. Proc Natl Acad Sci USA 63:1389–1394
31. Rawlings ND, Barrett AJ (1993) Evolutionary families of peptidases. Biochem J 290(Pt 1): 205–218
32. Vallee BL, Auld DS (1990) Zinc coordination, function, and structure of zinc enzymes and other proteins. Biochemistry 29:5647–5659
33. Bignell C, Thomas CM (2001) The bacterial ParA-ParB partitioning proteins. J Biotechnol 91:1–34
34. Schumacher MA (2007) Structural biology of plasmid segregation proteins. Curr Opin Struct Biol 17:103–109
35. Khare D, Ziegelin G, Lanka E et al (2004) Sequence-specific DNA binding determined by contacts outside the helix-turn-helix motif of the ParB homolog KorB. Nat Struct Mol Biol 11:656–663
36. Jones DT (1999) Protein secondary structure prediction based on position-specific scoring matrices. J Mol Biol 292:195–202
37. Oliva G, Fontes MR, Garratt RC et al (1995) Structure and catalytic mechanism of glucosamine 6-phosphate deaminase from *Escherichia coli* at 2.1 A resolution. Structure 3:1323–1332
38. Calcagno M, Campos PJ, Mulliert G et al (1984) Purification, molecular and kinetic properties of glucosamine-6-phosphate isomerase (deaminase) from *Escherichia coli*. Biochim Biophys Acta 787:165–173
39. Natarajan K, Datta A (1993) Molecular cloning and analysis of the NAG1 cDNA coding for glucosamine-6-phosphate deaminase from *Candida albicans*. J Biol Chem 268: 9206–9214
40. Lara-Lemus R, Libreros-Minotta CA, Altamirano MM et al (1992) Purification and characterization of glucosamine-6-phosphate deaminase from dog kidney cortex. Arch Biochem Biophys 297:213–220
41. Rogers MJ, Ohgi T, Plumbridge J et al (1988) Nucleotide sequences of the *Escherichia coli nagE* and *nagB* genes: the structural genes for the *N*-acetylglucosamine transport protein of the bacterial phosphoenolpyruvate: sugar phosphotransferase system and for glucosamine-6-phosphate deaminase. Gene 62:197–207
42. Montero-Moran GM, Lara-Gonzalez S, Alvarez-Anorve LI et al (2001) On the multiple functional roles of the active site histidine in catalysis and allosteric regulation of *Escherichia coli* glucosamine 6-phosphate deaminase. Biochemistry 40:10187–10196
43. Das AK, Helps NR, Cohen PT et al (1996) Crystal structure of the protein serine/threonine phosphatase 2C at 2.0 A resolution. EMBO J 15:6798–6809
44. Moore F, Weekes J, Hardie DG (1991) Evidence that AMP triggers phosphorylation as well as direct allosteric activation of rat liver AMP-activated protein kinase. A sensitive mechanism to protect the cell against ATP depletion. Eur J Biochem 199:691–697
45. Stone JM, Collinge MA, Smith RD et al (1994) Interaction of a protein phosphatase with an *Arabidopsis* serine-threonine receptor kinase. Science 266:793–795
46. Maeda T, Tsai AY, Saito H (1993) Mutations in a protein tyrosine phosphatase gene (PTP2) and a protein serine/threonine phosphatase gene (PTC1) cause a synthetic growth defect in *Saccharomyces cerevisiae*. Mol Cell Biol 13:5408–5417
47. Allen JW, Leach N, Ferguson SJ (2005) The histidine of the c-type cytochrome CXXCH haem-binding motif is essential for haem attachment by the *Escherichia coli* cytochrome c maturation (Ccm) apparatus. Biochem J 389:587–592
48. Stevens JM, Daltrop O, Allen JW et al (2004) C-type cytochrome formation: chemical and biological enigmas. Acc Chem Res 37:999–1007
49. Thony-Meyer L (2000) Haem-polypeptide interactions during cytochrome c maturation. Biochim Biophys Acta 1459:316–324
50. Praefcke GJ, McMahon HT (2004) The dynamin superfamily: universal membrane tubulation and fission molecules? Nat Rev Mol Cell Biol 5:133–147
51. Obar RA, Collins CA, Hammarback JA et al (1990) Molecular cloning of the microtubule-associated mechanochemical enzyme dynamin reveals homology with a new family of GTP-binding proteins. Nature 347:256–261
52. Shpetner HS, Vallee RB (1989) Identification of dynamin, a novel mechanochemical enzyme that mediates interactions between microtubules. Cell 59:421–432
53. Robinson PJ, Hauptschein R, Lovenberg W et al (1987) Dephosphorylation of synaptosomal

proteins P96 and P139 is regulated by both depolarization and calcium, but not by a rise in cytosolic calcium alone. J Neurochem 48:187–195
54. van der Bliek AM (1999) Functional diversity in the dynamin family. Trends Cell Biol 9:96–102
55. Low HH, Lowe J (2006) A bacterial dynamin-like protein. Nature 444:766–769
56. Dever TE, Glynias MJ, Merrick WC (1987) GTP-binding domain: three consensus sequence elements with distinct spacing. Proc Natl Acad Sci USA 84:1814–1818
57. Nar H, Huber R, Heizmann CW et al (1994) Three-dimensional structure of 6-pyruvoyl tetrahydropterin synthase, an enzyme involved in tetrahydrobiopterin biosynthesis. EMBO J 13:1255–1262
58. Makarova KS, Grishin NV (1999) The Zn-peptidase superfamily: functional convergence after evolutionary divergence. J Mol Biol 292:11–17
59. Kaul R, Gao GP, Balamurugan K et al (1993) Cloning of the human aspartoacylase cDNA and a common missense mutation in Canavan disease. Nat Genet 5:118–123
60. Kanehisa M, Araki M, Goto S et al (2008) KEGG for linking genomes to life and the environment. Nucleic Acids Res 36:D480–D484
61. Lauster R (1989) Evolution of type II DNA methyltransferases. A gene duplication model. J Mol Biol 206:313–321
62. Narva KE, Van Etten JL, Slatko BE et al (1988) The amino acid sequence of the eukaryotic DNA [N6-adenine]methyltransferase, M. CviBIII, has regions of similarity with the prokaryotic isoschizomer M.TaqI and other DNA [N6-adenine] methyltransferases. Gene 74:253–259
63. Timinskas A, Butkus V, Janulaitis A (1995) Sequence motifs characteristic for DNA [cytosine-N4] and DNA [adenine-N6] methyltransferases. Classification of all DNA methyltransferases. Gene 157:3–11
64. Bork P, Sander C, Valencia A (1992) An ATPase domain common to prokaryotic cell cycle proteins, sugar kinases, actin, and hsp70 heat shock proteins. Proc Natl Acad Sci USA 89: 7290–7294
65. Bork P, Sander C, Valencia A (1993) Convergent evolution of similar enzymatic function on different protein folds: the hexokinase, ribokinase, and galactokinase families of sugar kinases. Protein Sci 2:31–40
66. Andreassi JL 2nd, Leyh TS (2004) Molecular functions of conserved aspects of the GHMP kinase family. Biochemistry 43: 14594–14601
67. Zhou T, Daugherty M, Grishin NV et al (2000) Structure and mechanism of homoserine kinase: prototype for the GHMP kinase superfamily. Structure 8:1247–1257

Chapter 12

Identification of Novel Anthrax Toxin Countermeasures Using In Silico Methods

Ting-Lan Chiu, Kimberly M. Maize, and Elizabeth A. Amin

Abstract

Anthrax is an acute infectious disease caused by the spore-forming, gram-positive, rod-shaped bacterium *Bacillus anthracis*. The anthrax toxin lethal factor (LF) is the primary anthrax toxin component responsible for cytotoxicity and host death and has been a heavily researched target for design of postexposure therapeutics in the event of a bioterror attack. Various computer-aided drug design methodologies have proven useful for pinpointing new antianthrax drug scaffolds, optimizing existing leads and probes, and elucidating key mechanisms of action. We present a selection of in silico virtual screening protocols incorporating docking and scoring, shape-based searching, and pharmacophore mapping techniques to identify and prioritize small molecules with potential biological activity against LF. We also recommend screening parameters that have been shown to increase the accuracy and reliability of these computational results.

Key words Anthrax, Lethal factor, Metalloproteinases, Virtual screening, Docking and scoring, Pharmacophore mapping, Computer-aided drug design

1 Introduction

Molecular modeling techniques, including docking and scoring, shape-based searching, and pharmacophore mapping, are widely used in the drug discovery process to identify new molecular scaffolds, elucidate mechanisms of action, and prioritize particular compounds and/or series for experimental screening or synthesis (1–5). Incorporating virtual screening (VS) methodologies such as docking and topomeric searching into therapeutics design has had a positive impact on compound hit rates (1) and has led to better prediction of binding modes (1, 3, 4). However, the reliability of VS simulations varies broadly among the available docking algorithms, scoring functions, parameters, and descriptors (Maize and Amin, unpublished observation, 2012, (6–8)), which must be chosen carefully on the basis of ligand structure(s) and key receptor characteristics such as hydrophobicity and hydrogen-bonding environment.

Sandhya Kortagere (ed.), *In Silico Models for Drug Discovery*, Methods in Molecular Biology, vol. 993,
DOI 10.1007/978-1-62703-342-8_12, © Springer Science+Business Media, LLC 2013

Docking and scoring in particular are often conducted using default settings and parameters available in the VS package on hand; however, variations in binding-site structure and features, as well as applied tolerance thresholds for steric and electrostatic interactions, strongly influence the quality of results. Validation studies should be conducted whenever possible to assess the capability of VS protocols to accurately reproduce experimental bound configurations to the target of interest and to prioritize known active compounds within a dataset. Metalloenzymes, such as the anthrax toxin lethal factor (LF), which is primarily responsible for anthrax-related toxicity (9–13), have attracted particular attention as drug targets in various disease modalities. However, they pose significant challenges in VS (and in other molecular modeling techniques as well) owing to the presence of catalytic transition metals that are often poorly represented in molecular mechanics-based force fields used to construct scoring functions. We outline a series of experimentally validated protocols and recommended parameter sets for docking and scoring, shape-based (topomeric) searching, and pharmacophore perception that have been shown to increase experimental hit rates for small molecules targeting LF and that can be implemented to rank existing compounds for further prosecution and to search databases for new LF inhibitor scaffolds.

2 Materials

1. SciTegic Pipeline Pilot 8.0, Accelrys, Inc. (San Diego, CA)
2. SYBYL-X 1.0 expert molecular modeling environment, Tripos, Inc. (St. Louis, MO)
3. Surflex-Dock 2.1 virtual screening package (14–17) and CScore consensus scoring module (18), Tripos, Inc.
4. Topomer Search (19–22) three-dimensional (3D) shape-based searching module, Tripos, Inc.
5. MOE (Molecular Operating Environment) 2010.10, Chemical Computing Group, Inc. (Montreal, Quebec, Canada)
6. GALAHAD (Genetic Algorithm with Linear Assignment of Hypermolecular Alignment of Datasets) (23) pharmacophore hypothesis module, Tripos, Inc.

3 Methods

1. Small molecules of interest can be prepared for LF screening via a variety of mechanisms. For a relatively small library, compounds can be sketched individually and then geometry optimized in SYBYL-X or MOE using an appropriate force

field for drug-like small molecules, such as Tripos (24) or MMFF94s (Merck Molecular Force Field) (25), in order to obtain 3D coordinates for subsequent simulations. For a wider-ranging virtual screen (e.g., to identify new scaffolds in less explored areas of chemistry space), small-molecule compound libraries can be obtained from a variety of sources, including DrugBank (26, 27) (http://www.drugbank.ca/), the National Institutes of Health (NIH) Molecular Libraries Small Molecule Repository (MLSMR) (28), the National Cancer Institute (29), and eMolecules (http://www.emolecules.com/). For these larger compound sets, we recommend generating 3D configurations using SciTegic Pipeline Pilot via the "SD Reader," "3D Coordinates," "Add Hydrogens," "Minimize Molecule," and "SD Writer" components, in that order. This protocol breaks each compound into ring and chain fragments, generates 3D structures for the fragments, reassembles the compound, conducts a preliminary geometry optimization on the reassembled structure, adds hydrogen atoms, carries out a more thorough energy minimization via the Clean force field (30), and writes a new 3D SD file as output that can subsequently be used as input for a variety of software packages and modeling techniques.

2. For receptor-based procedures including docking and scoring, five experimental X-ray structures of LF-ligand complexes are currently available in the Protein Data Bank (http://www.rcsb.org/pdb/download/download.do): 1YQY (31), 1ZXV (32), 1PWP (33), 1PWQ (34), and 1PWU (34). Cocrystallized inhibitors in these complexes include a sulfonamide hydroxamate, MK-702/LF-1B, the most active LF inhibitor designed to date (half-maximal inhibitory concentration [IC_{50}] = 0.054 μM, 1YQY), rhodanine derivative BI-MFM3 (IC_{50} = 1.7 μM, 1ZXV), the *N*,*N′*-di-quinoline urea analogue NSC 12155 (K_i = 0.5 μM, 1PWP), and two peptidic hydroxamates, thioacetyl-Tyr-Pro-Met amide and GM6001 (K_i^{app} = 2.1 μM and 11 μM, 1PWQ and 1PWU, respectively). We recommend 1YQY.pdb (31) for VS, because it is a truncated LF structure comprising the three key domains (II–IV) that form the enzyme active site. Protein and ligand preparation can be done in MOE 2010.10: remove all cocrystallized water molecules; add hydrogens, check ligand bonds/protonation state(s) and edit if necessary; examine residue bonds/protonation states within 4.5 Å of the cocrystallized ligand and correct if necessary; fix heavy atoms in space and then energy minimize the complex using an initial gradient of 0.05.

3. Docking and scoring can be carried out using Surflex-Dock (14–17) and CScore (18) in SYBYL-X 1.0. In detailed evaluations (1) of various docking and scoring programs for the LF

system, Surflex-Dock was found to be superior in terms of reproducing cocrystallized LF-inhibitor bound configurations, within root-mean-square-deviation (RMSD) values of only 0.54 Å. In this VS environment, the target area on the receptor for small-molecule docking is rendered by a protomol, which is a representation of an "ideal" ligand located in the active site of the protein, constructed by modeling specific interactions of small-molecule, fragment-based probes within that area (14–17). Probes include a variety of hydrophobic and hydrophilic fragments as well as hydrogen-bond donor and acceptor groupings. In the screening procedure, the targeted area in the active site can be obtained from an automatic detection algorithm, the location of the cocrystallized ligand, or manual selection of relevant residues using the SYBYL interface. Extensive validation studies (Maize and Amin, unpublished observation, 2012) on experimental LF X-ray structures and their cocrystallized ligands indicated that ligand-based protomol generation *without* hydrogens added to the ligand yielded greater screening accuracy, in terms of RMSD between predicted and experimental bound configurations as well as docking enrichment factor (35), compared with automatic site detection or manual residue selection.

4. Choosing acceptable threshold and bloat values is critical to maximizing LF docking precision. In Surflex-Dock, the threshold value determines how much of the protomol may be "buried" within the protein, while the bloat parameter allows for protomol expansion in order to reach into crevices or longer binding channels (especially those with open ends). Evaluating a variety of threshold values from 0.01 to 1 for docking the five available cocrystallized LF inhibitors into their respective crystal structures (31–34) led to an optimal threshold value of 0.74 and a bloat value of zero (Amin and Maize, unpublished observation, 2012). These studies also pinpointed two key user-defined variables that impact LF docking outcomes: ring flexibility and search density (14–16). Including ring flexibility (a binary operator) and implementing the highest available density of search level ($d=9$) increase computation time but significantly improve the quality of results and are therefore recommended for all LF VS runs. The maximum number of conformations per compound fragment and the maximum number of poses per ligand should be set to 20, with the maximum number of rotatable bonds per small molecule set to 100. Postdock minimizations are recommended for each bound configuration and, given the significant amount of steric and electrostatic variation in the LF active site, all four consensus scoring functions in CScore (G_SCORE, PMF_SCORE, D_SCORE, and CHEMSCORE) should be implemented in order to represent the broadest possible selection of ligand–receptor

interactions taking place in the sterically and electrostatically diverse LF binding site.

5. Shape-based, "topomeric" searching can be done in order to find new LF inhibitor scaffolds occupying un- or underexplored regions in chemical space. In this procedure, one or more proven active compounds are utilized to search collections of molecules for matches that exhibit similar 3D shapes, as represented by conformationally independent topomeric fields (19–22). We have found this similarity searching protocol to be highly useful for identifying active LF inhibitors within data collections of various sizes and diversity levels as well as for selecting compounds for subsequent experimental evaluation via in vitro screening (1). Structures pinpointed by topomeric searching are often significantly dissimilar in terms of usual two-dimensional (2D) structural fingerprints, meaning that they are more likely to be located in less extensively explored chemical space than those identified by traditional 2D similarity searching (19–22). One often uses a highly active but pharmacokinetically "compromised" compound as the topomeric search template to "lead-hop" to new compounds that have similar 3D shapes (and, ostensibly, similar ability to bind to the desired target) but different chemical functionalities, in order to retain biological activity while avoiding impediments such as toxicity and metabolic instability. Effective searching using an active LF inhibitor template can be done using the Topomer Search module in SYBYL-X 1.0, using a "maximum distance considered hit" parameter of 185, with all weighting factors (steric, aromatic, positive/negative, donor/acceptor) set to 1,000.

6. Accurate and validated pharmacophore hypotheses have proven useful for identifying new LF inhibitor scaffolds via database searching on the basis of ligand–receptor interactions observed for one or more series of active compounds (33, 36–38). Several LF inhibitor pharmacophore hypotheses have been outlined in the literature (33, 36–38); however, these models were developed from relatively small training sets that occupy only one or two subsites of the LF active site and therefore do not necessarily represent the majority of key interactions that are essential for ligand binding. We recently reported (2) a new comprehensive pharmacophore map based on experimentally determined bound configurations for active compounds; this new hypothesis covers all three subsites (S1′, S1-S2, and S2′) of the LF active site and selectively identifies inhibitors with biological activity against LF in the nanomolar range. As reported by Chiu and Amin (2), for accurate and useful pharmacophore mapping based on active LF inhibitors, we recommend a genetic algorithm approach incorporating Pareto scoring, as

implemented in the GALAHAD pharmacophore perception module (23) (see Note 1), together with ligand–receptor interaction analysis based on experimental structural biology (see Note 2).

7. If experimental bound configurations (i.e., cocrystallized inhibitors) are to be used for pharmacophore perception, we recommend aligning all structures in Cartesian space by optimizing the sum of all pairwise alignment scores using the Homology/Align module in MOE 2010.10, basing the alignment on protein coordinates. Structures of additional small molecules can be subjected to geometry optimization by energy minimization within the LF X-ray structure of choice (we suggest 1YQY.pdb) in order to approach a putative bound configuration as closely as possible. Optimization can be done in MOE 2010.10 using the MMFF94s force field (25) with a convergence criterion of 0.05 kcal/mol·Å, with the receptor held rigid. Larger sets of molecules used for hypothesis validation and/or database searching can be prepared and optimized using MMFF94s and then docked into the LF active site using Surflex-Dock and CScore as described above, with the protomol defined to encompass all three LF binding area subsites and threshold and bloat parameters set to the optimal values of 0.74 and zero.

4 Notes

1. All genetic algorithm–based pharmacophore hypotheses should be subjected to multiple iterative refinement, with accuracy assessed by two criteria: (1) an overall Pareto score and (2) a rank sum value that includes the GALAHAD parameters of steric overlap, pharmacophoric concordance, and agreement between the query tuplet and the pharmacophoric tuplets for the ligands used to create the model (which is essentially a similarity value between the query and the ligand set) (23). Generally, if all Pareto scores are equal, the models are ordered by the rank sum value, with any remaining "ties" broken by a total strain energy term where lower energy is considered more favorable. Recommended user-specified parameters for LF inhibitors (to be fine-tuned based on the training set used) include a population size of 25–35; a maximum number of 90 generations; three to five molecules that must hit the query in order for a model to be retained; and a "keep best *n* models" value of 15–20.

2. The presence of a given chemical functionality in more than one compound used to generate a pharmacophore hypothesis does not guarantee a significant contribution to activity. It is

therefore helpful to examine structural biology data for ligand–receptor complexes whenever possible and to model experimentally observed interactions as 2D ligand–receptor interaction maps (in MOE 2010.10). Pharmacophoric features that either do not parallel experimental interactions or represent those interactions inaccurately (e.g., incorrect hydrogen-bonding directionality) can then be removed from the hypotheses of interest. Because hydrophobic interactions are not well rendered in MOE 2D interaction maps, supplemental PoseView (http://www.zbh.uni-hamburg.de/poseview) (39) 2D diagrams can be generated in cases such as LF where hydrophobic interactions demonstrate a significant contribution to compound activity.

Acknowledgments

This work was supported by NIH R01 AI083234 to E.A.A.; the Minnesota Supercomputing Institute for Advanced Computational Research (MSI); and the University of Minnesota Institute for Therapeutics Discovery and Development (ITDD).

References

1. Chiu T, Solberg J, Patil S et al (2009) Identification of novel non-hydroxamate anthrax toxin lethal factor inhibitors by topomeric searching, docking and scoring, and in vitro screening. J Chem Inf Model 49:2726–2734
2. Chiu TL, Amin EA (2012) Development of a comprehensive, validated pharmacophore hypothesis for anthrax toxin lethal factor (LF) inhibitors using genetic algorithms, Pareto scoring, and structural biology. J Chem Inf Model, 52:1886–1897
3. Amin EA, Welsh WJ (2006) Highly predictive CoMFA and CoMSIA models for two series of stromelysin-1 (MMP-3) inhibitors elucidate S1′ and S1-S2′ binding modes. J Chem Inf Model 46:1775–1783
4. Jia Y, Chiu TL, Amin EA et al (2010) Design, synthesis and evaluation of analogs of initiation factor 4E (eIF4E) cap-binding antagonist Bn7-GMP. Eur J Med Chem 45:1304–1313
5. Ambrose Amin E, Welsh WJ (2001) Three-dimensional quantitative structure-activity relationship (3D-QSAR) models for a novel class of piperazine-based stromelysin-1 (MMP-3) inhibitors: applying a "divide and conquer" strategy. J Med Chem 44:3849–3855
6. Moustakas DT, Lang PT, Pegg S et al (2006) Development and validation of a modular, extensible docking program: DOCK 5. J Comput Aided Mol Des 20:601–619
7. Hartshorn MJ, Verdonk ML, Chessari G et al (2007) Diverse, high-quality test set for the validation of protein-ligand docking performance. J Med Chem 50:726–741
8. Corbeil CR, Englebienne P, Moitessier N (2007) Docking ligands into flexible and solvated macromolecules. 1. Development and validation of FITTED 1.0. J Chem Inf Model 47:435–449
9. Pezard C, Berche P, Mock M (1991) Contribution of individual toxin components to virulence of Bacillus anthracis. Infect Immun 59:3472–3477
10. Chopra AP, Boone S, Liang X et al (2003) Anthrax lethal factor proteolysis and inactivation of MAPK kinase. J Biol Chem 278: 9402–9406
11. Vitale G, Bernardi L, Napolitani G et al (2000) Susceptibility of mitogen-activated protein kinase kinase family members to proteolysis by anthrax lethal factor. Biochem J 352:739–745
12. Moayeri M, Leppla SH (2004) The roles of anthrax toxin in pathogenesis. Curr Opin Microbiol 7:19–24
13. Warfel JM, Steele AD, D'Agnillo F (2005) Anthrax lethal toxin induces endothelial barrier dysfunction. Am J Pathol 166:1871–1881

14. Jain AN (2007) Surflex-Dock 2.1: robust performance from ligand energetic modeling, ring flexibility, and knowledge-based search. J Comput Aided Mol Des 21:281–306
15. Pham T, Jain AN (2006) Parameter estimation for scoring protein-ligand interactions using negative training data. J Med Chem 49: 5856–5868
16. Jain AN (2004) Virtual screening in lead discovery and optimization. Curr Opin Drug Discov Devel 7:396–403
17. Jain AN (2003) Surflex: fully automatic flexible molecular docking using a molecular similarity-based search engine. J Med Chem 46: 499–511
18. Meng EC, Shoichet BK, Kuntz ID (1992) Automated docking with grid-based energy evaluation. J Comput Chem 13:505–524
19. Cramer RD, Poss MA, Hermsmeier MA et al (1999) Prospective identification of biologically active structures by topomer shape similarity searching. J Med Chem 42:3919–3933
20. Cramer RD, Jilek RJ, Guessregen S et al (2004) Lead hopping. Validation of topomer similarity as a superior predictor of similar biological activities. J Med Chem 47:6777–6791
21. Cramer RD (2006) Leadhopping - and beyond. Expert Opin Drug Discov 1: 311–321
22. Jilek RJ, Cramer RD (2004) Topomers: a validated protocol for their self-consistent generation. J Chem Inf Comput Sci 44:1221–1227
23. Richmond NJ, Abrams CA, Wolohan PR et al (2006) GALAHAD: 1. pharmacophore identification by hypermolecular alignment of ligands in 3D. J Comput Aided Mol Des 20:567–587
24. Clark M, Cramer RD III, Van Opdenbosch N (1989) Validation of the general purpose tripos 5.2 force field. J Comput Chem 10:982–1012
25. Halgren TA (1999) MMFF VII. Characterization of MMFF94, MMFF94s, and other widely available force fields for conformational energies and for intermolecular-interaction energies and geometries. J Comput Chem 20:730–748
26. Wishart DS, Knox C, Guo AC et al (2008) DrugBank: a knowledgebase for drugs, drug actions and drug targets. Nucleic Acids Res 36:D901–D906
27. Wishart DS, Knox C, Guo AC et al (2006) DrugBank: a comprehensive resource for in silico drug discovery and exploration. Nucleic Acids Res 34:D668–D672
28. Austin CP, Brady LS, Insel TR et al (2004) NIH molecular libraries initiative. Science 306:1138–1139
29. National Cancer Institute [updated 16 Mar 2010]. http://cactus.nci.nih.gov/download/nci/. Accessed 29 Mar 2012
30. Hahn M (1995) Receptor surface models. 1. Definition and construction. J Med Chem 38: 2080–2090
31. Shoop WL, Xiong Y, Wiltsie J et al (2005) Anthrax lethal factor inhibition. Proc Natl Acad Sci USA 102:7958–7963
32. Forino M, Johnson S, Wong TY et al (2005) Efficient synthetic inhibitors of anthrax lethal factor. Proc Natl Acad Sci USA 102:9499–9504
33. Panchal RG, Hermone AR, Nguyen TL et al (2004) Identification of small molecule inhibitors of anthrax lethal factor. Nat Struct Mol Biol 11:67–72
34. Turk BE, Wong TY, Schwarzenbacher R et al (2004) The structural basis for substrate and inhibitor selectivity of the anthrax lethal factor. Nat Struct Mol Biol 11:60–66
35. Huang N, Shoichet BK, Irwin JJ (2006) Benchmarking sets for molecular docking. J Med Chem 49:6789–6801
36. Yuan H, Johnson SL, Chen LH et al (2010) A novel pharmacophore model for the design of anthrax lethal factor inhibitors. Chem Biol Drug Des 76:263–268
37. Agrawal A, de Oliveira CA, Cheng Y et al (2009) Thioamide hydroxypyrothiones supersede amide hydroxypyrothiones in potency against anthrax lethal factor. J Med Chem 52: 1063–1074
38. Roy J, Kumar UC, Machiraju PK et al (2010) In silico studies on anthrax lethal factor inhibitors: pharmacophore modeling and virtual screening approaches towards designing of novel inhibitors for a killer. J Mol Graph Model 29:256–265
39. Stierand K, Maass PC, Rarey M (2006) Molecular complexes at a glance: automated generation of two-dimensional complex diagrams. Bioinformatics 22:1710–1716

Chapter 13

Rational Design of HIV-1 Entry Inhibitors

Asim K. Debnath

Abstract

This chapter reviews studies that have used in silico techniques to design or identify potential HIV-1 entry inhibitors targeting cellular receptors CD4, CCR5, and CXCR4 and envelope glycoproteins, gp120 and gp41 of HIV-1. Both structure- and ligand-based design techniques have been used in those studies by applying diverse modeling techniques such as quantitative structure–activity relationship analysis, conformational analysis, molecular dynamics, pharmacophore generation, docking, virtual screening (using docking software and also shape-based ROCS techniques), and fragment-based design.

Key words HIV-1, Entry inhibitor, gp120, CD4, gp41, CCR5, CXCR4

1 Introduction

Entry of human immunodeficiency virus type 1 (HIV-1) into cells is obligatory to initiate the infection process, which involves a sequence of steps. The first step requires the interaction between gp120, the HIV-1 exterior envelope glycoprotein, and CD4, the primary receptor on the target cell. This interaction creates a high-affinity binding site on gp120 for specific chemokine receptors (CXCR4 or CCR5), which act as second receptors for HIV-1. CCR5 is the principal coreceptor involved in natural infection and transmission of HIV-1. The binding of the gp120–CD4 complex to CCR5 presumably triggers subsequent conformational changes in the envelope glycoproteins, followed by the dissociation of gp120 from gp41 and insertion of the fusion domain of gp41 into the cell membrane. Subsequently the viral and target cell membranes fuse and eventually the virus enters the cells. Each step of this entry process has been considered as a potential target for developing inhibitors, which are thus termed *entry inhibitors*. Two entry inhibitors currently have been approved by the US FDA: Fuzeon (Enfuvirtide, T-20; Roche, Indianapolis, IN) is targeted to gp41 and Maraviroc (Selzentry; Pfizer, New York, NY) is targeted

Sandhya Kortagere (ed.), *In Silico Models for Drug Discovery*, Methods in Molecular Biology, vol. 993,
DOI 10.1007/978-1-62703-342-8_13,

to the CCR5 coreceptor. However, numerous other efforts are underway (1–7) to develop small-molecule and peptide-based HIV-1 entry inhibitors. Review of those studies is beyond the scope of this chapter. Therefore, this chapter focuses on studies that have used in silico techniques to design or identify potential entry inhibitors.

2 gp120 as Target

The first successful report on structure-based rational design of miniproteins as CD4 mimics was based on transferring the critical binding sites of CD4 and the CDR2-like CD4 domain, especially amino acids Phe43 and Arg59, to a scorpion toxin scaffold, scyllatoxin (8). However, the CD4 mimics engineered by this approach, CD4M8 and CD4M9, only showed micromolar binding with $gp120_{HXB2}$ and $gp120_{JRFL}$. The major breakthrough came after the X-ray crystal structure of HIV-1 gp120 bound to CD4 and a monoclonal antibody 17b was solved (9). The nuclear magnetic resonance (NMR) structure of CD4M9 indicated that the functional unit of CD4 on this miniprotein was a very good mimic, with a root mean square deviation (RMSD) of only 0.7 Å. However, when this miniprotein mimic was modeled using the X-ray structure, steric clash was observed at the N-terminal part of the mimic. Based on the modeled structure, CD4M9 was modified and the binding affinity of the resulting mimic, CD4M32, improved by ~40- to ~80-fold to the nanomolar level. Further mutations of additional residues to protect the disulfide bonds in the scaffold structure and to stabilize the helical region resulted in the best miniprotein mimic, CD4M33, reported so far (10). CD4M33 also inhibited diverse strains of HIV-1 with both X4 and R5 coreceptor tropism as well as dual tropism with nanomolar potency (Table 1).

Table 1
Sequences and antiviral activity of CD4 mimetics

Peptide	Sequences	IC_{50} (nM)[a]
sCD4 (CDR2)	QIKILGNQGSFLTKGP	–
Scyllatoxin	AFCNLRMCQLSCRSLGLLGKCIGDKCECVKH	–
CD4M8	Ac-CNLARCQLSCKSLGLKGGCAGSFCACG-NH_2	–
CD4M9	Ac-CNLARCQLRCKSLGLLGKCAGSFCACGP-NH_2	1.6
CD4M32	Tpa-NLARCQLRCKSLGLLGKCAGSBCACV-NH_2	37.5
CD4M33	Tpa-NLHFCQLRCKSLGLLGKCAGSBCACV-NH_2	7.5

B Bip (biphenylalanine), *Tpa* thiopropionic acid
[a]50% inhibitory concentration obtained in competition ELISA between sCD4 and $gp120_{HXB2}$

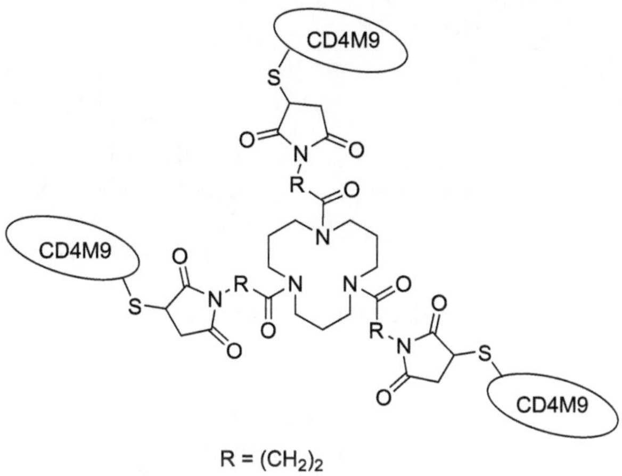

Fig. 1 Structure of trivalent CD4M9 miniprotein

Later a trivalent version of CD4M9 miniprotein (Fig. 1) was synthesized and was reported to have ~140-fold enhanced anti-HIV-1 activity compared with the monovalent miniprotein (11).

In 2005, our group first reported the discovery of two inhibitors of CD4–gp120 interaction, NBD-556 and 557, through screening a commercially available library of compounds using an HIV-1 syncytium formation assay (12). Later it was conclusively shown that NBD-556 bound to the Phe43 cavity of gp120 using a mutated (S375W) gp120 (13). This mutation fills the cavity and prevents these compounds from binding. Subsequently, a Japanese group confirmed in an escape mutation study that the binding site indeed was the Phe43 cavity (14). Our group (unpublished) and others have attempted to optimize these leads (13, 15, 16); however, we have failed so far to improve the antiviral potency appreciably. This failure was partly attributed to the lack of structural information regarding the binding of these inhibitors to gp120. However, recently, Kwong's group at the Vaccine Research Center at the NIH, in association with our group, was successful in crystallizing both NBD-556 and 557 with gp120 and an antibody that recognizes the CD4-bound gp120 conformation (16a).

Our group and others have observed that there is very little room for manipulation of regions I and II of the NBD-556 chemotype (Fig. 2). The X-ray structure of NBD-557 with gp120 revealed that the tetramethyl-piperidine moiety was in proximity to a negatively charged amino acid, D368, which is very important in forming an ionic interaction with R59 of CD4. A hydrophobic residue, V430, was also in proximity. Therefore, in an attempt to determine structure–activity relationships in region III and to identify new analogues, LaLonde et al. (15) used two orthogonal virtual screening techniques, one based on docking using GOLD software (CCDC, Cambridge, UK) (17, 18) and the other based on ligand-based

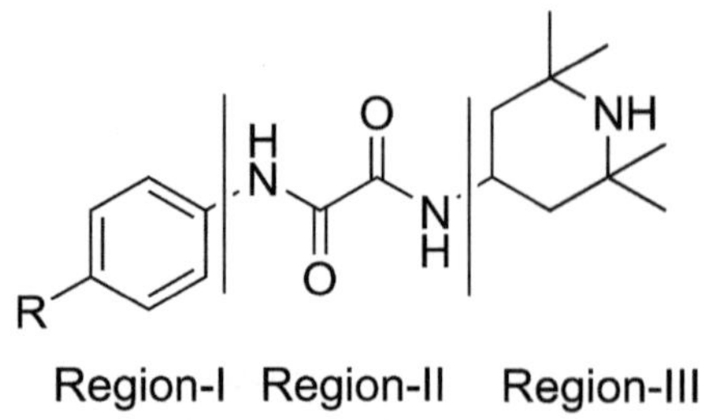

Fig. 2 Different regions of NBD-556 (*R* = Cl)

shape similarity matching, termed rapid overlay of chemical structure (ROCS) (OpenEye Scientific Software, Santa Fe, NM) (19). For the docking simulation, a database of more than 300 primary amines was selected from commercial sources, attached in silico to the oxalamide core of the lead scaffold, and then docked on the gp120 structure (PDB Code: 1G9M). The compounds were selected for synthesis based on commercial availability, docking score, and a slogp of less than 4.5. However, none of these compounds showed improved binding affinity in isothermal titration calorimetry experiments despite the fact that they showed inhibitory activity in a CD4–gp120 binding assay as well as in a virus infectivity assay.

In a further attempt to identify inhibitors of gp120–CD4 interaction, LaLonde et al. (15) used the docked pose of NBD-556 as a query in ROCS to search a Zinc database consisting of more than two million drug-like molecules. However, among several compounds tested only one compound showed inhibition of gp120 binding to CD4. Subsequently, the investigators used the unpublished X-ray structure of NBD-557 bound to gp120 and used the NBD-556 conformation as the search query. However, none of the 16 compounds that were purchased based on this study showed antiviral activity. Thus, only the tetramethyl-amino-piperidine was used as the query for ROCS to identify analogues from the Zinc database. Several amines were thus identified. As mentioned previously, the same approach was used to form the molecules in silico using the selected amines and docked in the gp120 pocket. These docked conformations were used as follow-up queries in the subsequent ROCS run, and several new compounds were identified that showed modest antiviral activity. However, utilization of these techniques did not produce compounds with improved activity over the original lead, NBD-556.

In a similar attempt to improve the antiviral activity, we have used a large database of commercially available primary amines and created molecules in silico by replacing the tetramethyl-piperidine ring in NBD-556 with the amines. We have used the crystal structure information of NBD-557 bound to gp120 (16a) to dock these molecules using GLIDE (20, 21) docking software (Schrödinger

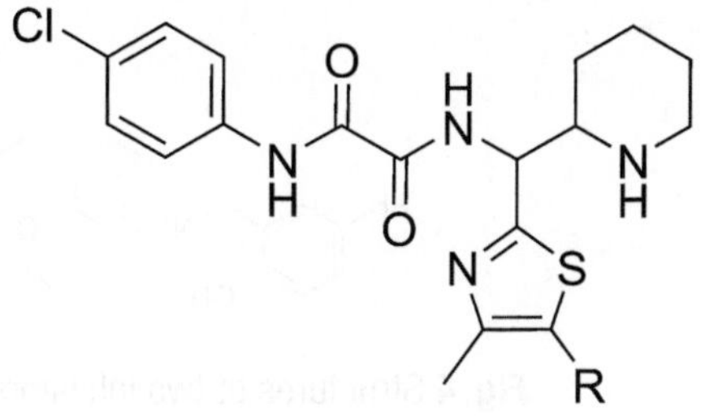

R = CH_2OH (**NBD09027**)

R = CH_2CH_2OH (**NBD-10007**)

Fig. 3 Structures of NBD-09027 and NBD-10007

LLC, New York, NY). We have selected a few compounds with top-scored dock poses for synthesis. Two of these new-generation inhibitors, NBD-09027 and NBD-10007 (Fig. 3), inhibited CD4–gp120 interaction and showed more potent antiviral activity against primary isolates than NBD-556 or 557 (unpublished).

In another approach, Caporuscio et al. (22) used a protocol combining molecular dynamics (MD), pharmacophore analysis, and docking-based screening to identify inhibitors of gp120–CD4 interaction by targeting the Phe43 cavity, the binding site of CD4. MD simulations (1 ns) were run with the Amber 9 (23) software (University of California, San Francisco) on the gp120 core structure extracted from a complex of gp120, CD4, and X5 antibody (PDB Code: 2B4C) to account for protein flexibility. Next, molecular interaction fields were computed using GRID (24) software (Molecular Discovery, Ltd, Pinner, Middlesex, UK) on a shell of residues in gp120 within 10 Å from the CD4 residues. Pharmacophores were generated using CATALYST (Accelrys, Inc., San Diego, CA) pharmacophore generation software. A four-feature pharmacophore with three hydrophobic (HYD1–3) and one hydrogen bond acceptor with 10 excluded volumes was selected to virtually screen a database from Asinex Corp. (the Asinex Gold collection; Moscow, Russia) consisting of 200,000 small molecules. Subsequently a limited set of data was selected from the pharmacophore-based screening run by removing any compounds with a fit value of lower than 1.5, more than 10 rotatable bonds, and more than one chiral center. In addition, Lipinski's rule of five and other drug-likeness criteria were applied to select a much smaller set of 729 compounds for use in a GLIDE-based docking protocol. The binding modes of the top 10% of hits in the Phe43 cavity were visually inspected and five compounds were purchased based on commercial availability. Two of these hits (Fig. 4) showed modest inhibitory activity (22 and 9 μM, respectively) on the HIV-1-induced cytopathic effect in MT4 cells infected with HIV-1 NL4-3. However, their toxicity was high, making their selectivity

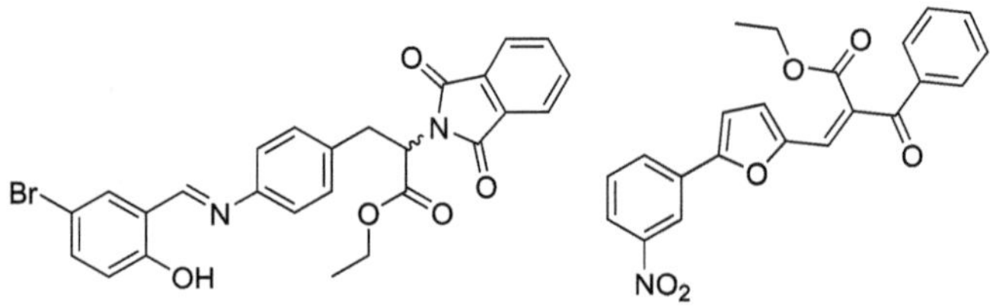

Fig. 4 Structures of two inhibitors targeted to gp120

index very low. Nevertheless, these compounds belong to two unique scaffolds and may be useful as leads for further optimization.

Cyanovirin-N (CVN), an 11-kDa lectin, was reported to have potent antiviral activity because it binds irreversibly to high-mannose oligosaccharides on the surface of the HIV-1 envelope glycoprotein gp120 (25–27). CVN has been proposed as a potential topical microbicide. However, its modest stability and tendency to form domain-swapped dimers restrict its use in the clinic. To improve the stability of CVN with regard to chemical and thermal unfolding while preserving its high-affinity binding property to glycans, Patsalo et al. designed stabilized variants of CVN using a computation-based rational design strategy (28). Structure-based observation of CVN leads to the identification of three buried polar residues: Ser11, Ser20, and Thr61. The investigators performed a Poisson–Boltzmann continuum electrostatic free energy calculation using the ICE software package (Massachusetts Institute of Technology, Cambridge, MA) to determine the contribution of these residues to the stability of CVN. They observed that these three residues carry a substantial solvation penalty. To improve the stability by mutating these residues, they resorted to the same calculation to determine the effect of such mutation on the stability, and they designed several mutants based on their initially designed stable protein, Pro51Gly (P51G). Double mutants (e.g., S11V. S20A, S11I.S20A, and S11A.S20A) provided the most stable variants. These mutants adopt the wild-type fold. They also retained the binding specificity with regard to glycans containing the Manα(1$\rightarrow$2)Man linkage.

3 CD4 as Target

Most of the inhibitors of gp120–CD4 interaction reported so far were targeted to gp120. In 2006 Neffe and Meyer reported an optimization strategy using a lead CD4-binding peptidomimetic and employing computer-based docking and saturation transfer difference (STD) NMR study (29). The peptidomimetic was based on their original discovery of a lead decapeptide NMWQKVGTPL (30). In their initial optimization strategy they considered

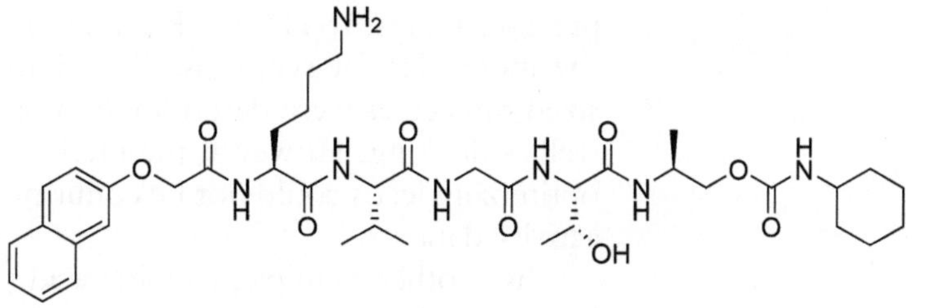

Fig. 5 Structure of a peptidomimetic targeted to CD4

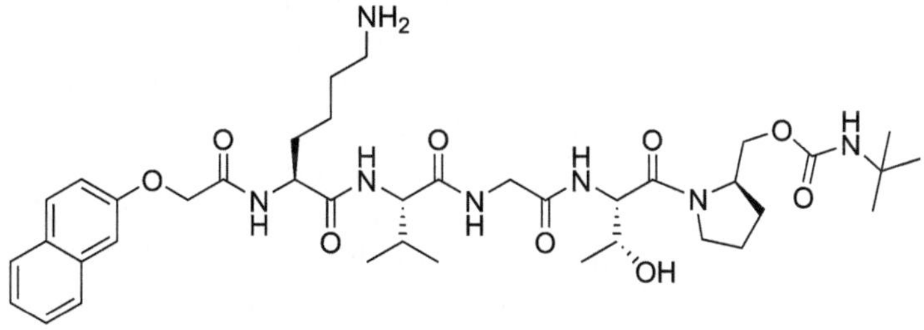

Fig. 6 Structural analogue of a peptidomimetic targeted to CD4

modifying only the N-terminal aromatic residue, the core peptide, and the amino acid mimetics forming the core and the linkers. Several aromatic substitutions at the N-terminal were selected by similarity search in the available chemical directory and docked to improve interactions; however, the original β-naphthyl ring provided the best interaction with CD4. The investigators also varied the linkers of the peptidomimetics that connect the aromatic ring with the core peptide. The oxyacetic acid and two slightly larger linkers provided the best interactions; however, the selection of the final linker was based on the commercial availability of the reagent. The STD-NMR study played a critical role in identifying the important amino acids for substitution in the docking study. The authors based the selection of the amino acids in the peptide core for synthesis on the calculated binding energies of the ligand–protein complex derived from the docking simulations. A set of 12 peptidomimetics were synthesized and their theoretical binding energy with CD4 was calculated using the FlexiDock docking program within the Sybyl 6.9 software package (Tripos, St. Louis, MO). The binding affinity (K_D) of the peptidomimetics to CD4 was determined using surface plasmon resonance. The best peptidomimetic (Fig. 5) showed a K_D of 10 μM, which was further analyzed in detail using STD-NMR and the K_D value agreed very well (9 μM).

In a subsequent study, the same group optimized the C terminus of the CD4 binding peptidomimetics (31), which yielded an even higher binding affinity (6 μM) compound (Fig. 6) than reported previously. Again STD-NMR was used as the guiding tool to design the molecules for synthesis. The binding affinities of the synthesized molecules were determined by surface plasmon resonance as well as docking. However, the merit of these design techniques in optimizing leads could not be confirmed in the absence of antiviral activity data.

In another example, palmitic acid (PA) was recently shown to inhibit gp120–CD4 interaction and HIV-1 infection of both R5- and X4-tropic viruses by targeting CD4 (32). The binding affinity of PA to CD4 was estimated to be only in the low micromolar range (K_D ~ 1.5 μM). Nevertheless, PA has been shown to inhibit attachment between gp120 and CD4 with a K_i of ~2.53 μM and to prevent R5-tropic HIV-1 infection in a cervical explant model of human vagina (33). Therefore, although PA showed potential as a possible lead for entry inhibitor, it requires improvement in binding and antiviral activity to be a viable drug candidate. In an attempt to improve the activity of this novel small molecule, the investigators used a standard protocol in Autodock automated docking software 4.0 (Scripps Research Institute, La Jolla, CA) (34) to dock this molecule to the known X-ray structure of CD4 to gain insight into the mechanism of binding of PA. They observed that PA binds tightly to a hydrophobic cavity in CD4 formed by Phe53, Ile60, Ile62, Leu63, and Leu70, indicating that extensive modification of the aliphatic chain was not possible, but the modeling study showed the scope of modifications at the PA carboxylic group as well as the methylene groups close to the carboxyl end. Using this information from in silico modeling, the investigators searched a database of chemical compounds and three compounds—2-bromohexadecanoic acid (2-BP), 6-*O*-palmitoyl-l-ascorbic acid, and sucrose palmitate—were selected for further study. Tryptophan fluorescence of soluble CD4 was used to estimate the binding affinity of these PA analogues. The estimated K_D was in the low nanomolar range (~74–364 nM). One of the PA analogues, 2-BP, did not form micelles and was shown to have 1:1 stoichiometry when it binds to CD4. These compounds also inhibited the binding between gp120 and CD4 (K_i ~ 122–1,486 nM) and they were nontoxic in a cell-based assay.

4 gp41 as Target

The major breakthrough in our structural knowledge of the HIV-1 envelope glycoprotein came in 1997 when the X-ray structure of the fusion-active form of gp41 was reported (35–37). Because it is inherently difficult to crystallize envelope glycoproteins, a protein-dissection approach was applied to derive substructures from the N- and C-terminal regions of the ectodomain termed N-36 and C-34. The structure indicates that the N-36/C-34 complex forms a hexahelical bundle consisting of a parallel trimeric coil–coil formed by the inner N-36 peptides and three C-34 helical peptides packed in an antiparallel orientation to the hydrophobic grooves formed by the inner N-36 trimer (35). Each groove in the trimer has a large hydrophobic cavity that accommodates three highly conserved hydrophobic residues (Trp628, Trp631, and Ile635 of C-34). Structure–activity analysis suggests that mutation of these residues drastically reduces the association between N-36 and C-34 peptides, indicating the importance of these residues in the complex formation. Therefore, these cavities have been suggested as the potential target for small-molecule inhibitors. It has been hypothesized that small molecules that bind to the cavity with high affinity may prevent the formation of the six-helix bundle, a necessary step for fusion of HIV-1 with cell membrane. In addition to these cavities, a critical salt-bridge was also identified at the periphery of the cavity formed by Asp632 of C-34 with Lys574 of N-36 (35).

We utilized this critical structure information and initiated a systematic study to identify small-molecule lead compounds targeted to the cavity (38). We screened a commercial database of 20,000 compounds from ComGenex, Inc. (Budapest, Hungary) using the automated molecular docking software DOCK3.5 (39, 40). To create the cavity on the N-36 trimer for docking the compounds, we removed one of the C-34 peptides from the X-ray structure (PDB Code: 1AIK). Based on crystallographic information, we selected all residues within an 8 Å radius surrounding Trp628, an indole-based hydrophobic amino acid that binds to the hydrophobic pocket, to create a negative image of the cavity. The 3D coordinates of all 20,000 molecules were generated using CONCORD software (Tripos), docked into the cavity, and their binding evaluated using a force-field scoring function. We selected the 200 top-scoring compounds from the docking run for further analysis by visual inspection and narrowed them to 20 compounds with the best fit and maximum possible interactions irrespective of their score. Of these 20 compounds, 16 were available for purchase. These compounds were tested by enzyme-linked immunosorbent assay (ELISA) for inhibitory activity in forming the complex N-36/C-34, which is recognized by the monoclonal antibody NC-1 (41) and by HIV-1-mediated cell fusion and cytopathic effects. The compounds were also tested for their in vitro cytotoxicity. Two of the tested compounds, ADS-J1 and ADS-J2 (Fig. 7), showed promising inhibitory activities; however, ADS-J1 had the better selectivity index. The docking of ADS-J1 indicated that hydrophobic groups (phenyl and naphthalene) interacted with the hydrophobic residues Leu568, Val570, and Trp571 in the cavity. The sulfonic acid group also formed a salt-bridge with Lys574 similar to that identified in the X-ray structure between Asp632 and Lys574. Although ADS-J1 lacked drug-like properties and could not be further optimized to a drug, it was the first proof of concept that the X-ray-identified cavity in the gp41 structure is a

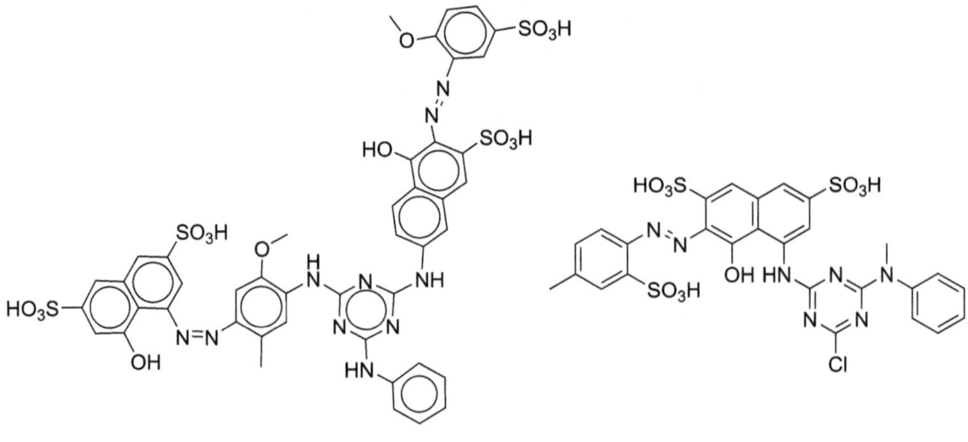

Fig. 7 Structures of ADS-J1 and ADS-J2 targeted to gp41

legitimate target for identifying small-molecule inhibitors. Later our group identified two N-substitute pyrroles, NB-2 and NB-64, by screening a chemical library consisting of drug-like small molecules from ChemBridge Corp. (San Diego, CA) (42). Computer-aided docking analysis also confirmed that these two inhibitors bind to the hydrophobic pocket identified in the X-ray structure and that their COOH group forms a salt-bridge with Lys574, as was observed with ADS-J1.

Recently, investigators from Ansaris Bio (a division of Locus Pharmaceuticals, Blue Bell, PA) reported the use of fragment-based discovery technique in creating ligands that bind to the same hydrophobic cavity described above (43, 44). The method includes systematic, rigorous sampling of fragments in the translational and rotational space in a protein target site. The fragments are then assembled to molecules and the binding free energy is calculated by statistical mechanics. Liu et al. applied this technique in designing ligands in the hydrophobic cavity of the gp41 structure formed by N-36 and C-34 peptides, as mentioned earlier (45). In essence, one of the C-34 peptides was removed from the crystal structure as described earlier. The fragments were sampled on the hydrophobic cavity formed by the N-36 trimer core and their binding was calculated and rank ordered. A high-affinity binding site was identified near the Lys574 that extended through a narrow channel to Arg579. One of the compounds generated by this technique contains a biphenyl group linked to a naphthalene ring by aliphatic ether. The compound (Fig. 8) was synthesized and tested for its ability to inhibit six-helix bundle formation in a binding assay based on size exclusion chromatography. The six-helix bundle was formed by mixing N-36 peptide with C-34. The compound showed a marginal inhibitory activity with a median inhibitory concentration (IC_{50}) of 31 μM.

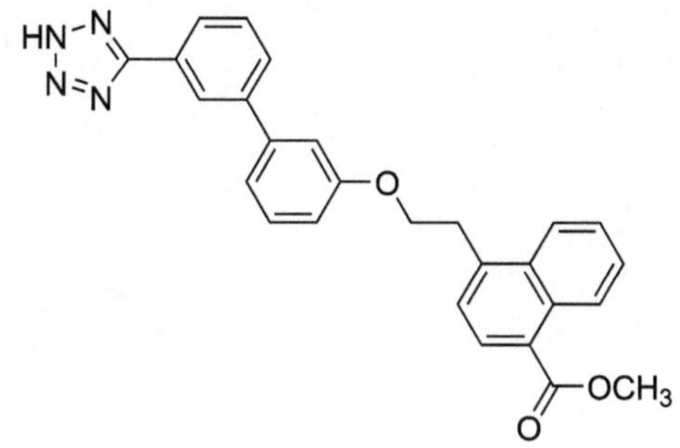

Fig. 8 Structure of a naphthalene-based inhibitor targeted to gp41

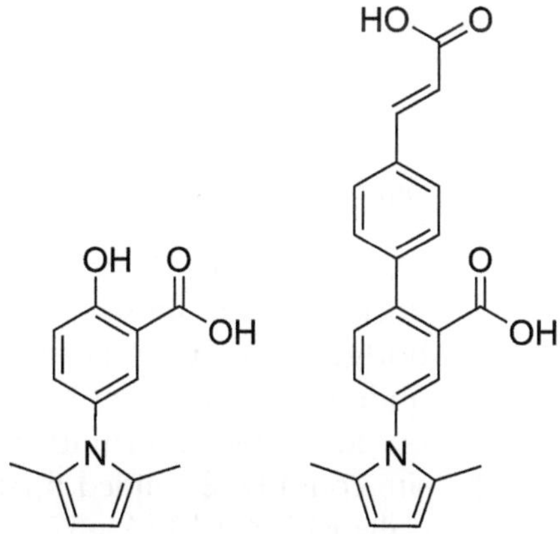

A12 **GLS_22**

Fig. 9 Structures of two inhibitors targeted to gp41

Wang et al. recently reported (46) the use of one of the most active inhibitors of gp41 hexahelical bundle formation as well as p24 production, discovered by Xie and Jiang's laboratory (45), as a starting point to design inhibitors. They used a proprietary structure-based de novo design technique termed GeometryFit developed by GeometryLifeSci (www.geometrylifesci.com). However, a detailed description of the technique was not available in the literature. The investigators used the IQN17 X-ray crystal structure (PDB ID: 2R5D) representing the gp41 N-trimer pocket to dock the original inhibitor A12 (Fig. 9) using Autodock 4.0 (34, 47). IQN17 was originally designed to overcome the aggregation problem of N-peptide and improve solubility. It was designed by fusing the portion of the N-peptide representing the gp41 pocket with a soluble trimeric coiled-coil (GCN4-pI$_q$I) (48). The docking study identified a different binding mode of A12 than reported previously. The most notable difference from the previous docking study is in the orientation of the A12 in the gp41 hydrophobic pocket, where its acid group forms a salt-bridge with Arg579

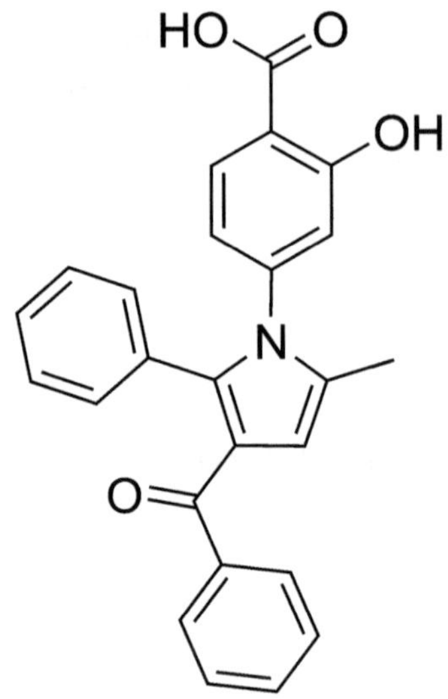

Fig. 10 Structure of a fusion inhibitor targeted to gp41

(Arg43 in 2R5D structure). The current Autodock 4.0-based docking study showed that the acid group of A12 forms a salt-bridge with Lys38 (Lys574 in 1AIK structure). Another major finding was that the phenolic group of A12 in this study showed no noticeable contribution to the overall binding affinity, but this site could be extended further to make other major contact, especially with Arg43 (Arg579 in 1AIK structure). This information led to the design of five new inhibitors showing better inhibitory activity than the parent compound A12. One of these five inhibitors, GLS_22 (Fig. 9), showed the most potent activity in inhibiting p24 production ($IC_{50}=4.91$ μM) and cell fusion ($IC_{50}=3.60$ μM).

In 2011 Tan et al. (49) reported the use of the detailed binding mode of NB-2 with the five-helix bundle of gp41 from their earlier reported studies using a combination of multiconformation docking, MD simulation, and the Poisson–Boltzmann/surface area and generalized Born/surface area molecular mechanics techniques (MM-PBSA and MM-GBSA) (50, 51). Several key interactions were revealed, which they utilized in LeapFrog software (Tripos) to design novel inhibitors. Autodock 4.0 (34) was used to dock these inhibitors in the binding site, and a cluster analysis was then performed. Two criteria (lowest binding energy and larger number of clusters) were used to select six molecules. However, this design technique did not consider the feasibility of synthesizing the molecules and the investigators decided not to pursue these molecules further. Instead, they have used NB-2 (42) structure information as the reference in the CombiLibMaker software from Tripos to create 160 molecules. They used the same docking techniques and compound selection criteria as above to select six compounds for synthesis and antiviral assay. Only one compound (Fig. 10) showed modest fusion inhibition ($IC_{50}=41.1$ μg/mL).

5 CCR5 as Target

In 2007, Kellenberger et al. reported the identification of CCR5 receptor agonists using structure-based virtual screening (52). They generated a 3D homology model of the receptor using bovine rhodopsin as the template. They developed the antagonist binding site by manually docking a set of well-characterized antagonists in the receptor cavity and refined it using an energy minimization technique. A preliminary validation study using the docking software GOLD (CCDC, Cambridge, UK) (17, 18) and Surflex-Dock (53) (Tripos) with a set of 1,000 compounds, including seven known antagonists, demonstrated that the CCR5 model was capable of discriminating known antagonists from a randomly chosen set of drug-like decoys. This model was used in in silico screening of a library of 1.6 million compounds available commercially. This large library consisting of compounds from several vendors was filtered to select only drug-like molecules. A subset of 44,524 compounds was extracted that satisfied a simple pharmacophore model generated from known CCR5 antagonists. This set was docked using high-throughput screening using both GOLD (17, 18) and Surflex docking protocols. A thorough analysis followed by visual inspection of the predicted binding mode resulted in 77 virtual hits; however, only 59 were available for purchase. These compounds were tested for their binding affinity to CCR5 expressed in CHO-K1 cells and their functional activity was tested by measuring the release of intracellular calcium in an aequorin-based functional assay. Of 59 molecules, only 10 showed detectable binding to CCR5 and -6 had functional activity. However, the majority of these compounds showed the agonist characteristics in the functional assay, while three had modest affinities in the high micromolar range. A follow-up similarity search based on these molecules identified three novel CCR5 agonists with enhanced binding affinity. However, one of the earlier three agonists (Fig. 11) was able to induce CCR5 receptor internalization, which is considered one of the strategies to prevent HIV-1 infection. This compound was considered a possible hit for further optimization.

It has been reported that the N-terminal CCR5, the coreceptor for binding of CD4–gp120 complex, is tyrosine rich and the tyrosines are posttranslationally converted to sulfated tyrosines (54–56). The NMR, X-ray, and docking-based structure of the

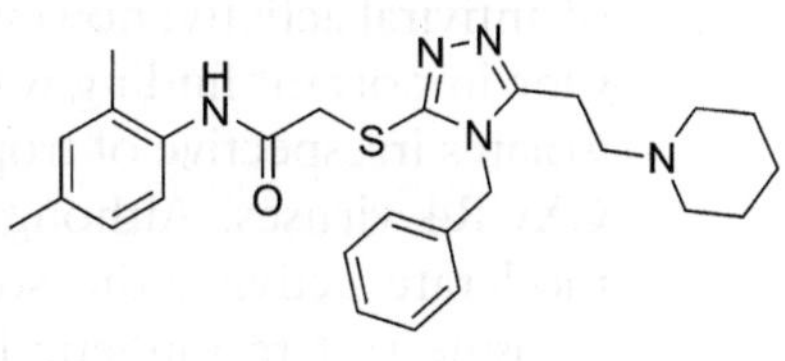

Fig. 11 Structure of an agonist targeted to CCR5

N-terminal CCR5 bound to gp120 confirmed that two sulfated tyrosines at position 10 and 14, respectively, are important in gp120 binding and play a critical role in HIV entry (57). Therefore, the sulfated tyrosine binding site in gp120 has been considered as a target for rational anti-HIV drug design. Acharya et al. recently reported the use of in silico virtual screening techniques to identify tyrosine sulfate mimetics from libraries of small molecules (58). They used two complementary screening techniques. First, they used the shape- and electrostatics-matching routine of ROCS (19) and GOLD-based docking analysis (17, 18) to search the ZINC V7 database, which consists of three million compounds. The chemical features of tyrosine sulfates in the 412d CDRH3 and CCR5 N-terminus were used as shape queries. To mimic the sulfated tyrosines, primary attention was paid to compounds with a sulfur–oxygen bond. Although the investigators found very few phenyl sulfates in the database, a large number of phenyl sulfonates were available. Thirty compounds were selected from this screening and tested in an ELISA-based inhibition assay. One of the highest scored compounds identified from both queries showed the best inhibition and was used as a query in a follow-up ROCS-based search to identify additional hits.

In a second approach, the investigators used a GLIDE-based docking protocol (20, 21) (Schrödinger) to screen a combined database from 21 vendors. Drug-like compounds were filtered using QikProp V3.0 (Schrödinger) software for ADME prediction (absorption, distribution, metabolism, and excretion). The X-ray crystal structure of antibody 412d (PDB ID: 2QAD) and a model of gp120 docked to a CCR5 N-terminus peptide were used to generate models. Four such models were used to screen the database. Standard precision scoring was used in the GLIDE-based docking simulations. Compounds containing an SO_3 group were selected for visual inspection and commercial availability. ELISA-based screening of 60 such compounds resulted in additional hits.

This combined approach identified two classes of seven tyrosine sulfate mimetics containing a "phenylsulfonate-linker-aromatic" motif, which inhibited the binding of gp120 to the CCR5 N-terminus. Two compounds (Fig. 12) showed the highest affinity to gp120 in the CD4-bound conformation, with a K_D of 49 and 94 nM, respectively. Four of these selected compounds were tested for inhibition of HIV-1 entry in a panel of 27 primary isolates. Only compounds that had high affinity showed a greater breadth of antiviral activity; however, the IC_{50} values were 50 μM or higher. One important finding was that these compounds inhibited HIV-1 isolates irrespective of tropism; i.e., they inhibited both CCR5 and CXCR4 viruses. Although the identified sulfated derivatives had moderate activity, the study represents a proof of concept that tyrosine sulfate mimetic hits may be optimized further as potent anti-HIV-1 agents.

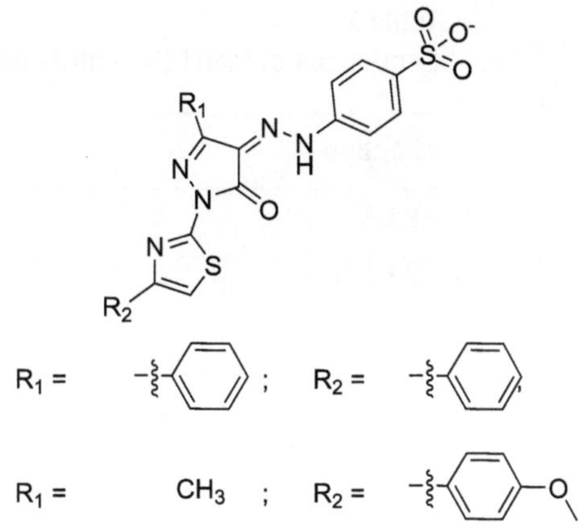

Fig. 12 Structures of two phenylsulfonate-based inhibitors targeted to CCR5

Chemokine receptors such as CCR5 and CXCR4 play important roles in HIV-1 infection, and chemokine ligands such as RANTES (regulated upon activation, normal T-cell expressed and secreted), MIP-1α (macrophage inhibitory protein-1α), and MIP-1β of CCR5 act as natural inhibitors of HIV-1 infection. Lusso's group identified the critical structural determinants in RANTES for binding to CCR5 and inhibiting HIV-1 infection through peptide scanning, alanine mutagenesis, and NMR structure-guided modeling (59, 60). These critical sites were located at the N-loop and β1-strand regions of RANTES. The surface potential map of the dimeric structure of RANTES indicated the location of large and solvent-exposed hydrophobic patches within the β1-strand of RANTES, which was shown to be responsible for antiviral activity. Lusso et al. used this information in structure-guided designing of peptide mimetics from the N-loop and β1-strand regions of RANTES, which yielded CCR5-specific inhibitors active against a broad spectrum of CCR5 isolates with potent antiviral activity (61). One such prototype inhibitor, R11–29, containing amino acids (aa) 11–29 of mature RANTES, was observed to have two hydrophobic clusters, one in the N-loop and the other in the β1-strand region connected by a ten-residue amphiphilic linker (aa 17–26). The R11–29 region in the full-length RANTES is structured and those two hydrophobic patches play significant roles, but this peptide in solution was found to be unstructured, which was confirmed by NMR. In an attempt to stabilize the N-terminal regions containing the hydrophobic clusters, the authors observed that the mutation of Ala13Pro substantially improved the inhibitory activity against envelope-mediated cell fusion, indicating that the rigidity introduced by proline contributed to the stabilization of the structure. This was confirmed when Ala13Val mutation resulted in loss of activity. The investigators then concentrated on optimizing the length of the amphiphilic linker amino acids. They observed that

Table 2
Sequences of RANTES peptide mimetics

Peptide	Sequence
R1.5	CFPYIARPLPIKEYFY
R1.5G	CFPYIARP**G**PIKEYFY
R1.5G3	CFPYITRP**G**PIKEY**X**Y
R2.0	CFPYITRP**GTYHD**Y**X**Y**O**

X 1Nal; *O* Orn

truncation of aa Leu19-Glu26 was tolerable, and one of the peptides, R1.5, with truncation of aa 21–23, had the highest cell fusion inhibitory activity (Table 2). This peptide was used as a template and mutated further to improve activity. The peptide, R1.5G3, with dual mutation at position 16 where Ala was replaced by Thr and Phe28, was replaced by a hydrophobic nonnatural amino acid 1Nal, yielding slightly improved antiviral activity. The 3D structure of this peptide was determined by NMR and a helical motif was observed at the C-terminal region. Further in silico optimization of the linker region of this peptide using the de novo protein-modeling software Rosetta 2.3.0 (Rosetta Design Group, University of Washington) resulted in peptide R2.0, which had significantly improved antiviral activity (104 nM) over R1.5G3 (403 nM) against HIV-1 BaL. The NMR structure of R2.0 showed remarkable similarity to the Rosetta-predicted model, with an average Cα-RMSD of 0.37 Å.

6 CXCR4 as Target

Because structure information on the CXCR4 coreceptor was not available until recently, most attempts to design entry inhibitors targeted to CXCR4 have involved building homology models of this coreceptor. Few attempts have been made to use rational design in identifying entry inhibitors targeted to CXCR4. Perez-Nueno et al. reported the use of multiple virtual screening techniques, both ligand-based and receptor-based, such as quantitative structure–activity relationship analysis, pharmacophore mapping, docking, and shape matching to identify novel compounds from a virtual library, which were generated based on the structure of the most active CXCR4 inhibitors reported to date, i.e., AMD3100 (62). Before using prospective virtual screening protocols, the investigators validated these protocols by identifying known inhibitors from a set of diverse compounds in retrospective studies. For this purpose, they selected a set of 248 known CXCR4 antagonists from the literature with activity value lower than 100 μM. These antagonists belong to seven representative families of compounds such as AMD3100 analogues, macrocycles, KRH1636 analogues, dipicolil amine zinc(II) complexes, tetrahydroquinolinamines, cyclic peptides, and the most active CXCR4 inhibitors from the in-house combinatorial library. For the prospective screening study, they used the same inactive compounds along with 34 virtual compounds belonging to amino/hydrazono-amine/hydrazone, hydrazono/amino-aldehyde, and cyclam-hydrazone/amine.

Different docking techniques such as Autodock 3.0 (34), GOLD 3.0.1 (18), FRED 2.2.1 (Fast Rigid Exhaustive Docking; OpenEye Scientific Software) (63), and *Hex* 4.8 were used in combination with several scoring systems. FRED consensus, Consensus scoring (Autodock energy, GOLD Goldscore, and ChemScore), and CemScore yielded the best enrichments.

For the retrospective study, using ligands to generate pharmacophore, a subset of compounds was selected from the original database consisting of AMD3100 analogues, KRH1636 analogues, dipicolil amine zinc(II) complexes, and the most active compounds from the combinatorial library. The training set consisted of the most active compounds from the database. Pharmacophores were generated using both MOE software (Chemical Computing Group, Montreal, Canada) and Discovery Studio software (Accelrys). For the MOE-based pharmacophore generation, the polarity-charge-hydrophobicity (PCH) scheme was used, and for the Discovery Studio-based study, hydrogen bond acceptor, hydrogen bond donor, hydrophobic, ionizable positive, and charged positive pharmacophore features were used. However, the most reliable results in the retrospective study were obtained using a consensus pharmacophore model from MOE software, which was used in the prospective screening study.

For the shape-based virtual screening study, again very diverse protocols were used, consisting of PARAFIT08 Shape Tanimoto (Cepos InSilico Ltd, Kempston, Bedford, UK), ROCS2.2 Combo score and Shape Tanimoto, and *Hex* 4.8 Shape Tanimoto scores. In the absence of crystallographic conformation, a docking-based conformation of AMD3100 was used as the shape query. For the retrospective study, a consensus query was built using known CXCR4 inhibitors from analogues of AMD3100 and KRH1636 and macrocycles. The virtual library was then used as before for the prospective screening study.

The shape-based virtual screening protocols performed better than the docking protocols overall. Finally, a consensus "rank-by-vote" selection technique was used and the five best compounds were selected from the hits for synthesis and for determining antiviral activity. Remarkably, two of the five compounds (Fig. 13) showed activity at low micromolar ranges (0.022 μg/mL and 0.058 μg/mL, respectively).

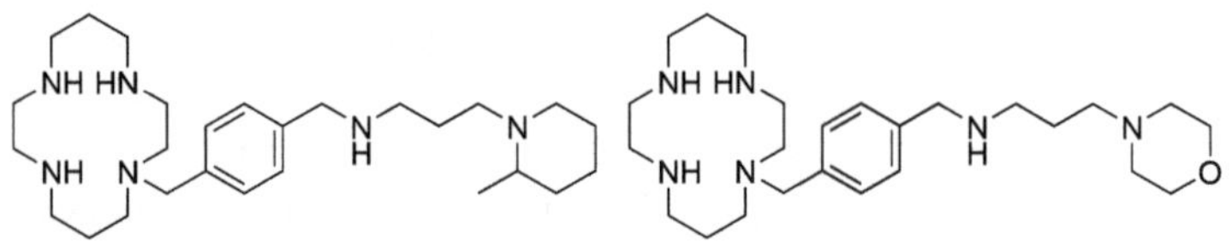

Fig. 13 Structures of two cyclam-based inhibitors targeted to CXCR4

7 Conclusion

Analyses of the in silico studies described in this chapter indicate that both structure-based and ligand-based design techniques have been used by applying diverse modeling techniques such as quantitative structure–activity relationship analysis, conformational analysis, molecular dynamics, pharmacophore generation, docking, virtual screening (using docking software and also shape-based ROCS techniques), and fragment-based design. Docking-based methods dominated, however, especially in virtual screening-based inhibitor identification. Peptide/protein-based inhibitors designed using structure-based techniques yielded the most active CD4 mimics with nanomolar potency. Although many small-molecule inhibitors have been identified using in silico approaches, none have shown potency at the nanomolar level. However, this is probably not expected from initial hit-identification protocols. Future studies will show whether any of these micromolar hits/leads can be optimized to drugs with nanomolar potency.

References

1. Biscone MJ, Pierson TC, Doms RW (2002) Opportunities and challenges in targeting HIV entry. Curr Opin Pharmacol 2:529–533
2. Starr-Spires LD, Collman RG (2002) HIV-1 entry and entry inhibitors as therapeutic agents. Clin Lab Med 22:681–701
3. Hertje M, Zhou M, Dietrich U (2010) Inhibition of HIV-1 entry: multiple keys to close the door. ChemMedChem 5:1825–1835
4. Sodroski JG (1999) HIV-1 entry inhibitors in the side pocket. Cell 99:243–246
5. Caffrey M (2011) HIV envelope: challenges and opportunities for development of entry inhibitors. Trends Microbiol 19:191–197
6. Jiang S, Debnath AK (2000) Development of HIV entry inhibitors targeted to the coiled coil regions of gp41. Biochem Biophys Res Commun 269:641–646
7. Jiang S, Zhao Q, Debnath AK (2002) Peptide and non-peptide HIV fusion inhibitors. Curr Pharm Des 8:563–580
8. Vita C, Drakopoulou E, Vizzavona J et al (1999) Rational engineering of a miniprotein that reproduces the core of the CD4 site interacting with HIV-1 envelope glycoprotein. Proc Natl Acad Sci USA 96:13091–13096
9. Wyatt R, Kwong PD, Desjardins E et al (1998) The antigenic structure of the HIV gp120 envelope glycoprotein. Nature 393: 705–711
10. Martin L, Stricher F, Misse D et al (2003) Rational design of a CD4 mimic that inhibits HIV-1 entry and exposes cryptic neutralization epitopes. Nat Biotechnol 21:71–76
11. Li H, Guan Y, Szczepanska A et al (2007) Synthesis and anti-HIV activity of trivalent CD4-mimetic miniproteins. Bioorg Med Chem 15:4220–4228
12. Zhao Q, Ma L, Jiang S et al (2005) Identification of N-phenyl-N′-(2,2,6,6-tetramethyl-piperidin-4-yl)-oxalamides as a new class of HIV-1 entry inhibitors that prevent gp120 binding to CD4. Virology 339:213–225
13. Madani N, Schon A, Princiotto AM et al (2008) Small-molecule CD4 mimics interact

with a highly conserved pocket on HIV-1 gp120. Structure 16:1689–1701

14. Yoshimura K, Harada S, Shibata J et al (2010) Enhanced exposure of human immunodeficiency virus type 1 primary isolate neutralization epitopes through binding of CD4 mimetic compounds. J Virol 84:7558–7568
15. Lalonde JM, Elban MA, Courter JR et al (2011) Design, synthesis and biological evaluation of small molecule inhibitors of CD4-gp120 binding based on virtual screening. Bioorg Med Chem 19:91–101
16. Yamada Y, Ochiai C, Yoshimura K et al (2010) CD4 mimics targeting the mechanism of HIV entry. Bioorg Med Chem Lett 20:354–358

16a. Kwon YD, Finzi A, Wu X et al (2012) Unliganded HIV-1 gp120 core structures assume the CD4-bound conformation with regulation by quaternary interactions and variable loops. Proc Natl Acad Sci USA 109:5663–5668

17. Jones G, Willett P, Glen RC (1995) Molecular recognition of receptor sites using a genetic algorithm with a description of desolvation. J Mol Biol 245:43–53
18. Verdonk ML, Cole JC, Hartshorn MJ et al (2003) Improved protein-ligand docking using GOLD. Proteins 52:609–623
19. Grant JA, Gallardo MA, Pickup BT (1996) A fast method of molecular shape comparison: a simple application of a Gaussian description of molecular shape. J Comput Chem 17:1653–1666
20. Friesner RA, Banks JL, Murphy RB et al (2004) Glide: a new approach for rapid, accurate docking and scoring. 1. Method and assessment of docking accuracy. J Med Chem 47:1739–1749
21. Halgren TA, Murphy RB, Friesner RA et al (2004) Glide: a new approach for rapid, accurate docking and scoring. 2. Enrichment factors in database screening. J Med Chem 47:1750–1759
22. Caporuscio F, Tafi A, Gonzalez E et al (2009) A dynamic target-based pharmacophoric model mapping the CD4 binding site on HIV-1 gp120 to identify new inhibitors of gp120-CD4 protein-protein interactions. Bioorg Med Chem Lett 19:6087–6091
23. Cornell WD, Cieplak P, Bayly CI et al (1995) A second generation force field for the simulation of proteins, nucleic acids, and organic molecules. J Am Chem Soc 117:5179–5197
24. Goodford PJ (1985) A computational procedure for determining energetically favorable binding sites on biologically important macromolecules. J Med Chem 28:849–857
25. Boyd MR, Gustafson KR, McMahon JB et al (1997) Discovery of cyanovirin-N, a novel human immunodeficiency virus-inactivating protein that binds viral surface envelope glycoprotein gp120: potential applications to microbicide development. Antimicrob Agents Chemother 41:1521–1530
26. Dey B, Lerner DL, Lusso P et al (2000) Multiple antiviral activities of cyanovirin-N: blocking of human immunodeficiency virus type 1 gp120 interaction with CD4 and coreceptor and inhibition of diverse enveloped viruses. J Virol 74:4562–4569
27. Esser MT, Mori T, Mondor I et al (1999) Cyanovirin-N binds to gp120 to interfere with CD4-dependent human immunodeficiency virus type 1 virion binding, fusion, and infectivity but does not affect the CD4 binding site on gp120 or soluble CD4-induced conformational changes in gp120. J Virol 73:4360–4371
28. Patsalo V, Raleigh DP, Green DF (2011) Rational and computational design of stabilized variants of cyanovirin-N that retain affinity and specificity for glycan ligands. Biochemistry 50:10698–10712
29. Neffe AT, Meyer B (2004) A peptidomimetic HIV-entry inhibitor directed against the CD4 binding site of the viral glycoprotein gp120. Angew Chem Int Ed Engl 43:2937–2940
30. Wulfken J (2000) Development of CD4 binding peptides as inhibitors of HIV infection. Ph.D. Thesis, University of Hamburg, Germany
31. Neffe AT, Bilang M, Gruneberg I, Meyer B (2007) Rational optimization of the binding affinity of CD4 targeting peptidomimetics with potential anti HIV activity. J Med Chem 50:3482–3488
32. Lee DY, Lin X, Paskaleva EE et al (2009) Palmitic acid is a novel CD4 fusion inhibitor that blocks HIV entry and infection. AIDS Res Hum Retroviruses 25:1231–1241
33. Lin X, Paskaleva EE, Chang W et al (2011) Inhibition of HIV-1 infection in ex vivo cervical tissue model of human vagina by palmitic acid: implications for a microbicide development. PLoS One 6:e24803
34. Morris GM, Goodsell DS, Halliday RS et al (1998) Automated docking using Lamarckian genetic algorithm and empirical binding free energy function. J Comput Chem 19:1639–1662
35. Chan DC, Fass D, Berger JM, Kim PS (1997) Core structure of gp41 from the HIV envelope glycoprotein. Cell 89:263–273
36. Weissenhorn W, Dessen A, Harrison SC et al (1997) Atomic structure of the ectodomain from HIV-1 gp41. Nature 387:426–430
37. Lu M, Blacklow SC, Kim PS (1995) A trimeric structural domain of the HIV-1 transmembrane glycoprotein. Nat Struct Biol 2:1075–1082

38. Debnath AK, Radigan L, Jiang S (1999) Structure-based identification of small molecule antiviral compounds targeted to the gp41 core structure of the human immunodeficiency virus type 1. J Med Chem 42:3203–3209
39. Good AC, Ewing TJ, Gschwend DA, Kuntz ID (1995) New molecular shape descriptors: application in database screening. J Comput Aided Mol Des 9:1–12
40. Shoichet BK, Bodian DL, Kuntz ID (1992) Molecular docking using shape descriptors. J Comput Chem 13:380–397
41. Jiang S, Lin K, Lu M (1998) A conformation-specific monoclonal antibody reacting with fusion-active gp41 from the HIV-1 envelope glycoprotein. J Virol 72:10213–10217
42. Jiang S, Lu H, Liu S et al (2004) N-substituted pyrrole derivatives as novel human immunodeficiency virus type 1 entry inhibitors that interfere with the gp41 six-helix bundle formation and block virus fusion. Antimicrob Agents Chemother 48:4349–4359
43. Liu B, Joseph RW, Dorsey BD et al (2009) Structure-based design of substituted biphenyl ethylene ethers as ligands binding in the hydrophobic pocket of gp41 and blocking the helical bundle formation. Bioorg Med Chem Lett 19:5693–5697
44. Clark M, Meshkat S, Talbot GT et al (2009) Fragment-based computation of binding free energies by systematic sampling. J Chem Inf Model 49:1901–1913
45. Liu K, Lu H, Hou L et al (2008) Design, synthesis, and biological evaluation of N-carboxyphenylpyrrole derivatives as potent HIV fusion inhibitors targeting gp41. J Med Chem 51:7843–7854
46. Wang Y, Lu H, Zhu Q et al (2010) Structure-based design, synthesis and biological evaluation of new N-carboxyphenylpyrrole derivatives as HIV fusion inhibitors targeting gp41. Bioorg Med Chem Lett 20:189–192
47. Welch BD, VanDemark AP, Heroux A et al (2007) Potent D-peptide inhibitors of HIV-1 entry. Proc Natl Acad Sci USA 104: 16828–16833
48. Eckert DM, Malashkevich VN, Hong LH et al (1999) Inhibiting HIV-1 entry: discovery of D-peptide inhibitors that target the gp41 coiled-coil pocket. Cell 99:103–115
49. Tan JJ, Zhang B, Cong XJ et al (2011) Computer-aided design, synthesis, and biological activity evaluation of potent fusion inhibitors targeting HIV-1 gp41. Med Chem 7:309–316
50. Cong XJ, Tan JJ, Liu M et al (2010) Computational study of binding mode of N-substituted pyrrole derivatives to HIV-1 gp41. Prog Biochem Biophys 37:904–915
51. Wang CX, Cong XJ, Kong R et al (2010) Binding mode of HIV-1 gp41 with its inhibitor NB-2. J Beijing Univ Technol 36: 1118–1123
52. Kellenberger E, Springael JY, Parmentier M et al (2007) Identification of nonpeptide CCR5 receptor agonists by structure-based virtual screening. J Med Chem 50: 1294–1303
53. Jain AN (2003) Surflex: fully automatic flexible molecular docking using a molecular similarity-based search engine. J Med Chem 46:499–511
54. Farzan M, Choe H, Vaca L et al (1998) A tyrosine-rich region in the N terminus of CCR5 is important for human immunodeficiency virus type 1 entry and mediates an association between gp120 and CCR5. J Virol 72:1160–1164
55. Farzan M, Vasilieva N, Schnitzler CE et al (2000) A tyrosine-sulfated peptide based on the N-terminus of CCR5 interacts with a CD4-enhanced epitope of the HIV-1 gp120 envelope glycoprotein and inhibits HIV-1 entry. J Biol Chem 275:33516–33521
56. Farzan M, Mirzabekov T, Kolchinsky P et al (1999) Tyrosine sulfation of the amino terminus of CCR5 facilitates HIV-1 entry. Cell 96: 667–676
57. Huang CC, Lam SN, Acharya P et al (2007) Structures of the CCR5 N terminus and of a tyrosine-sulfated antibody with HIV-1 gp120 and CD4. Science 317:1930–1934
58. Acharya P, Dogo-Isonagie C, Lalonde JM et al (2011) Structure-based identification and neutralization mechanism of tyrosine sulfate mimetics that inhibit HIV-1 entry. ACS Chem Biol 6:1069–1077
59. Nardese V, Longhi R, Polo S et al (2001) Structural determinants of CCR5 recognition and HIV-1 blockade in RANTES. Nat Struct Biol 8:611–615
60. Vangelista L, Longhi R, Sironi F et al (2006) Critical role of the N-loop and beta1-strand hydrophobic clusters of RANTES-derived peptides in anti-HIV activity. Biochem Biophys Res Commun 351:664–668
61. Lusso P, Vangelista L, Cimbro R et al (2011) Molecular engineering of RANTES peptide mimetics with potent anti-HIV-1 activity. FASEB J 25:1230–1243
62. Perez-Nueno VI, Pettersson S, Ritchie DW et al (2009) Discovery of novel HIV entry inhibitors for the CXCR4 receptor by prospective virtual screening. J Chem Inf Model 49:810–823
63. McGann M (2011) FRED pose prediction and virtual screening accuracy. J Chem Inf Model 51:578–596

Chapter 14

Malarial Kinases: Novel Targets for In Silico Approaches to Drug Discovery

Kristen M. Bullard, Robert Kirk DeLisle, and Susan M. Keenan

Abstract

Malaria, the disease caused by infection with protozoan parasites from the genus *Plasmodium*, claims the lives of nearly 1 million people annually. Developing nations, particularly in the African Region, bear the brunt of this malaria burden. Alarmingly, the most dangerous etiologic agent of malaria, *Plasmodium falciparum*, is becoming increasingly resistant to current first-line antimalarials. In light of the widespread devastation caused by malaria, the emergence of drug-resistant *P. falciparum* strains, and the projected decrease in funding for malaria eradication that may occur over the next decade, the identification of promising new targets for antimalarial drug design is imperative. *P. falciparum* kinases have been proposed as ideal drug targets for antimalarial drug design because they mediate critical cellular processes within the parasite and are, in many cases, structurally and mechanistically divergent when compared with kinases from humans. Identifying a molecule capable of inhibiting the activity of a target enzyme is generally an arduous and expensive process that can be greatly aided by utilizing in silico drug design techniques. Such methods have been extensively applied to human kinases, but as yet have not been fully exploited for the exploration and characterization of antimalarial kinase targets. This review focuses on in silico methods that have been used for the evaluation of potential antimalarials and the *Plasmodium* kinases that could be explored using these techniques.

Key words *Plasmodium falciparum*, Kinase, Drug, Quantitative structure–activity relationships, Pharmacophore, Docking, Antimalarials

1 Introduction

Malaria is a devastating disease caused by protozoan parasites from the genus *Plasmodium*. While reported cases of malaria decreased by 50% between 2000 and 2010, 216 million cases and almost 700,000 malaria-related deaths were reported in 2010 (1). In addition, 86% of these malaria deaths claimed the lives of children under 5 years of age. The African Region continues to bear the brunt of the malaria burden with an estimated 81% of all reported malaria cases. In general, tropical and subtropical regions have the greatest rates of malaria transmission as the climate in these regions supports development of the mosquito vector.

Sandhya Kortagere (ed.), *In Silico Models for Drug Discovery*, Methods in Molecular Biology, vol. 993, DOI 10.1007/978-1-62703-342-8_14,

Unfortunately, developing nations where malaria is endemic are also plagued with other endemic diseases and economic burdens (2). In malarious regions, both the growth of income per capita between 1965 and 1990 and the purchasing power parity gross domestic product were five times lower than in countries where malaria was not present (3). Malaria is intimately tied to poverty and affects countries that are least able to support the costs of eradication efforts. Furthermore, while funding for malaria control efforts was substantial in 2010, it fell well short of funding goals and is projected to stagnate if not decrease in the immediate future (1).

In light of the increasing parasite resistance to all currently used therapeutics and the projected scarcity of funds for malaria control efforts, it has become necessary to identify not only new drugs to treat resistant strains of *Plasmodium* but also novel targets for drug design. Here we focus on malarial protein kinases as a family of enzymes that are collectively worthy targets for inhibitor discovery efforts. We discuss in silico techniques that have been applied to date and further suggest rational design approaches that could purposefully and efficiently direct future drug discovery efforts.

1.1 Plasmodium *Species*

Four principal species from the genus *Plasmodium* cause natural human infection: *Plasmodium vivax*, *Plasmodium ovale*, *Plasmodium malariae*, and *Plasmodium falciparum*. *P. falciparum* is the most lethal as it causes approximately 90% of malaria-related deaths (1). An additional species, *Plasmodium knowlesi*, which generally infects macaques, has also been shown increasingly to infect humans as well (4). As more sophisticated diagnostic tests are now able to easily distinguish one species of *Plasmodium* from another, it is thought that infection with *P. knowlesi* has heretofore been underreported because this species morphologically resembles other *Plasmodium* species in blood smears (5).

The life cycle of *Plasmodium* requires two vectors for completion. The definitive hosts are female mosquitoes from the genus *Anopheles*. *Anopheles gambiae* and *Anopheles funestus* are two such definitive hosts in the African Region, while *Anopheles darlingi* transmits malaria in South and Central America (6–8). The second host is a vertebrate that may be a bird, reptile, or small mammal. The malaria species listed above are all able to utilize humans as a vertebrate host.

The intrinsic life cycle begins when a female mosquito takes a blood meal. At this time, sporozoites that have matured in the mosquito's salivary glands are inoculated into the blood stream. These sporozoites migrate to the liver where they infect hepatocytes and progress from early trophozoites to mature schizonts. This process is referred to as the exoerythrocytic cycle. Schizonts in the liver contain thousands of merozoites that are able to infect red blood cells (RBCs) when they are released from the hepatocyte. Merozoites flow through the blood stream until they come into contact with an RBC. The merozoite then begins the process of invasion by loosely attaching to the RBC membrane, orienting its apical end towards the membrane, and finally forming a tight junction with the membrane itself (9–12). The parasite then moves into the RBC and is surrounded by a parasitophorous vacuole, an invagination of the RBC membrane that has pinched off around the parasite during invasion. The intraerythrocytic cycle takes place inside the RBC as the parasite progresses from an early trophozoite to a mature schizont. The number of merozoites that are produced during schizont formation (schizogony) is characteristic of each *Plasmodium* species. *P. falciparum* forms 8–32 merozoites per schizont. When the schizont bursts, merozoites are released into the blood stream to infect new RBCs. Microgametocytes and macrogametocytes, which develop in a smaller number of infected RBCs, are infectious to the mosquito vector and migrate to the periphery to be taken up by a mosquito during a blood meal.

The extrinsic life cycle takes place within the mosquito vector. When a female anopheline mosquito takes a blood meal and ingests microgametocytes and macrogametocytes, these cells must first mature into microgametes and macrogametes. For microgametes, the next step in the process is exflagellation where the microgamete develops into eight flagellated cells, one of which fuses with a macrogamete. The product of this fusion is an ookinete, which embeds in the wall of the mosquito midgut and develops into an oocyst. Mature oocysts contain hundreds of sporozoites, which then migrate to the mosquito's salivary glands to complete maturation. Mature sporozoites are able to infect a new vertebrate host when the mosquito next takes a blood meal.

1.2 Drug Resistance

Drug-resistant malaria strains first became a global health issue in the 1950s when countries in South America and Southeast Asia noted the reduced efficacy of chloroquine (13). Over the following decades, chloroquine-resistant *P. falciparum* became so common that there are few countries where malaria is endemic that are not inhabited by chloroquine-resistant parasites. Resistance to the drug mefloquine was seen only 10 years after its introduction and pyrimethamine-sulfadoxine resistance in *P. falciparum* was seen within the same year of its introduction (14–17). Because of the clear capability of this parasite to develop drug resistance, the World Health Organization (WHO) recommends that all uncomplicated cases of *P. falciparum* malaria be treated with the last weapon in the drug resistance arsenal, artemisinin combination therapy (ACT) (1). Artemisinin derivatives are the class of drugs that currently show the most rapid therapeutic benefits and parasite clearance (18). The WHO implores countries not to administer artemisinin as a monotherapy because the likelihood of resistance development increases when these drugs are used alone. However, the World Malaria Report 2011 showed that of 106 countries where malaria was endemic, 25 countries still allowed the marketing of artemisinin monotherapies and 28 pharmaceutical companies were still marketing such products.

The administration of artemisinin derivatives as monotherapies, the selling of black market low-dose or fake antimalarials, the cost of drugs, noncompliance with therapeutic regimens, and the propensity of *P. falciparum* parasites to acquire drug resistance are just some of the many factors that have created an environment for the emergence of drug resistance to artemisinin. In fact, resistance to artemisinin was first observed in 2008 along the Thai-Cambodian border and since that time has been reported in other areas of Southeast Asia (18, 19). If widespread resistance to ACTs is seen in the near future, there will be no effective way to treat resistant *P. falciparum* malaria and a potentially devastating scenario will ensue as there are no first-line drug treatments that are now capable of replacing ACTs (20).

1.3 Kinases as Drug Targets

Drug resistance and the lack of replacements for ACTs as a first-line drug treatment against resistant *P. falciparum* malaria have necessitated the identification of promising new drug targets for antimalarial drug discovery. One such class of drug targets is the protein kinases. Kinases mediate various critical cellular processes such as homeostasis, apoptosis, and cell division, and while protein kinases are just beginning to be exploited as target enzymes in the war against malaria, they continue to be targeted for many other diseases including cancer, diabetes, inflammation, cardiovascular disease, autoimmune diseases, and neurological disorders; see, for example, Antoniou et al. (21), Burgess and Echeverria (22), Cohen, S., Fleischmann (23), Ding et al. (24), Fabbro et al. (25), Gálvez (26), Moriguchi (27), and Satoh et al. (28).

Protein kinase activity is ubiquitous in the cell, and regardless of the specific target(s), all kinases function by binding adenosine triphosphate (ATP) and facilitating the transfer of the γ-ATP phosphate to an acceptor residue. The very nature of this reaction was initially the cause for concern regarding kinases as potential drug targets. In order to bind the same cofactor (ATP), there must be overall sequence (and structural) homology of ATP binding site residues among kinases, suggesting that the development of a selective ATP-competitive inhibitor might not be easy. Furthermore, the level of ATP in the intracellular environment (approximately 2 mM) is far greater than the concentrations for affinity for most kinases (generally in the micromolar range), suggesting that it would also be difficult for a small-molecule inhibitor to effectively compete for the ATP binding site.

Much of the skepticism as to the drugability of kinases was assuaged by the development of imatinib (Gleevec; Novartis, Basel,

Switzerland) (29). Imatinib, a selective inhibitor for the nonreceptor tyrosine kinase ABL, was rationally designed and was the first Food and Drug Administration (FDA)-approved protein kinase inhibitor. Imatinib was commercialized in 2002 for the treatment of chronic myelogenous leukemia (30). Since 2002 at least 11 kinase inhibitors have been approved for clinical use and an estimated 150 clinical trials for kinase-targeted drugs are currently ongoing. Along with imatinib, dasatinib (Bristol-Myers Squibb; Princeton, NJ) and nilotinib (Novartis) target ABL1-2, platelet-derived growth factor receptor (PDGFR), and KIT. Dasatinib additionally targets SRC. Gefitinib (AstraZeneca, Wilmington, DE), erlotinib (Roche, Indianapolis, IN), and lapatinib (GlaxoSmithKline, Philadelphia, PA) selectively target epidermal growth factor receptor (EGFR), and sunitinib (Pfizer, New York, NY), sorafenib (Onyx Pharmaceuticals, South San Francisco, CA, and Bayer Pharmaceuticals, Pittsburgh, PA), and pazopanib (GlaxoSmithKline) are considered inhibitors of vascular endothelial growth factor receptor isoforms. All of these inhibitors target, at least in part, the ATP binding site (25). Two other FDA-approved drugs, everolimus (Novartis) and temsirolimus (Wyeth), target the mammalian target of rapamycin (25). Ruxolitinib (Cephalon/Teva, Jerusalem), the first small-molecule inhibitor of Janus kinase (JAK), was approved late in 2011 for the treatment of myeloproliferative neoplasms (31).

A plethora of mammalian kinase structures are available, and analyses of these structures suggest a number of structural determinants for kinase inhibitor selectivity that can be used for the rational design of inhibitors competing for the ATP binding site. Residue identities in the hinge region, the hydrophobic pocket (including the gatekeeper residue), and the pocket floor of the ATP binding site can all affect inhibitor affinity. Moreover, a lysine and an acidic group within the binding site coordinate the ATP phosphates in conjunction with the divalent metal ion. The orientation of the phenylalanine (of the DFG motif)—occluding the ATP binding site (Phe-out) or oriented to the hydrophobic pocket (Phe-in)—can also affect inhibition profiles of a small-molecule compound.

That kinase inhibitors have successfully made their way through the drug discovery pipeline and been approved for clinical use suggests that kinase inhibition would also be a successful strategy for the development of novel antimalarial drugs (32, 33).

1.4 Malaria Kinome

Much of the promise of kinase inhibitor antimalarials derives from the availability of structural and sequence data for this organism, the wealth of functional activity data that has been gathered over the past several decades regarding *Plasmodium* proteins, and the unique characteristics of the *Plasmodium* kinome (34–36). *Plasmodium* parasites are single-celled eukaryotic organisms. Cellular pathways are shared between malaria parasites and their human hosts and the mechanisms of action of the proteins that comprise these pathways are similar. Therefore, in order to combat the growing resistance of these parasitic organisms to traditional antimalarials, it is necessary to identify divergences between host and parasite pathways so promising drug targets for malaria drug discovery can be pursued (37). Kinases are one such class of promising targets for antimalarial drug discovery as there are structural and mechanistic differences between parasitic and host kinases that can be exploited (36, 38, 39).

The malaria kinome consists of 86–99 proteins and 65 of these kinases are related to the eukaryotic protein kinase (ePK) superfamily (40). The ePK superfamily can be subdivided into seven main groups: cyclin-dependent kinase, mitogen-activated protein kinase, glycogen synthase kinase, and cdc-like kinase (CMGC) group; the calcium signaling kinases (CamK); the casein kinase 1 (Ck1) group; the tyrosine kinase (TyrK) group; the cyclic-nucleotide-and calcium/phospholipid-dependent kinase (AGC) group; the yeast sterile mutant (STE) group; and the tyrosine kinase-like (TKL) group, as reviewed by Hanks (41). *P. falciparum* has more ePKs than any other *Plasmodium* species but the size of the *Plasmodium* kinome is smaller than that of other sequenced Apicomplexans (42). When compared as a percentage of the total genome, however, the malaria kinome comprises a percentage of the genome similar to the human kinome (2%). In general, *Plasmodium spp.* have fewer overall kinases compared with other Apicomplexans, with the notable exception of *P. falciparum.*

The kinase group that is most highly represented within the *P. falciparum* kinome is the CMGC group (40, 43). Eighteen *P. falciparum* kinases cluster within this group. The number of kinases within the CMGC group underlies the importance of cell cycle regulation to parasite survival and proliferation. The complexity of the *Plasmodium* life cycle as it possesses separate extrinsic, exoerythrocytic, and intraerythrocytic stages must be tightly regulated by the parasite in order for all the required events to progress in the correct order, and protein members from this group orchestrate these transitions.

Kinases that cluster within the CamK group are also highly represented within the kinome, thus pointing to the vital nature of calcium signaling during cellular processes within the parasite (43, 44). Calcium is an important second messenger in many eukaryotic cellular pathways and within the parasite calcium signaling has been shown to be crucial for parasite motility as well as the cellular invasion processes (45). The ability of parasites to find and invade RBCs is crucial for parasite survival. Interestingly, a particular subset of CamKs, the calcium-dependent protein kinases (CDPKs), are more closely related to kinases from plants and protists, and six proteins belonging to this kinase group are included in the *P. falciparum* kinome (40). These proteins have been shown to be involved in activities such as ookinete penetration into epithelial cells of the mosquito midgut and parasite egress from erythrocytes (44, 46).

There is one member of the Ck1 group and five members of the TKL groups within the *P. falciparum* kinome (40). In addition, there are several unique aspects of the *P. falciparum* kinome. STE and TyrK group members are absent from the *P. falciparum* kinome, and there are a reduced number of AGC kinases compared with other eukaryotic organisms. STE kinases generally function in mitogen-activated protein (MAP) kinase signaling pathways (47). The canonical three-component MAP kinase signaling pathway is absent in *P. falciparum* and to date no MAPKK homologue has been identified (40). Two MAP kinases are represented in the kinome; however, at least one of these proteins has been shown to be activated not by a MAPKK, but instead by a NIMA (never-in-mitosis/*Aspergillus*)-related kinase, Pfnek-1 (48, 49). Pfnek-1 in this case acts as a MAPKK, which, considered with the absence of STE kinases and the lack of a traditional MAPKK homologue, points to a unique MAP kinase signaling pathway operating in these parasites. Another unique feature of the *P. falciparum* kinome is the large quantity of kinases that fail to cluster within any ePK group (40). Termed orphan kinases, these proteins represent an additional opportunity to selectively target *P. falciparum* proteins for the development of antimalarials, as these proteins are absent in host cells.

A group of kinases that are unique to Apicomplexans was first identified during the characterization of the *P. falciparum* kinome, and the members were named for a four-character motif that is conserved in the kinases of this group (40). The FIKK group of proteins, which are targeted to the parasitophorous vacuolar membrane and eventually the host cell membrane, has been shown to allow trafficking of substances between the parasite and the host membranes (50). The average Apicomplexan kinome possesses one FIKK kinase. In contrast, the *P. falciparum* kinome contains 20 such proteins (51).

The characterization of the *P. falciparum* kinome along with functional analysis of a growing number of protein kinases from *P. falciparum* has enabled the identification of unique aspects of these parasitic proteins that can allow the identification of promising drug targets for antimalarial drug discovery. The unique aspects of MAP kinase signaling, the presence of numerous orphan kinases that are structurally distinct from host kinases, and the presences of the FIKK kinases, an Apicomplexan-specific group of kinases, will inform future drug discovery efforts.

2 In Silico Methods

Numerous in silico studies have been undertaken with the focus of identifying or developing novel antimalarial drugs as well as understanding the structure–activity relationships (SARs) for sets of compounds with known activities. The primary targets of inhibitors subject to computational modeling work include dihydrofolate reductase (DHFR), farnesyltransferase, heme polymerization, enoyl-acyl carrier protein reductase, aspartyl protease, spermidine synthase, reductoisomerase, and cysteine protease. In addition, the general antimalarial activity of structurally similar sets of compounds has been evaluated computationally in order to illuminate important contributing or interfering chemical features. The results of these studies, whether conducted on targeted inhibitors or related sets of compounds lacking a defined target, can contribute to the development of new compounds through focused design or development of models for the purpose of virtual screening and prioritization of large collections of small molecules.

Strikingly, computational studies applied to analyzing malarial kinase inhibitors are nearly nonexistent within the literature, with only two found at the time of this writing. This is quite surprising given the extensive structure-based and ligand-based in silico studies that have been performed for human kinases, the prominent role that kinases play in regulation of the cell cycle, and the interest in kinase inhibition for treatment of human disease. Furthermore, the methodologies used to evaluate targeted or nontargeted antimalarial effects computationally are fully amenable to those specifically targeted to kinase inhibition. Here we describe various modeling techniques that have been applied to malarial drug targets or antimalarial activity in general and include select references that well illustrate the capabilities of these techniques.

2.1 Quantitative Structure–Activity Relationships

Development of quantitative structure–activity relationship (QSAR) models involves establishing a numerical representation for the chemical structures under study and subsequently using statistical modeling techniques to find a mathematical relationship between that representation and the desired activity. The number of available representations, or descriptors, is vast (52, 53) and includes simple counts of atom types, atom pairs, or specific substructural groups; binary indicators, or fingerprints representing the presence or absence of substructural features with the molecules; mathematically sophisticated representations deriving from graph theoretic operations applied to the chemical structure and its corresponding connectivity matrix; descriptors derived from information theory; and many others. Additionally, many descriptors are topological in nature and do not require the generation of a three-dimensional conformation, thus reducing the overall computational costs. The choice of which representation to use is often guided by the expected usage of the model as many chemical descriptors do not lend themselves to direct interpretation and application to the design of new chemical structures. Choosing a statistical modeling method is also influenced by computational costs and usage expectations. As with descriptor selection, the types of modeling methods span a wide range of complexity, with simple linear regression, linear discriminant analysis, decision trees, random forests, neural networks, support-vector machines, clustering, and self-organizing maps being most common. Furthermore, the process of developing a model must include various steps of data preprocessing (validation of literature-derived data, detection of outliers, etc.), model validation using properly constructed training and test sets, and model evaluation using a previously unseen test set of structures. The likelihood of overtraining a model and overestimating its usefulness is far too high to ignore any of these steps.

Mahmoudi et al. (54) compiled a dataset of 395 antimalarial chemicals from literature sources, paying very close attention to the quality of the data and including only compounds that met specific criteria: publication between 1996 and 2003; antimalarial activity assessed by in vitro radioisotope assay; only chloroquine-resistant parasite strains used for drug assessment; specific chemical structure illustrated; median inhibitory concentration (IC_{50}) expressed in micromolar concentration; and parasitic cultures synchronized for testing. The compounds were then assigned to one of three groups: highly active ($IC_{50} < 0.06$ μM), active (0.06 μM $\leq IC_{50} \leq 5$ μM), and inactive ($IC_{50} > 5$ μM) with a linear discriminant analysis performed in multiple steps. First, a classifier was developed to distinguish the inactive group from the other two. Next, structures predicted as active in the first step were used to develop a three-class classifier, yielding good separation of the three groups of compounds (95% accurate prediction of a held out test set of structures). These equations were then used to screen 2,000 structures from the Merck Index, leading to the selection of 22 compounds for experimental testing. Upon in vitro testing, 16 of these compounds were found to have antimalarial activity below 5 μM, two of which had subnanomolar IC_{50}, and seven of which were found to have submicromolar activity.

2.2 Three-Dimensional Quantitative Structure–Activity Relationships

Much like the QSAR model discussed above, three-dimensional QSAR models attempt to relate a numerical representation of the molecules of interest to a given activity value (55). The primary difference is that reasonable, low-energy three-dimensional conformations of each structure are required, and these structures must be aligned to each other in a manner that is consistent with their expected mode of action. Once the structures are aligned in a common three-dimensional space, various energetic parameters can be calculated for each structure at regularly spaced grid points surrounding the entire set. Commonly used parameters include electrostatic, steric, hydrophobic, and hydrogen bonding fields, and the array of resulting values are analyzed by statistical modeling methods. The resulting three-dimensional models can be used to predict activities of novel structures and can be interpreted by superimposing structures of interest onto visualizations of those regions of space containing fields highly correlated with the endpoint. The result is identification of regions on the molecules where certain characteristics are expected to be favored or disfavored, e.g., regions where steric bulk is or is not tolerated, positive/negative electrostatic preferences, and hydrophobicity requirements.

Using a set of 40 naphthyl-isoquinoline alkaloids, Bringmann and Rummey (56) compared two different methods of generating alignments automatically. Typically, data are divided into a training set used to optimize the parameters of the model and a test set that does not influence the model in any way but is used for evaluation of the predictive abilities of the final model. In this study, all available compounds were used for model development, and given that the researchers synthesized and tested new compounds on a continuous basis, the test set consisted of newly synthesized chemical compounds. When seven newly synthesized compounds were evaluated and predictions were made, a poor correlation was seen between actual and predicted IC_{50} values ($r^2 = 0.279$). If instead, the results are evaluated as a classification of "active" or "inactive," five of the seven structures are predicted correctly.

2.3 Pharmacophores

Pharmacophore modeling (57) is a technique in which actual molecular features are represented by generalized features, typically but not exclusively hydrogen bond donors, hydrogen bond acceptors, aromatic rings, hydrophobic regions, and charge centers. These generalized features, or pharmacophores, can be positioned relative to each other in three dimensions based on the actual conformation of the structure they are being used to represent. Often, a number of acceptable, low-energy conformations of structures with known activities are developed and pharmacophore features are mapped to each one, resulting in a collection of possible pharmacophore arrangements for a single structure. Statistical modeling and data mining techniques are then applied to the entire library of structures and associated conformations in order to develop classification or regression models of the desired property. These models can then in turn be used to search conformational pharmacophore libraries of structures in an effort to identify new chemical matter that can be tested for activity. Alternatively, a pharmacophore model may be developed from the known binding mode of a small molecule to its protein target and the interactions identified between the two.

Using a set of 15 known kinase inhibitors having a range of activities against Pfmrk, Bhattacharjee et al. (58) developed a pharmacophore model consisting of four functional groups: two hydrogen bond acceptors, one hydrophobic group, and one planar aromatic group. When this model was used to evaluate an independent set of 15 additional known inhibitors, a correlation of $r^2 = 0.7$ was found between the predicted activities and actual activities, providing strong validation of the modeling procedure. The model was then used to search a database of 290,000 compounds and 16 were selected for further testing, revealing compounds with inhibitory activities as low as 2.5 μM.

Gupta et al. (59) developed a pharmacophore model for a series of 88 trioxanes with literature-reported antimalarial activities, being careful to include only consistent biological data. A pharmacophore consisting of two hydrogen bond acceptors, an aromatic hydrophobic feature, and two aliphatic hydrophobic features was found to be weakly predictive on a set of 43 trioxanes not used during model development ($r^2=0.51$, $r^2=0.61$ with one outlier removed). Interestingly, the authors chose to develop a virtual library of structures based on the trioxane core with various pendant groups attached combinatorially for the purpose of pharmacophore screening and subsequent synthesis of the identified hits. The authors report five novel structures that were synthesized and found to have antimalarial activity with IC_{50} values of less than 100 nM.

2.4 Docking Studies

Molecular docking studies (60) require a homology model or crystal structure of the protein target of interest. This structure is prepared with the docking software by cleaning the structure of aberrant molecules (e.g., solvents, buffers, or ions), adjusting the amino acid residue charges for the desired pH, and reducing the structure to the desired binding site and some surrounding residues in order to reduce the overall computational costs required. Small molecules can be docked rigidly or flexibly, with flexible docking using either a predefined conformational library or in-process flexibility during docking. Each putative docking pose is given a score taking into account factors such as electrostatic and steric interactions with the binding site and molecular strain for the particular compound's conformation. While it is well known that docking scores do not necessarily correlate well with actual binding affinities, particularly for closely related analogues (61), the scores can be used in larger virtual screening campaigns to prioritize groups of structures for wet lab evaluation.

Singh et al. (62) describe in detail the development of a reductoisomerase homology model, identifying the numerous critical steps involved in the construction of a valid model. The predicted structure was then used to dock a series of fosmidomycin analogues, and it was found that the predicted binding scores and actual pIC_{50} values correlated with $r^2=0.84$. While virtual screening was not performed to identify novel inhibitors, in the absence of X-ray crystallographic information, the generation of a reductoisomerase homology model provides a tool that can be used not only for virtual screening of structure databases but also in structure-based rational design.

Focusing upon Pfmrk, Peng et al. (63) developed a homology model of Pfmrk based on the known crystal structure of human CDK2. This model was then used to computationally dock a set of oxindole-based inhibitors, resulting in a correlation of $r^2 = 0.67$ between docking scores and actual inhibitory pIC_{50} values, providing another method to conduct virtual screening in order to identify novel malarial kinase inhibitors.

2.5 Combination Approach: Pharmacophores and Docking

Rastelli et al. (64) combined pharmacophore modeling and docking in an effort to identify novel inhibitors of *P. falciparum* dihydrofolate reductase (DHFR). Six pharmacophore models were developed based on the proposed binding modes of cycloguanil, pyrimethamine, and WR99210 taking into account interactions that were hypothesized to be important from previous studies as well as interactions identified in crystal structures of similar inhibitors bound to DHFRs. An overall molecular volume limit was established based on cycloguanil in order to prevent the identification of molecules that, while meeting the pharmacophoric interaction requirements, would be too large for the DHFR binding site. Further, an excluded volume region was included in order to select for molecules that would not be prone to a known cycloguanil resistance mechanism consisting of the mutation of alanine 16 to valine. A database of 230,000 compounds was screened with these models and reduced to a set of 4,061 compounds that scored well. This reduced set was then docked to a *P. falciparum* DHFR homology model and the best scoring compounds from each identified structural family were chosen for biological evaluation. Of the 24 compounds selected, 12 were found to inhibit wild-type and relevant mutants of DHFR with IC_{50} ranging from 0.6 to ~100 μM.

Similarly, Jacobsson et al. (65) developed pharmacophore models based on the structure of AdoDATO (*S*-adenosyl-1,8-diamo-3-thiooctane) bound to spermidine synthase. Two models were developed, one capturing interaction details of the adenosine moiety of AdoDATO and the other capturing those details of the amine substrate. Pharmacophore screening reduced an initial library of 2.6 million structures to 7,355 structures that were then docked into the spermidine synthase structure. As the docking process allowed conformational exploration of the structures of interest, the results were reevaluated with the pharmacophore models to eliminate any conformations not consistent with the pharmacophore constraints. From these, a second round of more elaborate and computationally expensive docking was performed. Ultimately, 28 compounds were selected for experimental evaluation, seven of which were found by nuclear magnetic resonance studies to be capable of reversible binding to the target enzyme's active site.

2.6 Malarial Kinases Applicable for Rational Design Approaches

Many of the computational approaches described above utilize structural information about the target enzyme. In the following section and summarized in Table 1, we describe the plasmodium kinases with solved kinase domains. We include known activities of

Table 1
***Plasmodium falciparum* kinases with solved kinase domains**

Kinase	Structures	Molecules cocrystallized
P. falciparum thymidylate kinase (PfTMK)	2WWG (71)	dGMP and ADP
	2WWI (71)	AZTMP and ADP
	2WWH (71)	AP5DT
	2WWF (71)	TMP and ADP
P. falciparum choline kinase (PfCK)	3FI8 (125)	ADP, phosphoric acid mono-(2-amino-ethyl) ester, and magnesium ion
P. falciparum pyruvate kinase (PfPyrK)	3KHD (126)	None
P. falciparum mitogen-activated protein kinase 2 (Pfmap-2)	3NIE (127)	ANP
P. falciparum phosphoglycerate kinase (PfPGK)	3OZA (128)	Glycerol and sulfate ion
	3OZ7 (128)	Sulfate ion
	1LTK (open conformation) (128)	AMP, glycerol, sulfate ion
P. falciparum adenylate kinase (PfAK)	3TLX (129)	ADP, AMP, ATP, and magnesium ion
P. falciparum protein kinase 5 (PfPK5)	1OB3 (98)	None
	1V0B (T198A mutant) (98)	None
	1V0O (98)	Indirubin-5-sulfonate
	1V0P (98)	Purvalanol B
P. falciparum nucleoside diphosphate kinase B (PfNDK)	1XIQ (130)	None
P. falciparum guanylate kinase (PfGK)	1Z6G (131)	EPE and sulfate ion
P. falciparum protein kinase 7 (PfPK7)	2PML (105)	ANP and manganese (II) ion
	2PMO (105)	HMD
	2PMN (105)	K51

ADP adenosine diphosphate, *ANP* atrial natriuretic peptide, *AP5DT* P1-(5′-adenosyl)P5-(5′-thymidyl)pentaphosphate, *ATP* adenosine triphosphate, *AZTMP* 3′-azido-3′-deoxythymidine monophosphate, *dGMP* deoxyguanosine monophosphate, *EPE* 4-(2-hydroxyethyl)-1-piperazine ethanesulfonic acid, *HMD* hymenialdisine, *TMP* thymidine monophosphate

each kinase along with interesting features or data that suggest the kinase as a good target for inhibitor design.

P. falciparum Thymidylate Kinase (*PfTMK*). PfTMK is involved in nucleotide synthesis in the parasite. As nucleotide synthesis pathways are indispensable for most organisms, it has been proposed that this enzyme may be a good target for antimalarial drug design. This enzyme is a stable homodimer and phosphorylates thymidine monophosphate (dTMP) and deoxyuridine monophosphate (dUMP) as well as larger purines, which indicates that this enzyme has a broader range of substrates than other thymidylate kinases (66). Both purine and pyrimidine nucleosides can inhibit the enzymatic activity of PfTMK, which may allow for the design of more selective inhibitors (67–69). This protein can also phosphorylate deoxyguanosine monophosphate (dGMP) with a specificity similar to that of the pyrimidines (70). While PfTMK is similar in sequence to the human type I TMK enzymes, the ability of the enzyme to phosphorylate AZTMP (3′-azido-3′-deoxythymidine monophosphate) was very robust and more similar to the actions of a type II TMK. Structural analysis revealed more differences between the *Plasmodium* and human enzymes in the lid region and P-loop that may be useful for the rational design of selective inhibitors (71).

P. falciparum Choline Kinase (PfCK). Biosynthesis of phosphocholine by choline kinase is a necessary step in the synthesis of phosphatidylcholine by the Kennedy pathway. Phosphatidylcholine is essential for parasite survival during intraerythrocytic development (72–74). PfCK is localized to the parasite cytoplasm and increases in expression as intraerythrocytic development progresses (75). The characterization of PfCK showed that recombinant PfCK is able to use choline in the presence of ATP to form phosphocholine (76). An inhibitor of PfCK, hexadecyltrimethylammonium bromide (HDTAB), was identified in the same study and was shown not only to inhibit recombinant PfCK but also to inhibit growth of *P. falciparum* in vitro and *Plasmodium yoelii* in vivo (77). PfCK has also been shown to be inhibited by the ammonium compound hemicholinium-3 and the choline analogue bis-thiazolium (T3), which at the time of this study was in clinical trials as an antimalarial (75). A screen of compounds from commercially available libraries revealed a 3-chlorobenzothiophene, CK-1, as well as two other analogues of this compound that are able to inhibit PfCK in the low micromolar range and possess a promising structure for chemical optimization (78). The essential nature of phospholipid biosynthesis in parasite development during erythrocytic stages and the recent discovery of potent inhibitors that may help further characterize PfCK make this kinase an attractive target for antimalarial drug design.

P. falciparum Pyruvate Kinase (PfPyrK). Glycolysis is the major energy utilization pathway for *Plasmodium* parasites and PfPyrK is one enzyme that contributes to this process and the breakdown of carbohydrates during intraerythrocytic development (79). Glycolytic processes have been shown to increase 50-fold in infected erythrocytes compared with uninfected erythrocytes, and this enzyme may be a crucial component of metabolic pathways for the parasite and thus an ideal drug target for antimalarial drug design (80–82). Recombinant PfPyrK is a 55.6-kDa protein that assumes a homotetramer in its active form and neither fructose-1,6-bisphosphate, which is an activating factor for the human enzyme, nor glucose-6-phosphate, which is an activating factor for the *Toxoplasma gondii* enzyme, is an activating factor for PfPyrK (79). The same study found that citrate and ATP are potent inhibitors of PfPyrK. A genome analysis revealed that there are two pyruvate kinase genes in the *Plasmodium* genome, *PK1* and *PK2*, that are both expressed during malaria blood stages (83, 84). *PK2* is restricted to the Apicomplexan lineage and localizes to the apicoplast. *PK2* may represent an ideal antimalarial drug target because is not found in humans.

P. falciparum MAPK2 (*Pfmap-2*). Pfmap-2 is in the CMGC group of kinases and is most likely involved in processes that regulate cellular proliferation. Pfmap-2 is expressed in gametocytes, is able to phosphorylate traditional MAPK substrates such as myelin basic protein, can autophosphorylate, and is able to be modulated by a MAPK-specific phosphatase (85). Pfmap-2 is specifically phosphorylated by Pfnek-1, a protein that shows maximal homology to the NIMA family of kinases that are involved in cell division of eukaryotes (48). Pfmap-2 is crucial for male gametogenesis in parasites, and parasites lacking a functional copy of Pfmap-2 cannot complete the process of exflagellation. Transmission of these mutants to the mosquito vector was also greatly reduced in *Plasmodium berghei* (86). Furthermore, Pfmap-2 is indispensable for intraerythrocytic development in *P. falciparum*, which makes it an attractive drug target (49).

P. falciparum Phosphoglycerate Kinase (*PfPGK*). Phosphoglycerate kinase is a protein that is necessary for glucose metabolism. PfPGK is an enzyme that was first found in parasite isolates (80). The gene for PfPGK encodes a 45.5-kDa protein that displays the most homology to the host enzyme when compared with other glycolytic enzymes from this organism (87). PGK-specific activity was found to be seven times higher in infected RBCs as compared with uninfected RBCs, which is partially explained by the later identification of two isoenzymes in *P. falciparum* (88). Suramin was found to inhibit PfPGK in the low micromolar range (89). The vital nature of glucose metabolism in malaria parasites points to this protein as a promising drug target for antimalarial drug development.

P. falciparum Adenylate Kinase (*PfAK*). PfAK helps the parasite meet the high demand for adenine nucleotide interconversion and energy homeostasis within the infected RBC. One study investigated the ATP concentrations in infected erythrocytes and found that ATP levels are kept constant throughout the host and parasite compartments and that this equalization of ATP concentration is accomplished in part by host and parasite adenylate kinase activities

(90). Recombinant PfAK was an enzymatically active 28.9-kDa protein that showed a mononucleotide binding preference for adenosine monophosphate and a trinucleotide binding preference for adenosine triphosphate (91).

P. falciparum Protein Kinase 5 (PfPK5). PfPK5 was first isolated using a partially redundant oligonucleotide, which was constructed based on conserved CDK and cdc2-like protein sequences (92). The same study found that when the resulting isolate was expressed, it was enzymatically active and was able to phosphorylate histone H1 and casein. PfPK5 is likely involved in either initiating or maintaining S-phase as blocking PfPK5 activity markedly reduced the parasite's ability to synthesize DNA and immunolocalization studies also suggest that PfPK5 is a mediator of DNA synthesis (93, 94). Human cyclin H and p25 both activate PfPK5 and this protein is able to autophosphorylate (95). PfPK5 is inhibited by CDK inhibitors, and a protein from *P. falciparum* (Pfcyc-1) that shows maximal homology to cyclin H is also able to activate PfPK5. This study, however, also found that Pfmrk, a putative cyclin-activating kinase (CAK), did not modulate PfPK5 activity. Human p21, a protein that negatively modulates human CDKs, was shown to inhibit PfPK5 as well (96). The crystal structure of PfPK5, which was resolved in 2003, was the first nonhuman CDK structure to be determined (97). Cocrystallization of inhibitors in the active site of PfPK5 has allowed for characterization of the active site, and a comparison of the ATP binding site of a PfPK5 homology model with the ATP binding site of human CDK2 has allowed for the identification of similarities and differences between the binding characteristics of human CDK and PfPK5 (98, 99). The characterization of the active site of this kinase along with its role in parasite cell cycle makes PfPK5 an attractive drug target.

P. falciparum Nucleoside Diphosphate Kinase (PfNDK). Nucleoside diphosphate kinases generally aid in maintaining concentrations of various nucleoside triphosphates, which modulates the regulation of energy metabolism. Tracking of recombinant PfNDK revealed the essentiality of $Mg^{(2+)}$ for PfNDK kinase activity but not necessarily for nucleotide binding (100). The specific activity of PfNDK was shown to be twice that of the human enzyme. The nucleotide binding site of PfNDK has more affinity for purine nucleotides and ribonucleotides than for pyrimidine nucleotides and deoxyribonucleotides, with nucleotide size determined to be the key element in the affinity of PfNDK for its ligand (101). The regulation of energy metabolism in *Plasmodium* is critical for maintaining cellular processes that contribute to parasite survival and this essential nature makes PfNDK an attractive target for antimalarial drug design.

P. falciparum Guanylate Kinase (PfGK). PfGK is involved in nucleic acid synthesis, specifically the conversion of dGMP to dGDP (deoxyguanosine diphosphate) (78). The expression and subsequent characterization of PfGK yielded important differences between this enzyme and other guanylate kinases, especially in three areas that are important for kinase activity and at helix 3, which is involved in domain movements (102). The catalytic efficiency of this enzyme was found to be much greater for GMP than for dGMP, which is another distinction between this protein and other known guanylate kinases (70, 103). The unique substrate binding preferences may make this enzyme an attractive drug target candidate; however, the essentiality of this kinase is still undetermined as its activities overlap with those of PfTMK (78).

P. falciparum Protein Kinase 7 (PfPK7). PfPK7 was first characterized in an effort to identify a MAPKK from *P. falciparum* and a classical MAP kinase pathway in this parasite (104). PfPK7 is an orphan kinase as it has no ePK homologue (105). This kinase is expressed during both extrinsic and intrinsic parasite life cycle stages and can phosphorylate various substrates (104). However, PfPK7 was not inhibited by protein kinase A (PKA) or MEK inhibitors and could not phosphorylate two MAPK homologues from *P. falciparum* in vitro. PfPK7(−) clones were less able to efficiently complete the processes of schizogony and sporogony, producing fewer merozoites in humans and fewer oocysts in the mosquito vector (106). Cocrystallization of PfPK7 with ATP-competitive inhibitors revealed structures in the ATP binding site that may be important for inhibitor design as well as promising drug scaffolds for chemical optimization (105). A high-throughput screen identified several imidazopyridazines as potent inhibitors of PfPK7 activity and synthesis of 1,4-disubstituted trizoles yielded promising results as two of these compounds showed inhibitory activity in the range of 10–20 μM (107, 108). It was also recently shown that PfPK7 may function in the melatonin transduction pathway as PfPK7(−) parasites failed to display the typical effects that are induced by melatonin introduction (e.g., increased schizont number and decreased ring stage populations) (109). The unique structural nature of this protein kinase makes it an attractive target for antimalarial drug design.

2.7 Additional Plasmodium *Kinase Targets with Unsolved Kinase Domains*

Structures for the following kinases have not yet been solved. Homology models based on human kinases have been developed, however.

P. falciparum MO15-Related Kinase (Pfmrk). When Pfmrk was first expressed and characterized in 1996, it was found to be predominately expressed in gametocytes and to share maximal sequence homology with human MO15 (CDK7), CAK (110). CDK7 has two primary roles in humans. The first is to act as a CAK, regulating the activity of various CDKs during the cell cycle, and the second is to regulate transcription by phosphorylating

RNA polymerase II carboxyl-terminal domain (CTD) (111, 112). Pfmrk is autophosphorylated in vitro and phosphorylates histone H1; however, this kinase is not inhibited by the traditional CDK inhibitors olomoucine and roscovitine, a finding that points to differences between human CDKs and Pfmrk (113). Pfmrk also does not activate PfPK5, a cdc-2-related protein kinase (95). The compound 3-phenyl-quinolinone was found to inhibit Pfmrk activity at 10 μM, and human p21, a negative modulator of human CDKs, is able to inhibit Pfmrk activity (96, 114). In addition, several oxindole-based compounds were found to inhibit this enzyme in the low micromolar range (IC_{50} = 1.5 μM) (115). A QSAR pharmacophore model containing two hydrogen bond acceptors and two hydrophobic sites was developed for Pfmrk using a set of 15 kinase inhibitors that displayed varying degrees of activity. The pharmacophore model was successfully able to select 16 potential inhibitors that were subsequently shown to be potent inhibitors of Pfmrk activity in an in vitro CDK assay (described more fully above) (58). In an effort to elucidate the role of Pfmrk within the parasite and to pinpoint the mechanisms involved in regulating this kinase, a *Plasmodium* homologue of the human effector protein MAT1, PfMAT1, was identified and found to stimulate Pfmrk activity and to allow phosphorylation of RNA polymerase II CTD. Even in the presence of PfMAT1, Pfmrk was not able to phosphorylate PfPK5, a putative homologue of human CDK1, which indicates Pfmrk may be acting as a regulator of transcription and not as a CAK (116). A screen of PKA inhibitors against Pfmrk showed that isoquinoline sulfonamides were potent inhibitors of Pfmrk and that the nitrogen in the isoquinoline ring was important for maintaining interactions between these small molecules and the protein binding site (117). Chalcone derivatives as well as thiophene and benzene sulfonamides have also been shown to inhibit Pfmrk activity in the low micromolar range (118, 119). These studies highlight the ability of small amino acid differences in the binding sites of various kinases to greatly affect the binding preferences of each protein, which indicates the promise of developing Pfmrk-selective inhibitors. It was recently shown that PfMAT1 confers substrate specificity to Pfmrk and that these two proteins colocalize to the nucleus (120). The same study identified two *Plasmodium* DNA replication proteins, PfRFC-5 and PfMCM-6, that interact with this protein.

P. falciparum Glycogen Synthase Kinase 3 (PfGSK-3). GSK-3 is a kinase involved in numerous cell signaling pathways and has been the focus of extensive research because of its proposed involvement in various disease processes. PfGSK-3 is a homologue of GSK-3β and a homology model that was developed based on the crystal structure of human GSK-3 suggested well-conserved regions of the protein (121). Recombinant PfGSK-3 was able to phosphorylate glycogen synthase, recombinant axin, microtubule-binding protein tau, and GS-1. PfGSK-3 is most highly expressed during the early trophozoite stage during the erythrocytic cycle and is localized to the cytoplasm before it associates with parasite vesicle-like structures. A screen of inhibitors against human GSK-3 and PfGSK-3 indicated the possibility of selectively targeting this enzyme. Several homology models have been developed from various GSK-3 crystal structures and have been extensively evaluated (122). The presence of numerous homology models of PfGSK-3 and the differential affinity of the binding sites of PfGSK-3 and human GSK-3 for inhibitors suggest that this protein could be selectively targeted for antimalarial drug design.

P. falciparum Protein Kinase 6 (PfPK6). PfPK6 was first isolated from parasite cDNA and was found to share sequence similarity with both CDKs and MAPKs (123). Western blot analysis suggested the upregulation of this protein in trophozoites and during trophozoite, schizont, and segmenter stages. PfPK6 was localized to both the parasite nucleus and cytoplasm. The same study found that histone and plasmodial ribonucleotide reductase were phosphorylated by recombinant PfPK6 and that traditional CDK inhibitors, olomoucine and roscovitine, inhibited this protein. However, PfPK6 was not affected by the heterologous CDK modulators p21(CIP1) or p16(INK4) (96). Structural models of PfPK6 complexed with either olomoucine or roscovitine have helped explain the difference between the inhibitory effect of these two molecules and may guide future efforts to design inhibitors for PfPK6 (124).

3 Conclusions

Malaria affects approximately 40% of the world's population. If resistance to artemisinin becomes widespread, its impact on the global community will become increasingly more devastating. Novel targets for the development of new drugs are urgently needed to replace artemisinin as the frontline therapy. As a family, protein kinases are exploited by the pharmaceutical industry to combat diseases as diverse as cancer, neurodegeneration, and inflammation. We (and others) contend that kinases are excellent targets for the design of antimalarial compounds. The numerous examples of solved *P. falciparum* kinase domains, coupled with the many unique features observed in the *Plasmodium* kinome, suggest these enzymes as good targets for inhibitor design. The development of pharmacophores, QSAR models, and molecular docking are a few examples of available in silico techniques that could be used to aid in the rational design of new antimalarial drugs targeting kinases.

References

1. *World malaria report* (2011) Geneva, Switzerland; World Health Organization. http://www.who.int/malaria/world_malaria_report_2011/9789241564403_eng.pdf. Accessed 02/20/2010.
2. Sachs J, Malaney P (2002) The economic and social burden of malaria. Nature 415(6872): 680–685
3. Gallup JL, Sachs JD (2001) The economic burden of malaria. Am J Trop Med Hyg 64(1–2 Suppl):85–96
4. Sermwittayawong N, Singh B, Nishibuchi M et al (2012) Human Plasmodium knowlesi infection in Ranong province, southwestern border of Thailand. Malar J 11(1):36
5. Lucchi NW, Poorak M, Oberstaller J et al (2012) A new single-step PCR assay for the detection of the zoonotic malaria parasite Plasmodium knowlesi. PLoS One 7(2): e31848
6. Dia I, Sagnon N, Guelbeogo MW, Diallo M (2011) Bionomics of sympatric chromosomal forms of Anopheles funestus (Diptera: Culicidae). J Vector Ecol 36(2):343–347
7. Chilaka N, Perkins E, Tripet F (2012) Visual and olfactory associative learning in the malaria vector Anopheles gambiae sensu stricto. Malar J 11:27
8. Girod R, Roux E, Berger F et al (2011) Unravelling the relationships between Anopheles darlingi (Diptera: Culicidae) densities, environmental factors and malaria incidence: understanding the variable patterns of malarial transmission in French Guiana (South America). Ann Trop Med Parasitol 105(2): 107–122
9. Dvorak JA, Miller LH, Whitehouse WC, Shiroishi T (1975) Invasion of erythrocytes by malaria merozoites. Science 187(4178): 748–750
10. Hossain ME, Dhawan S, Mohmmed A (2012) The cysteine-rich regions of Plasmodium falciparum RON2 bind with host erythrocyte and AMA1 during merozoite invasion. Parasitol Res 110(5):1711– 1721
11. Tonkin ML, Roques M, Lamarque MH et al (2011) Host cell invasion by apicomplexan parasites: insights from the co-structure of AMA1 with a RON2 peptide. Science 333(6041):463–467
12. Lamarque M, Besteiro S, Papoin J et al (2011) The RON2-AMA1 interaction is a critical step in moving junction-dependent invasion by apicomplexan parasites. PLoS Pathog 7(2):e1001276
13. Chan CW, Spathis R, Reiff DM et al (2012) Diversity of Plasmodium falciparum chloroquine resistance transporter (pfcrt) exon 2 haplotypes in the Pacific from 1959 to 1979. PLoS One 7(1):e30213
14. Brockelman CR, Monkolkeha S, Tanariya P (1981) Decrease in susceptibility of Plasmodium falciparum to mefloquine in continuous culture. Bull World Health Organ 59(2):249–252
15. Lambros C, Notsch JD (1984) Plasmodium falciparum: mefloquine resistance produced in vitro. Bull World Health Organ 62(3): 433–438
16. Nosten F, ter Kuile F, Chongsuphajaisiddhi T et al (1991) Mefloquine pharmacokinetics and resistance in children with acute falciparum malaria. Br J Clin Pharmacol 31(5):556–559
17. Black F, Bygbjerg I, Effersøe P et al (1981) Fansidar resistant falciparum malaria acquired in South East Asia. Trans R Soc Trop Med Hyg 75(5):715–716
18. White NJ (2009) Malaria. In: Cook GC, Zumia A (eds) Manson's tropical diseases, 22nd edn. Elsevier, Amsterdam, The Netherlands, pp 1201–1301
19. Noedl H, Se Y, Schaecher K et al (2008) Evidence of artemisinin-resistant malaria in western Cambodia. N Engl J Med 359(24): 2619–2620
20. O'Brien C, Henrich PP, Passi N, Fidock DA (2011) Recent clinical and molecular insights into emerging artemisinin resistance in Plasmodium falciparum. Curr Opin Infect Dis 24(6):570–577
21. Antoniou X, Falconi M, Di Marino D, Borsello T (2011) JNK3 as a therapeutic target for neurodegenerative diseases. J Alzheimers Dis 24(4):633–642
22. Burgess S, Echeverria V (2010) Raf inhibitors as therapeutic agents against neurodegenerative diseases. CNS Neurol Disord Drug Targets 9(1):120–127
23. Cohen S, Fleischmann R (2010) Kinase inhibitors: a new approach to rheumatoid arthritis treatment. Current Opin Rheumatol 22(3):330–335
24. Ding RQ, Tsao J, Chai H et al (2011) Therapeutic potential for protein kinase C inhibitor in vascular restenosis. J Cardiovasc Pharmacol Ther 16(2):160–167
25. Fabbro D, Cowan-Jacob SW, Möbitz H, Martiny-Baron G (2012) Targeting cancer with small-molecular-weight kinase inhibitors. Methods Mol Biol 795:1–34

26. Gálvez MI (2011) Protein kinase C inhibitors in the treatment of diabetic retinopathy. Review. Curr Pharm Biotechnol 12(3):386–391
27. Moriguchi S (2011) Pharmacological study on Alzheimer's drugs targeting calcium/calmodulin-dependent protein kinase II. J Pharmacol Sci 117(1):6–11
28. Satoh K, Fukumoto Y, Shimokawa H (2011) Rho-kinase: important new therapeutic target in cardiovascular diseases. Am J Physiol Heart Circ Physiol 301(2):H287–H296
29. Savage DG, Antman KH (2002) Imatinib mesylate—a new oral targeted therapy. N Engl J Med 346(9):683–693
30. Capdeville R, Buchdunger E, Zimmermann J, Matter A (2002) Glivec (STI571, imatinib), a rationally developed, targeted anticancer drug. Nat Rev Drug Discov 1(7):493–502
31. Seavey MM, Dobrzanski P (2012) The many faces of Janus kinase. Biochem Pharmacol 83(9):1136–1145
32. Kunimasa K, Yoshioka H, Iwasaku M et al (2012) Successful treatment of non-small cell lung cancer with gefitinib after severe erlotinib-related hepatotoxicity. Intern Med 51(4):431–434
33. Park BJ, Whichard ZL, Corey SJ (2012) Dasatinib synergizes with both cytotoxic and signal transduction inhibitors in heterogeneous breast cancer cell lines—lessons for design of combination targeted therapy. Cancer Lett 320(1):104–110
34. Doerig C, Abdi A, Bland N et al (2010) Malaria: targeting parasite and host cell kinomes. Biochim Biophys Acta 1804(3):604–612
35. Doerig C, Billker O, Pratt D, Endicott J (2005) Protein kinases as targets for antimalarial intervention: kinomics, structure-based design, transmission-blockade, and targeting host cell enzymes. Biochim Biophys Acta 1754(1–2):132–150
36. Doerig C (2004) Protein kinases as targets for anti-parasitic chemotherapy. Biochim Biophys Acta 1697(1–2):155–168
37. Hammarton TC, Mottram JC, Doerig C (2003) The cell cycle of parasitic protozoa: potential for chemotherapeutic exploitation. Prog Cell Cycle Res 5:91–101
38. Kappes B, Doerig CD, Graeser R (1999) An overview of Plasmodium protein kinases. Parasitol Today 15(11):449–454
39. Doerig C, Meijer L, Mottram JC (2002) Protein kinases as drug targets in parasitic protozoa. Trends Parasitol 18(8):366–371
40. Ward P, Equinet L, Packer J, Doerig C (2004) Protein kinases of the human malaria parasite Plasmodium falciparum: the kinome of a divergent eukaryote. BMC Genomics 5:79
41. Hanks SK (2003) Genomic analysis of the eukaryotic protein kinase superfamily: a perspective. Genome Biol 4(5):111
42. Talevich E, Mirza A, Kannan N (2011) Structural and evolutionary divergence of eukaryotic protein kinases in Apicomplexa. BMC Evol Biol 11:321
43. Anamika K, Srinivasan N (2007) Comparative kinomics of Plasmodium organisms: unity in diversity. Protein Pept Lett 14(6):509–517
44. Dvorin JD, Martyn DC, Patel SD et al (2010) A plant-like kinase in Plasmodium falciparum regulates parasite egress from erythrocytes. Science 328(5980):910–912
45. Billker O, Lourido S, Sibley LD (2009) Calcium-dependent signaling and kinases in apicomplexan parasites. Cell Host Microbe 5(6):612–622
46. Ishino T, Orito Y, Chinzei Y, Yuda M (2006) A calcium-dependent protein kinase regulates Plasmodium ookinete access to the midgut epithelial cell. Mol Microbiol 59(4):1175–1184
47. Ahn SH, Acurio A, Kron SJ (1999) Regulation of G2/M progression by the STE mitogen-activated protein kinase pathway in budding yeast filamentous growth. Mol Biol Cell 10(10):3301–3316
48. Dorin D, Le Roch K, Sallicandro P et al (2001) Pfnek-1, a NIMA-related kinase from the human malaria parasite Plasmodium falciparum. Biochemical properties and possible involvement in MAPK regulation. Eur J Biochem 268(9):2600–2608
49. Dorin-Semblat D, Quashie N, Halbert J et al (2007) Functional characterization of both MAP kinases of the human malaria parasite Plasmodium falciparum by reverse genetics. Mol Microbiol 65(5):1170–1180
50. Nunes MC, Okada M, Scheidig-Benatar C et al (2010) Plasmodium falciparum FIKK kinase members target distinct components of the erythrocyte membrane. PLoS One 5(7):e11747
51. Schneider AG, Mercereau-Puijalon O (2005) A new Apicomplexa-specific protein kinase family: multiple members in Plasmodium falciparum, all with an export signature. BMC Genomics 6:30
52. Tetko IV, Gasteiger J, Todeschini R et al (2005) Virtual computational chemistry laboratory—design and description. J Comput Aided Mol Des 19:453–463
53. Mannhold R, Kubinyi H, Timmerman H (eds) (2000) Handbook of molecular

descriptors. Methods and principles in medicinal chemistry, vol 11. Wiley-VCH, Weinheim
54. Mahmoudi N, de Julián-Ortiz JV, Ciceron L et al (2006) Identification of new antimalarial drugs by linear discriminant analysis and topological virtual screening. J Antimicrob Chemother 57(3):489–497
55. Verma J, Khedkar VM, Coutinho EC (2010) 3D-QSAR in drug design—a review. Curr Top Med Chem 10(1):95–115
56. Bringmann G, Rummey C (2003) 3D QSAR investigations on antimalarial naphthylisoquinoline alkaloids by comparative molecular similarity indices analysis (CoMSIA), based on different alignment approaches. J Chem Inf Comput Sci 43(1):304–316
57. Guner OF (2002) History and evolution of the pharmacophore concept in computer-aided drug design. Curr Top Med Chem 2(12):1321–1332
58. Bhattacharjee AK, Geyer JA, Woodard CL et al (2004) A three-dimensional in silico pharmacophore model for inhibition of Plasmodium falciparum cyclin-dependent kinases and discovery of different classes of novel Pfmrk specific inhibitors. J Med Chem 47(22):5418–5426
59. Gupta AK, Chakroborty S, Srivastava K et al (2010) Pharmacophore modeling of substituted 1,2,4-Trioxanes for quantitative prediction of their antimalarial activity. J Chem Inf Model 50(8):1510–1520
60. Kitchen DB, Decornez H, Furr JR, Bajorath J (2004) Docking and scoring in virtual screening for drug discovery: methods and applications. Nat Rev Drug Discov 3(11):935–949
61. Bissantz C, Folkers G, Rognan D (2000) Protein-based virtual screening of chemical databases. 1. Evaluation of different docking/scoring combinations. J Med Chem 43(25):4759–4767
62. Singh N, Chevé G, Avery MA, McCurdy CR (2006) Comparative protein modeling of 1-deoxy-D-xylulose-5-phosphate reductoisomerase enzyme from Plasmodium falciparum: a potential target for antimalarial drug discovery. J Chem Inf Model 46(3):1360–1370
63. Peng Y, Keenan SM, Welsh WJ (2005) Structural model of the Plasmodium CDK, Pfmrk, a novel target for malaria therapeutics. J Mol Graph Model 24:72–80
64. Rastelli G, Pacchioni S, Sirawaraporn W et al (2003) Docking and database screening reveal new classes of Plasmodium falciparum dihydrofolate reductase inhibitors. J Med Chem 46(14):2834–2845
65. Jacobsson M, Gäredal M, Schultz J, Karlén A (2008) Identification of Plasmodium falciparum spermidine synthase active site binders through structure-based virtual screening. J Med Chem 51(9):2777–2786
66. Kandeel M, Kitade Y (2008) Molecular characterization, heterologous expression and kinetic analysis of recombinant Plasmodium falciparum thymidylate kinase. J Biochem 144(2):245–250
67. Kandeel M, Ando T, Kitamura Y et al (2009) Mutational, inhibitory and microcalorimetric analyses of Plasmodium falciparum TMP kinase. Implications for drug discovery. Parasitology 136(1):11–25
68. Kandeel M, Kato A, Kitamura Y, Kitade Y (2009) Thymidylate kinase: the lost chemotherapeutic target. Nucleic Acids Symp Ser (Oxf) 53:283–284
69. Kandeel M, Kitade Y (2011) The substrate binding preferences of Plasmodium thymidylate kinase. Biol Pharm Bull 34(1):173–176
70. Kandeel M, Kitamura Y, Kitade Y (2009) The exceptional properties of Plasmodium deoxyguanylate pathways as a potential area for metabolic and drug discovery studies. Nucleic Acids Symp Ser (Oxf) 53:39–40
71. Whittingham JL, Carrero-Lerida J, Brannigan JA et al (2010) Structural basis for the efficient phosphorylation of AZT-MP (3′-azido-3′-deoxythymidine monophosphate) and dGMP by Plasmodium falciparum type I thymidylate kinase. Biochem J 428(3):499–509
72. Ancelin ML, Vial HJ (1986) Several lines of evidence demonstrating that Plasmodium falciparum, a parasitic organism, has distinct enzymes for the phosphorylation of choline and ethanolamine. FEBS Lett 202(2):217–223
73. Ancelin ML, Vial HJ (1986) Quaternary ammonium compounds efficiently inhibit Plasmodium falciparum growth in vitro by impairment of choline transport. Antimicrob Agents Chemother 29(5):814–820
74. Ancelin ML, Vial HJ, Philippot JR (1985) Inhibitors of choline transport into Plasmodium-infected erythrocytes are effective antiplasmodial compounds in vitro. Biochem Pharmacol 34(22):4068–4071
75. Alberge B, Gannoun-Zaki L, Bascunana C et al (2010) Comparison of the cellular and biochemical properties of Plasmodium falciparum choline and ethanolamine kinases. Biochem J 425(1):149–158
76. Choubey V, Guha M, Maity P et al (2006) Molecular characterization and localization of Plasmodium falciparum choline kinase. Biochim Biophys Acta 1760(7):1027–1038

77. Choubey V, Maity P, Guha M et al (2007) Inhibition of Plasmodium falciparum choline kinase by hexadecyltrimethylammonium bromide: a possible antimalarial mechanism. Antimicrob Agents Chemother 51(2): 696–706
78. Crowther GJ, Napuli AJ, Gilligan JH et al (2011) Identification of inhibitors for putative malaria drug targets among novel antimalarial compounds. Mol Biochem Parasitol 175(1):21–29
79. Roth E Jr (1990) Plasmodium falciparum carbohydrate metabolism: a connection between host cell and parasite. Blood Cells 16(2–3): 453–460, discussion 461–466
80. Roth E Jr, Joulin V, Miwa S et al (1988) The use of enzymopathic human red cells in the study of malarial parasite glucose metabolism. Blood 71(5):1408–1413
81. Roth EF Jr, Calvin MC, Max-Audit I et al (1988) The enzymes of the glycolytic pathway in erythrocytes infected with Plasmodium falciparum malaria parasites. Blood 72(6):1922–1925
82. Mehta M, Sonawat HM, Sharma S (2006) Glycolysis in Plasmodium falciparum results in modulation of host enzyme activities. J Vector Borne Dis 43(3):95–103
83. Chan M, Tan DS, Sim TS (2007) Plasmodium falciparum pyruvate kinase as a novel target for antimalarial drug-screening. Travel Med Infect Dis 5(2):125–131
84. Maeda T, Saito T, Harb OS et al (2009) Pyruvate kinase type-II isozyme in Plasmodium falciparum localizes to the apicoplast. Parasitol Int 58(1):101–105
85. Dorin D, Alano P, Boccaccio I et al (1999) An atypical mitogen-activated protein kinase (MAPK) homologue expressed in gametocytes of the human malaria parasite Plasmodium falciparum. Identification of a MAPK signature. J Biol Chem 274(42):29912–29920
86. Rangarajan R, Bei AK, Jethwaney D et al (2005) A mitogen-activated protein kinase regulates male gametogenesis and transmission of the malaria parasite Plasmodium berghei. EMBO Rep 6(5):464–469
87. Hicks KE, Read M, Holloway SP et al (1991) Glycolytic pathway of the human malaria parasite Plasmodium falciparum: primary sequence analysis of the gene encoding 3-phosphoglycerate kinase and chromosomal mapping studies. Gene 100:123–129
88. Grall M, Srivastava IK, Schmidt M et al (1992) Plasmodium falciparum: identification and purification of the phosphoglycerate kinase of the malaria parasite. Exp Parasitol 75(1): 10–18
89. Pal B, Pybus B, Muccio DD, Chattopadhyay D (2004) Biochemical characterization and crystallization of recombinant 3-phosphoglycerate kinase of Plasmodium falciparum. Biochim Biophys Acta 1699(1–2):277–280
90. Kanaani J, Ginsburg H (1989) Metabolic interconnection between the human malarial parasite Plasmodium falciparum and its host erythrocyte. Regulation of ATP levels by means of an adenylate translocator and adenylate kinase. J Biol Chem 264(6):3194–3199
91. Ulschmid JK, Rahlfs S, Schirmer RH, Becker K (2004) Adenylate kinase and GTP:AMP phosphotransferase of the malarial parasite Plasmodium falciparum. Central players in cellular energy metabolism. Mol Biochem Parasitol 136(2):211–220
92. Ross-Macdonald PB, Graeser R, Kappes B et al (1994) Isolation and expression of a gene specifying a cdc2-like protein kinase from the human malaria parasite Plasmodium falciparum. Eur J Biochem 220(3):693–701
93. Graeser R, Franklin RM, Kappes B (1996) Mechanisms of activation of the cdc2-related kinase PfPK5 from Plasmodium falciparum. Mol Biochem Parasitol 79(1):125–127
94. Graeser R, Wernli B, Franklin RM, Kappes B (1996) Plasmodium falciparum protein kinase 5 and the malarial nuclear division cycles. Mol Biochem Parasitol 82(1):37–49
95. Le Roch K, Sestier C, Dorin D et al (2000) Activation of a Plasmodium falciparum cdc2-related kinase by heterologous p25 and cyclin H. Functional characterization of a P. falciparum cyclin homologue. J Biol Chem 275(12):8952–8958
96. Li Z, Le Roch K, Geyer JA et al (2001) Influence of human p16(INK4) and p21(CIP1) on the in vitro activity of recombinant Plasmodium falciparum cyclin-dependent protein kinases. Biochem Biophys Res Commun 288(5):1207–1211
97. Brinen LS, Stout TJ (2003) Can mosquitoes be bitten? A new hope for anti-malarial drug design. Structure 11(11):1309–1310
98. Holton S, Merckx A, Burgess D et al (2003) Structures of P. falciparum PfPK5 test the CDK regulation paradigm and suggest mechanisms of small molecule inhibition. Structure 11(11):1329–1337
99. Keenan SM, Welsh WJ (2004) Characteristics of the Plasmodium falciparum PK5 ATP-binding site: implications for the design of novel antimalarial agents. J Mol Graph Model 22(3):241–247
100. Kandeel M, Miyamoto T, Kitade Y (2009) Bioinformatics, enzymologic properties, and comprehensive tracking of Plasmodium

falciparum nucleoside diphosphate kinase. Biol Pharm Bull 32(8):1321–1327
101. Kandeel M, Kitade Y (2010) Substrate specificity and nucleotides binding properties of NM23H2/nucleoside diphosphate kinase homolog from Plasmodium falciparum. J Bioenerg Biomembr 42(5):361–369
102. Kandeel M, Nakanishi M, Ando T et al (2008) Molecular cloning, expression, characterization and mutation of Plasmodium falciparum guanylate kinase. Mol Biochem Parasitol 159(2):130–133
103. Kandeel M, Kitade Y (2011) Binding dynamics and energetic insight into the molecular forces driving nucleotide binding by guanylate kinase. J Mol Recognit 24(2):322–332
104. Dorin D, Semblat JP, Poullet P et al (2005) PfPK7, an atypical MEK-related protein kinase, reflects the absence of classical three-component MAPK pathways in the human malaria parasite Plasmodium falciparum. Mol Microbiol 55(1):184–196
105. Merckx A, Echalier A, Langford K et al (2008) Structures of P. falciparum protein kinase 7 identify an activation motif and leads for inhibitor design. Structure 16(2):228–238
106. Dorin-Semblat D, Sicard A, Doerig C et al (2008) Disruption of the PfPK7 gene impairs schizogony and sporogony in the human malaria parasite Plasmodium falciparum. Eukaryot Cell 7(2):279–285
107. Bouloc N, Large JM, Smiljanic E et al (2008) Synthesis and in vitro evaluation of imidazopyridazines as novel inhibitors of the malarial kinase PfPK7. Bioorg Med Chem Lett 18(19):5294–5298
108. Klein M, Dinér P, Dorin-Semblat D et al (2009) Synthesis of 3-(1,2,3-triazol-1-yl)- and 3-(1,2,3-triazol-4-yl)-substituted pyrazolo[3,4-d]pyrimidin-4-amines via click chemistry: potential inhibitors of the Plasmodium falciparum PfPK7 protein kinase. Org Biomol Chem 7(17):3421–3429
109. Koyama FC, Ribeiro RY, Garcia JL et al (2012) Ubiquitin proteasome system and the atypical kinase PfPK7 are involved in melatonin signaling in Plasmodium falciparum. J Pineal Res Jan 30 [Epub ahead of print] doi: 10.1111/j.1600-079X.2012.00981.x
110. Li JL, Robson KJ, Chen JL et al (1996) Pfmrk, a MO15-related protein kinase from Plasmodium falciparum. Gene cloning, sequence, stage-specific expression and chromosome localization. Eur J Biochem 241(3):805–813
111. Fisher RP, Morgan DO (1994) A novel cyclin associates with MO15/CDK7 to form the CDK-activating kinase. Cell 78(4):713–724
112. Shiekhattar R, Mermelstein F, Fisher RP et al (1995) Cdk-activating kinase complex is a component of human transcription factor TFIIH. Nature 374(6519):283–287
113. Waters NC, Woodard CL, Prigge ST (2000) Cyclin H activation and drug susceptibility of the Pfmrk cyclin dependent protein kinase from Plasmodium falciparum. Mol Biochem Parasitol 107(1):45–55
114. Xiao Z, Waters NC, Woodard CL et al (2001) Design and synthesis of Pfmrk inhibitors as potential antimalarial agents. Bioorg Med Chem Lett 11(21):2875–2878
115. Woodard CL, Li Z, Kathcart AK et al (2003) Oxindole-based compounds are selective inhibitors of Plasmodium falciparum cyclin dependent protein kinases. J Med Chem 46(18):3877–3882
116. Chen Y, Jirage D, Caridha D et al (2006) Identification of an effector protein and gain-of-function mutants that activate Pfmrk, a malarial cyclin-dependent protein kinase. Mol Biochem Parasitol 149(1):48–57
117. Woodard CL, Keenan SM, Gerena L et al (2007) Evaluation of broad spectrum protein kinase inhibitors to probe the architecture of the malarial cyclin dependent protein kinase Pfmrk. Bioorg Med Chem Lett 17(17): 4961–4966
118. Caridha D, Kathcart AK, Jirage D, Waters NC (2010) Activity of substituted thiophene sulfonamides against malarial and mammalian cyclin dependent protein kinases. Bioorg Med Chem Lett 20(13):3863–3867
119. Geyer JA, Keenan SM, Woodard CL et al (2009) Selective inhibition of Pfmrk, a Plasmodium falciparum CDK, by antimalarial 1,3-diaryl-2-propenones. Bioorg Med Chem Lett 19(7):1982–1985
120. Jirage D, Chen Y, Caridha D et al (2010) The malarial CDK Pfmrk and its effector PfMAT1 phosphorylate DNA replication proteins and co-localize in the nucleus. Mol Biochem Parasitol 172(1):9–18
121. Droucheau E, Primot A, Thomas V et al (2004) Plasmodium falciparum glycogen synthase kinase-3: molecular model, expression, intracellular localisation and selective inhibitors. Biochim Biophys Acta 1697(1–2): 181–196
122. Kruggel S, Lemcke T (2009) Generation and evaluation of a homology model of PfGSK-3. Arch Pharm (Weinheim) 342(6):327–332
123. Bracchi-Ricard V, Barik S, Delvecchio C et al (2000) PfPK6, a novel cyclin-dependent kinase/mitogen-activated protein kinase-related protein kinase from Plasmodium falciparum. Biochem J 347(Pt 1):255–263

124. Manhani KK, Arcuri HA, da Silveira NJ et al (2005) Molecular models of protein kinase 6 from Plasmodium falciparum. J Mol Model 12(1):42–48
125. Wernimont AK, Pizarro JC, Artz JD et al. Crystal structure of choline kinase from Plasmodium Falciparum, PF14_0020.
126. Wernimont AK, Hutchinson A, Hassanali A et al. Crystal structure of PFF1300w.
127. Wernimont AK, Hutchinson A, Sullivan H et al. Crystal structure of PF11_0147 (CASP Target).
128. Smith CD, Chattopadhyay D, Pal B (2011) Crystal structure of Plasmodium falciparum phosphoglycerate kinase: evidence for anion binding in the basic patch. Biochem Biophys Res Commun 412(2):203–206
129. Wernimont AK, Loppnau P, Crombet L et al. Crystal structure of PF10_0086, adenylate kinase from plasmodium falciparum.
130. Robien MA, Bosch J, Hol WG Crystal structure of nucleoside diphosphate kinase B from Plasmodium falciparum.
131. Vedadi M, Lew J, Artz J et al (2007) Genome-scale protein expression and structural biology of Plasmodium falciparum and related Apicomplexan organisms. Mol Biochem Parasitol 151(1):100–110

Chapter 15

Designing Novel Inhibitors of *Trypanosoma brucei*

Özlem Demir and Rommie E. Amaro

Abstract

Computational simulations of essential biological systems in pathogenic organisms are increasingly being used to reveal structural and dynamical features for targets of interest. At the same time, increased research efforts, especially from academia, have been directed toward drug discovery for neglected tropical diseases. Although these diseases cripple large populations in less fortunate parts of the world, either very few new drugs are being developed or the available treatments for them have severe side effects, including death. This chapter walks readers through a computational investigation used to find novel inhibitors to target one of these neglected diseases, African sleeping sickness (human African trypanosomiasis). Such studies may suggest novel small-molecule compounds that could be considered as part of an early-stage drug discovery effort. As an example target protein of interest, we focus on the essential protein RNA-editing ligase 1 (REL1) in *Trypanosoma brucei*, the causative agent of human African trypanosomiasis.

Key words *Trypanosoma brucei*, RNA-editing ligase 1, REL1, Human African trypanosomiasis, African sleeping sickness, TbREL1, Editosome

1 Introduction

The trypanosome parasites responsible for many neglected tropical diseases, such as African sleeping sickness (human African trypanosomiasis or HAT), Chagas disease, and leishmaniasis, all go through a unique posttranscriptional mitochondrial RNA (mRNA)-editing process. This unique process has transformed the central dogma of biology, which limits information transfer in biological systems to a one-way direction from DNA to RNA, by introducing information transfer between different types of RNA. Using a multiprotein complex called "the editosome," trypanosomes add or delete single or multiple uridylates (Us) and transform premature mRNAs into mature mRNAs using guide RNAs (gRNA) as templates (1, 2).

The editosome consists of 16–20 proteins, and its composition is dynamic because the functional proteins equipped on the RNA-editing core complex (RECC)—consisting of structural proteins—differ to achieve different stages of editing (3–5). When a pre-mRNA

Sandhya Kortagere (ed.), *In Silico Models for Drug Discovery*, Methods in Molecular Biology, vol. 993,
DOI 10.1007/978-1-62703-342-8_15, © Springer Science+Business Media, LLC 2013

base pairs with a gRNA through a conserved "anchor sequence," the point of mismatch between these sequences determines where the editing process will be initiated through endonucleolytic cleavage of the pre-mRNA. Subsequently, depending on the exact type of mismatch between the gRNA and the pre-mRNA, Us are either inserted into, or deleted from, the pre-mRNA. Such reactions are carried out by the terminal uridylyl transferase (RET2) or a U-specific 3′-exoribonuclease, respectively. Finally, RNA-editing ligase 1 (REL1) or 2 (REL2) religates the processed mRNA fragments.

*Tb*REL1 has been shown to be required for the viability of the pathogen in both the insect and bloodstream forms (6, 7) and is thus considered a promising drug target. Like all DNA and RNA ligases, *Tb*REL1 achieves nick joining in RNA in three steps (8). In the first step, a lysine residue (Lys87) in the active site attacks the α-phosphate of adenosine triphosphate (ATP), forming an adenosine monophosphate (AMP)-bound enzyme intermediate and releasing pyrophosphate. In the second step, the 5′ phosphate group on the nicked RNA attacks the intermediate, releases the catalytic lysine residue, and forms a new AMP-bound RNA intermediate. In the final step, the 3′ OH group on the nicked RNA attacks the AMP-bound RNA intermediate, joins the two ends of the RNA, and releases AMP. This chapter outlines the critical system setup considerations for computational simulations and docking studies of *Tb*REL1.

2 Materials

Computational "starting materials" for simulations are crystal structures. To date, there is only one high-resolution *Tb*REL1 crystal structure deposited in the Protein Data Bank identified with PdbID:1XDN (9), which is the ATP-bound form of the adenylation domain of *Tb*REL1. The apo form of the enzyme could not be crystallized, likely owing to the high flexibility (10).

3 Methods

Modifications to the crystal structure of choice are necessary before starting any atomic-level simulations. The critical points to consider in order to prepare the best biophysical system with the *Tb*REL1 crystal structure are presented here (see Note 1).

The *Tb*REL1 crystal structure, which consists of residues 52–316 corresponding to the N-terminal domain, was resolved using the SeMet multiple-wavelength anomalous dispersion method (9). The selenomethionines used for this method should be replaced with methionines to obtain the original form of the protein for simulations (see Note 2).

*Tb*REL1 requires magnesium to function because a two-Mg^{+2} mechanism is suggested for *Tb*REL1 on the basis of experimental evidence on related systems (11). In the ATP-bound crystal structure of *Tb*REL1, the electron density clearly shows a single magnesium ion coordinated between the nonbridging oxygen atoms of the ATP β- and γ-phosphates (9). However, kinetic evidence for the relevant superfamily members suggests a two-Mg^{+2} mechanism for the nucleotide transfer step (11), and the crystal structures of PBCV-1 capping enzyme (from *Paramecium bursaria* Chlorella virus 1) and T4 RNA ligase 2 both have a divalent metal ion in the vicinity of α-phosphate after catalytic step 1 (12, 13). Thus, one needs to determine the number of magnesium ions to model depending on which step of the catalysis is being simulated. If the second magnesium needs to be placed into the crystal structure, one could be guided by the position of the magnesium ion (and its coordinating waters) in the crystal structures of PBCV-1 capping enzyme and T4 RNA ligase 2 (12, 13) or by the corresponding Ca^{+2} ion in crystal structure of T4 RNA ligase 1 (14) (see Note 3).

Apart from the catalytic magnesium ions, including the water molecules resolved in the crystal structure is an important detail. Including deeply buried water molecules in the binding site is especially important in order to prevent nonrealistic structural motions during the simulation of the protein. If one wants to simulate *Tb*REL1 in the apo form, the ATP molecule and Mg^{+2} ion in the binding pocket of the crystal structure should be deleted, and water molecules should be used to fill the empty space. For this purpose, the water molecules in crystal structures of closely related proteins can be used to homology model water molecules in the *Tb*REL1 binding site. As a second option, a water prediction program, such as DOWSER (15), can be used to predict the optimal positions of waters to replace the ATP molecule and the Mg^{+2} ion. Alternatively, one can restrain the protein during dynamics to prevent artificial rearrangements due to missing structural water molecules and let the bulk water reach and fill in the optimal water binding sites. However, this latter method is less preferable, because an optimal arrangement of the deeply buried waters may not occur in a reasonable simulation timescale.

Many researchers will wish to simulate a protein of interest with possible inhibitors to investigate the basis of molecular recognition. In these cases, use of a crystal structure with the relevant ligand bound to the protein of interest is best. In the cases for which there is no crystal structure resolved with a particular ligand, such as *Tb*REL1, a docking program can be used to predict the binding pose of a ligand (see Note 4). In ligand-bound protein simulations, the effect of using explicit or implicit solvent models should be carefully considered. As there are deeply buried water molecules in the *Tb*REL1 binding site as well as water molecules that coordinate the catalytic magnesium ion, employing implicit solvent models in *Tb*REL1 simulations will not be a good representation of the system and may cause artificial results.

Finally, protonation states of all titratable residues should be determined before any atomic-level simulation. Several programs, such as WHATIF (16) and PROPKA (17–19) (see Notes 5 and 6), exist to predict protonation states of standard residues.

Because numerous programs are available to pursue atomic-level simulations and docking, we do not review in detail the input parameters of each program here. We instead list the available programs and refer the reader to the manuals of each program for specific details. Additionally, we present the published atomic-level simulation and docking studies on *Tb*REL1 to provide the reader with specific examples.

All-atom molecular dynamics (MD) simulations of *Tb*REL1 in apo and ATP-bound forms with explicit solvent (10) have been performed using the NAMD2 program (20). *Tb*REL1 has also been simulated in complex with its double-stranded RNA substrate using NAMD2 (21). Various other MD simulation software packages, including AMBER (22), GROMACS (23), DESMOND (24), and GROMOS (25), can be used alternatively (see Note 7). To obtain the protein topology parameters, the Charm27 force field (26) was used for the apo and ATP-bound MD simulations, whereas the AmberFF99SB force field (27) was used for RNA-complexed *Tb*REL1 simulations. For a nonstandard residue in the RNA-complexed *Tb*REL1 simulation, which modeled the adenylated intermediate, additional parameters from the GAFF (general Amber force field) (28) were used. To simulate ATP, topology parameters developed by Meagher et al. (29) were utilized in both studies. For future studies, a new magnesium parameter set may be considered (30) instead of the standard parameters, in order to improve the simulation performance of otherwise troublesome divalent cations.

As *Tb*REL1 is a promising drug target for HAT, it has been used in early-stage receptor-based drug discovery studies that included docking. In Amaro et al. (31), the Autodock program (32) was used to dock ligands to the *Tb*REL1 active site (see Note 8). Alternative docking programs that could be used are Autodock Vina (33), Dock (34), FRED (35, 36), Gold (37), Glide (38, 39), Surflex-Dock (40), and Ligandfit (41) among others (see Notes 9 and 10).

Virtual screening (VS) is a widely used computational method to suggest potential inhibitors for a specific enzyme among a large database of ligands using either docking or pharmacophore algorithms (42). Integrating receptor flexibility into the VS protocol is an active area of research because different conformations of the active site affect the molecular recognition and binding affinity of ligands (43). The relaxed complex scheme (RCS) is a computational method that combines the strength of docking with dynamic

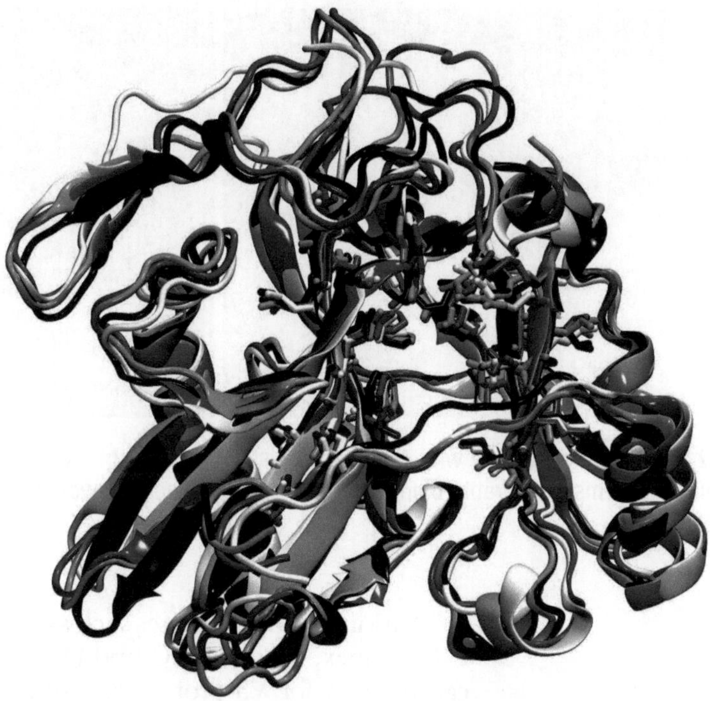

Fig. 1 *Tb*REL1 crystal structure (*in black*) and three snapshots (*in shades of gray*) from MD simulations. Active site residues are shown in sticks to depict their flexibility

structural information obtained by MD to fully account for both ligand and receptor flexibility (44, 45). While typical VS protocols include only one or a few static receptor structures, the RCS incorporates an ensemble of receptor structures and thus is able to take advantage of different conformational states and new binding pockets in or near the active site that are only revealed during MD simulations (Fig. 1).

For the RCS, one needs to determine which ensemble of receptor structures to incorporate into VS. Because using the entire ensemble of receptor structures generated by MD will be computationally too costly, the number of receptor structures must be distilled and reduced with methods such as root-mean-square deviation (RMSD)-based clustering (see Note 11) or QR factorization (see Note 12). Once the nonredundant set of receptor structures is determined, a so-called binding spectrum for each ligand in a compound database is computed by docking the compound into the crystal structure and the ensemble of structures extracted from the MD simulations. Either the mean or minimum of this binding spectrum could be used to rank-order the ligands. Some ligands that are ranked poorly in a crystal-structure VS protocol can rank highly using the RCS protocol as a result of incorporating receptor flexibility (Fig. 2).

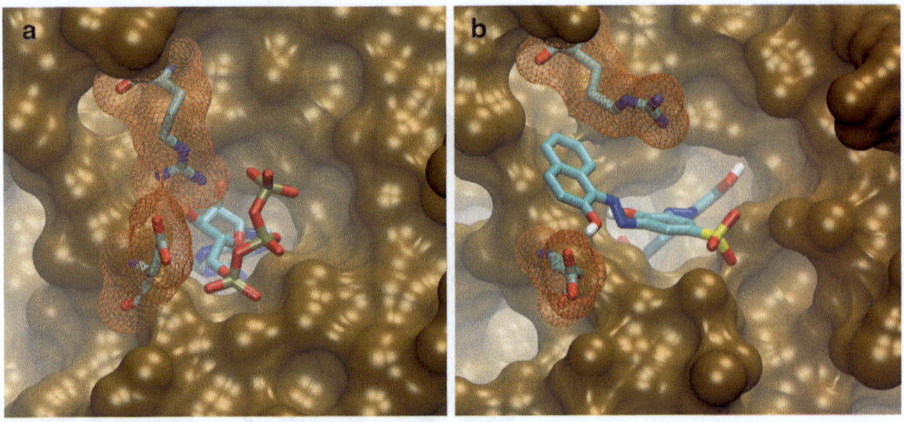

Fig. 2 (**a**) *Tb*REL1 active site with ATP in the crystal structure. (**b**) Motion of active site residues during MD simulations forms a favorable binding site for a compound that would not fit into the crystal structure in *panel* (**a**)

Various ligand databases such as the National Cancer Institute/DTP, Asinex, Otava, ZINC, and DrugBank databases (see Note 13) can be used for VS protocols. Using a combination of several of these may be preferred. An additional filtering to identify drug-like compounds in a database (or the set of compounds with high ranks in the RCS protocol) can be performed based on physicochemical properties or using a set of criteria such as Lipinski's rule of five or Jorgensen's rule of three (see Notes 14 and 15). These filters aim to remove molecules that are not likely to be used as drugs even if they are good inhibitors of the enzyme of interest.

The compounds with final high rank orders are then tested experimentally using inhibition assays. The number of compounds to test is generally limited by the monetary and time cost of the experimental assays. The assays can then be repeated in the presence of detergent, e.g., Triton X-100, as a first test to filter out promiscuous, aggregate-based inhibitors (46). The inhibitory effect of the active compounds can also be tested on similar enzymes to investigate the specificity of the compounds for the enzyme of interest (see Note 16).

If the experimental tests confirm some compounds as inhibitors at the protein level, it is generally necessary to modify these inhibitors to optimize important pharmacokinetic properties, such as membrane permeability, "drug-likeness," or cytotoxicity. For example, the five low-micromolar *Tb*REL1 inhibitors (V1, V3, V4, S1, and S5 in Fig. 3) identified by Amaro et al. (31) were effective against *Tb*REL1 in in vitro protein-level assays but were found to be ineffective against whole-cell *Trypanosoma brucei*, likely because of their low membrane permeability.

In an attempt to optimize the membrane permeability of one of the identified *Tb*REL1 inhibitor scaffolds, Durrant et al. (47)

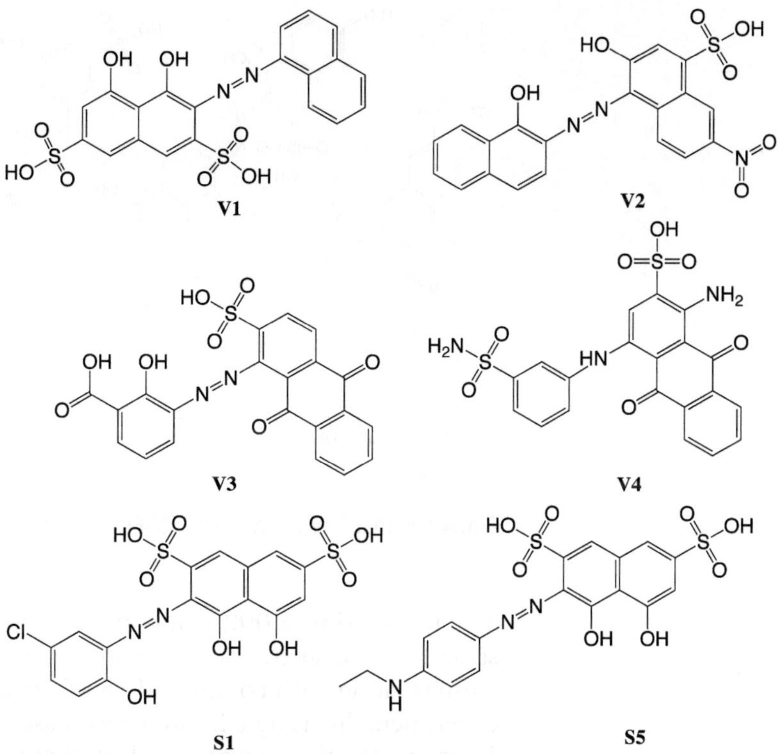

Fig. 3 The five low-micromolar *Tb*REL1 inhibitors in Amaro et al. (31)

performed a similarity search for these compounds (see Note 17) in commercially available databases, yielding 588 similar compounds. Among the top-ranked 100 compounds from a preliminary crystal-structure VS of these 588 compounds, 45 compounds had significant structural similarity to the best inhibitor (compound S5) identified by Amaro et al. (31). After reranking these 45 compounds with ensemble-based RCS scores, the top 12 compounds were tested experimentally, resulting in four compounds with low-micromolar *Tb*REL1 inhibitors (47) designated as compounds V1, V2, V3, and V4 in Fig. 4. Of these four compounds, only one would be placed in the top 12 compounds if all 588 compounds were ranked according to crystal-structure docking scores (47). Thus, three of these four compounds would not have been tested if the compounds had not been reranked based on RCS scores (47), indicating the importance of incorporating receptor flexibility into the VS protocols (see Note 18). One of these three compounds, V4, was even effective against whole-cell *T. brucei*, with a low-micromolar median effective concentration (EC_{50}) value (47).

An independent study suggested similar compounds as *Tb*REL1 inhibitors with a different protocol (48). Virtual screening of the entire 77,000-compound NCI library was performed against the *Tb*REL1 crystal structure using an in-house docking program.

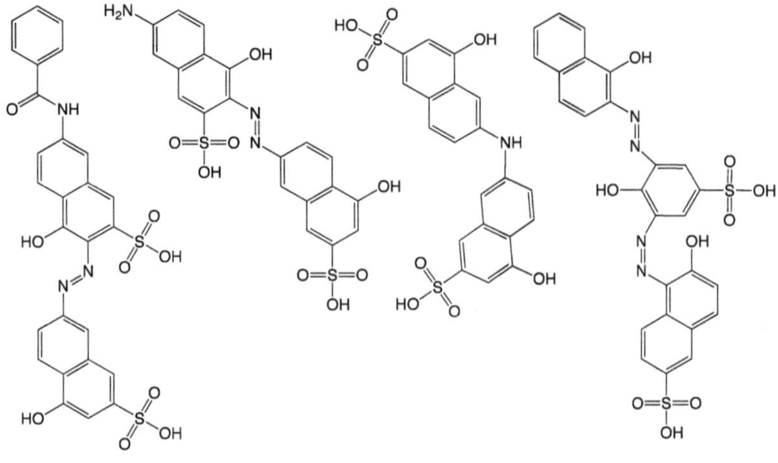

Fig. 4 The four low-micromolar *Tb*REL1 inhibitors in Durrant et al. (47)

The top-scoring 2,000 compounds were then clustered based on structural similarity, and the representatives of the top-ranking 12 clusters along with compound S5 of Amaro et al. (31) were tested experimentally using a fluorescence-based RNA-editing assay (49). Interestingly, the assays proved compounds S5 and C35—which is the same as compound V2 in Durrant et al. (47) (see Note 19)—as well as two other NCI compounds to be inhibitors.

Two other studies (50, 51) used fragment-based approaches to optimize membrane permeability of two of the inhibitors identified by Amaro et al. (31), S5 and V2. One of these studies used a program called Autogrow (52) (see Note 20) to add molecular fragments to compound S5 and improve its predicted binding affinity (50). The second study used an algorithm called CrystalDock (see Note 21), which analyzes the microenvironment of a binding site and identifies the best potential fragments to bind based on an analysis database of publicly available protein–ligand complexes (51). CrystalDock analysis of *Tb*REL1 predicted that forming a composite compound of V2 and toluene would increase the binding affinity of compound V2. A more computationally expensive and more accurate method called independent-trajectories thermodynamic integration (IT-TI) was then used to support the binding affinity improvement to V2 provided by the toluene substituent (51). The candidate inhibitors identified in these studies await experimental confirmation.

4 Notes

1. Although we present the methods using the specific protein *Tb*REL1, many of the strategies described here are directly applicable to other protein targets as well.

2. This corresponds to residues 115, 263, and 314 for the 1XDN structure.
3. The relevant PdbIDs of the proteins are 1CKN, 2HVQ, and 2C5U, respectively.
4. While introducing a new ligand to the ATP-bound *Tb*REL1 crystal structure, ATP and magnesium ion should be deleted. After introducing the new ligand into the binding site in the predicted binding pose of a docking software, one should delete the water molecules that have a steric clash with this new ligand. If the ligand occupies a smaller space than ATP, then it might be necessary to introduce water molecules, e.g., using DOWSER.
5. A Web service for PROPKA can be found at http://propka.ki.ku.dk. Also, a PROPKA graphical user interface (53) for VMD is available at http://propka.ki.ku.dk/~luca/wiki/index.php/GUI_Web that extends and validates the PROPKA approach.
6. There is a useful Web service at http://kryptonite.nbcr.net/pdb2pqr (54) hosted by the National Biomedical Computation Resource (NBCR) for users who do not have local resources for PDB2PQR (55, 56) computations that use the PROPKA approach to predict protonation states of titratable residues.
7. Among the listed MD software packages, NAMD2, DESMOND, and GROMACS are freely available to academic users.
8. The optimized Autodock parameters for *Tb*REL1 can be obtained from Amaro et al. (31).
9. Any molecular docking program can be chosen to study the system of choice as long as a control docking experiment with known inhibitors proves to be successful.
10. Among these, Autodock, Autodock Vina, Dock, and FRED are freely available for academic users.
11. In Durrant et al. (47), RMSD-based clustering was performed using an RMSD cutoff of 0.085 Å on a subset of residues that line the ATP binding site, which are residues 87–90, 155–162, 207–209, 283–287, and 305–308.
12. Amaro et al. (31) use QR factorization to distill the entire ensemble of 400 receptor structures to a nonredundant set of 33 structures (31).
13. Among these databases, several small-sized compound databases are freely available (excluding shipping costs) to academics from NCI/DTP (http://dtp.nci.nih.gov/branches/dscb/repo_open.html) upon project approval and could be a good starting point for different projects.
14. Many physicochemical properties can be predicted by programs like Schrödinger.

15. Lipinski's rule of five: The compound should have (a) at most five hydrogen-bond donors, (b) at most 10 hydrogen-bond acceptors, (c) at most 500 Da of molecular mass, (d) an octanol–water partition coefficient (logP) of not more than 5. Jorgensen's rule of three: The compound should have (a) logS greater than −5.7, (b) PCaco greater than 22 nm/s, and (c) fewer than seven metabolites. To filter compounds for drug-likeness, either or both criteria sets can be strictly enforced, or a looser filtering can be performed allowing up to a certain number of violations of these criteria.
16. In the case of *Tb*REL1, bacteriophage T4 RNA ligase 2 (T4Rnl2) (57, 58) and human DNA ligase IIIβ (HsLigIIIβ) (59) are used to check for selectivity of identified inhibitors.
17. The core scaffold of three of the previously identified inhibitors, V1, S1, and S5 (31), is 4,5-dihydroxynaphthalene-2,7-disulfonate—structure A in Durrant et al. (47). Similarity searches were performed using three structures similar to this core scaffold: naphthalene-2-sulfonic acid, 2-naphthoic acid, and 2-nitronaphthalene—structures B, C, and D in Fig. 1 of Durrant et al. (47), respectively.
18. Comparing the crystal structure with the best-scoring MD-generated receptor structures for each of the four low-micromolar inhibitors revealed that the crystallographic position of E60 in the active site was the reason for poor binding scores obtained for the crystal structure (47). During MD, conformation of residue E60 changes, opening up a new cleft that is entirely absent in the crystal structure. This new cleft, lined by residues I59-E60-I61-D62, has contacts with the identified inhibitors (47). The corresponding residues in human DNA ligase—(PdbID:1X9N) M543-L544-A545-H546—are significantly different and present an opportunity to design selective inhibitors for trypanosomes (47).
19. The binding pose of C35 (or V2) to the *Tb*REL1 crystal structure in Moshiri et al. (48) differs from the binding pose of V2 to the MD-generated frames in Durrant et al. (47) owing to the relative positions of E60 and R111.
20. The Autogrow program is freely available at http://autogrow.ucsd.edu.
21. The CrystalDock program is freely available at http://www.nbcr.net/crystaldock.

Acknowledgments

This work was funded in part by the National Institutes of Health through the NIH Director's New Innovator Award Program DP2-OD007237 to R.E.A.

References

1. Carnes J, Stuart K (2008) Working together: the RNA editing machinery in Trypanosoma brucei. In: Göringer HU (ed) RNA editing, vol 20. Nucleic acids and molecular biology (Gross HG, ed). Springer, Berlin/Heidelberg, pp 143–164. doi: 10.1007/978-3-540-73787-2_7
2. Ochsenreiter T, Hajduk S (2008) The function of RNA editing in trypanosomes. In: Göringer HU (ed), RNA editing, vol 20. Nucleic acids and molecular biology (Gross HG, ed). Springer, Berlin/Heidelberg, pp 181–197. doi: 10.1007/978-3-540-73787-2_9
3. Golas MM, Bohm C, Sander B et al (2009) Snapshots of the RNA editing machine in trypanosomes captured at different assembly stages in vivo. EMBO J 28:766–778. doi:10.1038/emboj.2009.19
4. Schnaufer A, Ernst NL, Palazzo SS et al (2003) Separate insertion and deletion subcomplexes of the Trypanosoma brucei RNA editing complex. Mol Cell 12:307–319. doi:S1097276503002867
5. Schnaufer A, Wu M, Park YJ et al (2010) A protein-protein interaction map of trypanosome 20S editosomes. J Biol Chem 285:5282–5295. doi:M109.059378
6. Rusché LN, Huang CE, Piller KJ et al (2001) The two RNA ligases of the Trypanosoma brucei RNA editing complex: cloning the essential band IV gene and identifying the band V gene. Mol Cell Biol 21:979–989. doi:10.1128/MCB.21.4.979-989.2001
7. Schnaufer A, Panigrahi AK, Panicucci B et al (2001) An RNA ligase essential for RNA editing and survival of the bloodstream form of Trypanosoma brucei. Science 291:2159–2162. doi:10.1126/science.1058955
8. Shuman S, Lima CD (2004) The polynucleotide ligase and RNA capping enzyme superfamily of covalent nucleotidyltransferases. Curr Opin Struct Biol 14:757–764. doi:10.1016/j.sbi.2004.10.006
9. Deng J, Schnaufer A, Salavati R et al (2004) High resolution crystal structure of a key editosome enzyme from Trypanosoma brucei: RNA editing ligase 1. J Mol Biol 343:601–613. doi:10.1016/j.jmb.2004.08.041
10. Amaro RE, Swift RV, McCammon JA (2007) Functional and structural insights revealed by molecular dynamics simulations of an essential RNA editing ligase in Trypanosoma brucei. PLoS Negl Trop Dis 1:e68. doi:10.1371/journal.pntd.0000068
11. Cherepanov AV, de Vries S (2002) Kinetic mechanism of the Mg2+-dependent nucleotidyl transfer catalyzed by T4 DNA and RNA ligases. J Biol Chem 277:1695–1704. doi:10.1074/jbc.M109616200
12. Håkansson K, Doherty AJ, Shuman S, Wigley DB (1997) X-ray crystallography reveals a large conformational change during guanyl transfer by mRNA capping enzymes. Cell 89:545–553. doi:10.1016/S0092-8674(00)80236-6
13. Nandakumar J, Shuman S, Lima CD (2006) RNA ligase structures reveal the basis for RNA specificity and conformational changes that drive ligation forward. Cell 127:71–84. doi:10.1016/j.cell.2006.08.038
14. El Omari K, Ren J, Bird LE et al (2006) Molecular architecture and ligand recognition determinants for T4 RNA ligase. J Biol Chem 281:1573–1579. doi:10.1074/jbc.M509658200
15. Zhang L, Hermans J (1996) Hydrophilicity of cavities in proteins. Proteins 24:433–438. doi:10.1002/(SICI)1097-0134(199604)24:4<433::AID-PROT3>3.0.CO;2-F
16. Vriend G (1990) WHAT IF: a molecular modeling and drug design program. J Mol Graph 8(52–56):29
17. Bas DC, Rogers DM, Jensen JH (2008) Very fast prediction and rationalization of pKa values for protein-ligand complexes. Proteins 73:765–783. doi:10.1002/prot.22102
18. Li H, Robertson AD, Jensen JH (2005) Very fast empirical prediction and rationalization of protein pKa values. Proteins 61:704–721. doi:10.1002/prot.20660
19. Olsson MH, Sondergaard CR, Rostkowski M, Jensen JH (2011) PROPKA3: consistent treatment of internal and surface residues in empirical pKa predictions. J Chem Theory Comput 7:525–537
20. Phillips JC, Braun R, Wang W et al (2005) Scalable molecular dynamics with NAMD. J Comput Chem 26:1781–1802. doi:10.1002/jcc.20289
21. Swift RV, Durrant J, Amaro RE, McCammon JA (2009) Toward understanding the conformational dynamics of RNA ligation. Biochemistry 48:709–719. doi:10.1021/bi8018114
22. Case DA, Cheatham TE 3rd, Darden T et al (2005) The Amber biomolecular simulation programs. J Comput Chem 26:1668–1688. doi:10.1002/jcc.20290
23. Hess B, Kutzner C, Van Der Spoel D, Lindahl E (2008) GROMACS 4: algorithms for highly efficient, load-balanced, and scalable molecular simulation. J Chem Theory Comput 4:435–447
24. Bowers KJ, Chow E, Xu H et al (2006) Scalable algorithms for molecular dynamics simulations on commodity clusters. In: International conference for high performance computing,

networking, storage and analysis (SC06), 11–17 Nov 2006, Tampa, FL
25. Christen M, Hunenberger PH, Bakowies D et al (2005) The GROMOS software for biomolecular simulation: GROMOS05. J Comput Chem 26:1719–1751. doi:10.1002/jcc.20303
26. MacKerell AD, Bashford D, Bellott M et al (1998) All-atom empirical potential for molecular modeling and dynamics studies of proteins. J Phys Chem B 102:3586–3616
27. Hornak V, Abel R, Okur A et al (2006) Comparison of multiple amber force fields and development of improved protein backbone parameters. Proteins 65:712–725. doi:10.1002/Prot.21123
28. Wang JM, Wolf RM, Caldwell JW et al (2004) Development and testing of a general amber force field. J Comput Chem 25:1157–1174
29. Meagher KL, Redman LT, Carlson HA (2003) Development of polyphosphate parameters for use with the AMBER force field. J Comput Chem 24:1016–1025. doi:10.1002/Jcc.10262
30. Oelschlaeger P, Klahn M, Beard WA et al (2007) Magnesium-cationic dummy atom molecules enhance representation of DNA polymerase beta in molecular dynamics simulations: improved accuracy in studies of structural features and mutational effects. J Mol Biol 366:687–701. doi:10.1016/J.Jmb.2006.10.095
31. Amaro RE, Schnaufer A, Interthal H et al (2008) Discovery of drug-like inhibitors of an essential RNA-editing ligase in Trypanosoma brucei. Proc Natl Acad Sci USA 105:17278–17283. doi:10.1073/pnas.0805820105
32. Morris GM, Huey R, Lindstrom W et al (2009) AutoDock4 and AutoDockTools4: Automated docking with selective receptor flexibility. J Comput Chem 30:2785–2791. doi:10.1002/jcc.21256
33. Trott O, Olson AJ (2010) AutoDock Vina: improving the speed and accuracy of docking with a new scoring function, efficient optimization, and multithreading. J Comput Chem 31:455–461. doi:10.1002/jcc.21334
34. Lang PT, Brozell SR, Mukherjee S et al (2009) DOCK 6: combining techniques to model RNA-small molecule complexes. RNA 15:1219–1230. doi:rna.1563609
35. OpenEye Scientific Software I (2010) OEChem, Santa Fe, NM
36. McGann MR, Almond HR, Nicholls A et al (2003) Gaussian docking functions. Biopolymers 68:76–90. doi:10.1002/Bip. 10207
37. Jones G, Willett P, Glen RC et al (1997) Development and validation of a genetic algorithm for flexible docking. J Mol Biol 267:727–748. doi:10.1006/jmbi.1996.0897
38. Halgren TA, Murphy RB, Friesner RA et al (2004) Glide: a new approach for rapid, accurate docking and scoring. 2. Enrichment factors in database screening. J Med Chem 47:1750–1759. doi: 10.1021/jm030644s
39. Friesner RA, Banks JL, Murphy RB et al (2004) Glide: a new approach for rapid, accurate docking and scoring. 1. Method and assessment of docking accuracy. J Med Chem 47:1739–1749. doi:10.1021/jm0306430
40. Jain AN (2007) Surflex-Dock 2.1: robust performance from ligand energetic modeling, ring flexibility, and knowledge-based search. J Comput Aided Mol Des 21:281–306. doi:10.1007/s10822-007-9114-2
41. Venkatachalam CM, Jiang X, Oldfield T, Waldman M (2003) LigandFit: a novel method for the shape-directed rapid docking of ligands to protein active sites. J Mol Graph Model 21:289–307. doi:10.1016/S1093-3263(02)00164-X
42. Oprea TI, Matter H (2004) Integrating virtual screening in lead discovery. Curr Opin Chem Biol 8:349–358. doi:10.1016/j.cbpa.2004.06.008
43. Carlson HA (2002) Protein flexibility and drug design: how to hit a moving target. Curr Opin Chem Biol 6:447–452. doi:S1367593102003411
44. Lin JH, Perryman AL, Schames JR, McCammon JA (2002) Computational drug design accommodating receptor flexibility: the relaxed complex scheme. J Am Chem Soc 124:5632–5633. doi:ja0260162
45. Amaro RE, Baron R, McCammon JA (2008) An improved relaxed complex scheme for receptor flexibility in computer-aided drug design. J Comput Aided Mol Des 22:693–705. doi:10.1007/s10822-007-9159-2
46. Ryan AJ, Gray NM, Lowe PN, Chung CW (2003) Effect of detergent on "promiscuous" inhibitors. J Med Chem 46(16):3448–3451. doi:10.1021/jm0340896
47. Durrant JD, Hall L, Swift RV et al (2010) Novel naphthalene-based inhibitors of Trypanosoma brucei RNA editing ligase 1. PLoS Negl Trop Dis 4:e803. doi:10.1371/journal.pntd.0000803
48. Moshiri H, Acoca S, Kala S et al (2011) Naphthalene-based RNA editing inhibitor blocks RNA editing activities and editosome assembly in Trypanosoma brucei. J Biol Chem 286:14178–14189. doi:10.1074/jbc.M110.199646
49. Moshiri H, Salavati R (2010) A fluorescence-based reporter substrate for monitoring RNA editing in trypanosomatid pathogens. Nucleic Acids Res 38:e138. doi:10.1093/nar/gkq333
50. Durrant JD, McCammon JA (2011) Towards the development of novel Trypanosoma brucei

RNA editing ligase 1 inhibitors. BMC Pharmacol 11:9. doi:10.1186/1471-2210-11-9

51. Durrant JD, Friedman AJ, McCammon JA (2011) CrystalDock: a novel approach to fragment-based drug design. J Chem Inf Model 51:2573–2580. doi:10.1021/ci200357y
52. Durrant JD, Amaro RE, McCammon JA (2009) Autogrow: a novel algorithm for protein inhibitor design. Chem Biol Drug Des 73:168–178. doi:10.1111/j.1747-0285.2008.00761.x
53. Rostkowski M, Olsson MH, Sondergaard CR, Jensen JH (2011) Graphical analysis of pH-dependent properties of proteins predicted using PROPKA. BMC Struct Biol 11:6. doi:10.1186/1472-6807-11-6
54. Unni S, Huang Y, Hanson RM et al (2011) Web servers and services for electrostatics calculations with APBS and PDB2PQR. J Comput Chem 32:1488–1491. doi:10.1002/jcc.21720
55. Dolinsky TJ, Czodrowski P, Li H et al (2007) PDB2PQR: expanding and upgrading automated preparation of biomolecular structures for molecular simulations. Nucleic Acids Res 35 (Web Server issue):W522–W525. doi: 10.1093/nar/gkm276
56. Dolinsky TJ, Nielsen JE, McCammon JA, Baker NA (2004) PDB2PQR: an automated pipeline for the setup of Poisson-Boltzmann electrostatics calculations. Nucleic Acids Res 32 (Web Server issue):W665–W667. doi: 10.1093/nar/gkh381
57. Ho CK, Wang LK, Lima CD, Shuman S (2004) Structure and mechanism of RNA ligase. Structure 12:327–339. doi:10.1016/j.str.2004.01.011
58. Yin S, Ho CK, Shuman S (2003) Structure-function analysis of T4 RNA ligase 2. J Biol Chem 278:17601–17608. doi:10.1074/jbc.M300817200
59. Tomkinson AE, Vijayakumar S, Pascal JM, Ellenberger T (2006) DNA ligases: structure, reaction mechanism, and function. Chem Rev 106:687–699. doi:10.1021/cr040498d

Chapter 16

Computational Models for Tuberculosis Drug Discovery

Sean Ekins and Joel S. Freundlich

Abstract

The search for small molecules with activity against *Mycobacterium tuberculosis* increasingly uses high-throughput screening and computational methods. Previously, we have analyzed recent studies in which computational tools were used for cheminformatics. We have now updated this analysis to illustrate how they may assist in finding desirable leads for tuberculosis drug discovery. We provide our thoughts on strategies for drug discovery efforts for neglected diseases.

Key words Bayesian models, Collaborative Drug Discovery Tuberculosis database, Docking, *Mycobacterium tuberculosis*, Quantitative structure–activity relationship, Tuberculosis

1 Introduction

Between 2007 and the end of 2011, only six new molecular entities were approved as antibiotics by the Food and Drug Administration in the United States (Table 1). On March 8, 2012, the Infectious Diseases Society of America (IDSA) proposed to a US House of Representatives subcommittee significant alterations in the drug approval process to address a dire healthcare need for new anti-infective agents (1). This follows other responses (3) seeking commitment for 10 new antibacterial drugs by 2020 (2) and other campaigns (3). However, the specter of drug resistance highlights the need for new therapeutic agents for which cross-resistance to existing treatments is nonexistent (4).

In order to deliver novel antibacterial agents, many strategies have been explored, despite the plummet in pharmaceutical research and less than optimal market opportunities (5). High-throughput screening (HTS) approaches focused on specific targets, informed by genomics, have yet to meet the need (6). Other efforts have involved structure-based design of inhibitors for a single target pathogen (7–10). Fischbach and Walsh highlighted the potential for novel chemical scaffolds in the world of antibacterials

Sandhya Kortagere (ed.), *In Silico Models for Drug Discovery*, Methods in Molecular Biology, vol. 993,
DOI 10.1007/978-1-62703-342-8_16, © Springer Science+Business Media, LLC 2013

Table 1
New molecular entities approved by the FDA since 2007

Antibiotic	Indication	Manufacturer	Date of FDA approval
Fidaxomicin	*Clostridium difficile*-associated diarrhea in adults (≥ 18 years of age)	Optimer Pharmaceuticals (San Diego, CA)	5/27/2011
Ceftaroline fosamil for injection	Acute bacterial skin and skin structure infections and community-acquired pneumonia	Cerexa, Forest Laboratories (Oakland, CA)	10/29/2010
Besifloxacin	Bacterial conjunctivitis	Bausch and Lomb (Rochester, NY)	5/28/2009
Telavancin	Complicated skin and skin structure infections	Theravance (South San Francisco, CA)	9/11/2009
Retapamulin	Treatment of impetigo	GlaxoSmithKline (Philadelphia, PA)	4/12/2007
Doripenem	Complicated intra-abdominal and urinary tract infections caused by susceptible isolates of the designated microorganisms	Johnson & Johnson (New Brunswick, NJ)	10/12/2007

FDA US Food and Drug Administration

to move beyond β-lactams and quinolones (11). These popular chemotypes have been optimized exhaustively, and while successful to some extent, the efforts have not afforded new classes of antibacterials with novel mechanisms of action essential to combat infections resistant to all current drug classes.

The study of *Mycobacterium tuberculosis* (Mtb), the causative agent of tuberculosis (TB), which infects approximately one third of the world's population and causes 1.7–1.8 million deaths annually (12), provides a clear example of our inability to discover new antibiotics. Agents active against Mtb are urgently needed to combat this global epidemic, which is heavily influenced by resistance to the available drugs, the lengthy treatment regimen, and coinfection with HIV (13, 14). A new antibiotic to address TB has not been developed in more than 40 years. We have seen in recent years massive phenotypic screening efforts of commercial vendor libraries looking for compounds that inhibit the growth of Mtb (14–17). These compounds predominantly broadly sample libraries of small molecules, and the hit rate of these screens tends to be in the low single digits (Table 2) (15–17) if not below 1% as seen elsewhere (18). Few such in vitro HTS screens use past knowledge of Mtb-active compounds to focus their screening. Leveraging such prior knowledge to produce computational models or rules for the virtual screening of compound libraries is an approach that is used elsewhere in the pharmaceutical industry in parallel or prior

Table 2
Published hit rates calculated from Southern Research Institute publications

Library size	Number of hits	Hit rate (%)	Notes	References
100,997	1,782	1.76	Diverse library	Ananthan et al. (15)
215,110	3,817	1.77	Diverse library	Maddry et al. (16)
25,671	1,329	5.18	Human kinase-focused library	Reynolds et al. (17)

to HTS to improve the efficiency of screening (19). Various computational and statistical analyses can provide insights into the physiochemical properties of small molecules that are important for activity against Mtb on the small-scale (20) as well as for very large datasets (21–23). Ligand-based (24) and protein-based (25) studies have used filters before or after a biological screening in order to identify molecules with an optimal set of physiochemical properties, e.g., hit-like, lead-like, or drug-like (26).

While there continue to be important reviews of bioinformatics as applied to tuberculosis, they tend to only briefly mention cheminformatics approaches (27). We suggest that cheminformatics can have a disproportionate impact if used widely in this and other neglected diseases. This chapter extends our recent review of computational tools for TB (28), by highlighting some of the recent ligand and structure-based studies performed for TB drug discovery, and describes how we see these approaches being used more widely for facilitating research on this and other neglected diseases.

2 Ligand-Based Methods

Ligand-based approaches towards TB drug discovery have consisted primarily of quantitative structure–activity relationship (QSAR), three-dimensional (3D)-QSAR, and pharmacophore models. Usually a computational model is generated using commercial software, and then testing is performed by leaving out one or more groups of compounds at random. Occasionally, an external test set generated after model building is used (Table 3) (29–39). Most such models are "local," focused on optimizing activity against a specific target or starting hit or lead (Table 3).

If molecules are found to interact with a specific protein, numerous analogues can be used along with their 3D conformations to generate 3D-QSAR (Table 4) (40–47). These usually consist of 21 to about 100 molecules for structurally related compounds; then external testing uses between <10 and <30 compounds. 3D-QSAR is rarely used for virtual screening studies. 3D-QSAR

Table 3
Quantitative structure–activity relationship studies

Compound types	Number of molecules in training set	Number of descriptors used	Algorithm used and testing	References
Pyrazinoate esters	32	43	Genetic function approximation models; c log P was a key descriptor; the model was tested with 11 external compounds	Fernandes et al. (29)
N-Benzylsalicylthioamides	29	177	Two MLR models for TB with the STATOO program[a]; c log P was a key descriptor, while there was no external testing	Dolezal et al. (30)
Ring-substituted-2/4-quinolinecarbaldehyde derivatives	24		PCA analysis; inclusion of log P did not improve model statistics. Actives appeared clustered in a small region of PCA plot	Nayyar et al. (31)
5-Aryl-2-thio-1,3,4-oxadiazoles	41	Topological descriptors	Neural networks ($q^2 = 0.8$), not tested externally	Macaev et al. (32)
Hydrazides	173	Abraham's descriptors; electronic, geometrical, or steric descriptors	MLR subsets were used for modeling. Hydrophobicity could not explain the biological response. For small subsets there were good correlations with test sets ($R^2 > 0.77$)	Ventura and Martins (33)

Isoniazid derivatives	91	HQSAR and DRAGON descriptors	Used HQSAR and generated a test set ($R^2 = 0.87$) for 24 compounds. The results were better than for PLS-QSAR with 2D descriptors from DRAGON ($R^2 = 0.72$)	Andrade et al. (34)
Chalcones and flavonoids	9–33	48	Genetic function approximation; internally cross-validated ($q^2 = 0.79–0.94$)	Sivakumar et al. (35)
Diverse	Three TAACF datasets from PubChem	179	Naïve Bayes, random forest, sequential minimal optimization, J48 decision tree. Used to create three models with different datasets. Naïve Bayes had external test set accuracy 73–82.7, random forest 60.7–82.7%, SMO 55.9–83.3, and J48 61.3–80%	Periwal et al. (36, 37)
2-(Substituted benzyl) sulfanyl benzimidazoles, benzoxazoles, benzothiazoles	29	9	MLR, NLM, and PCA were used to show derivatives with high polarizability and low lipophilicity, and electron-withdrawing substituents have higher activity. No cross-validation reported	Pytela and Klimesova (38)
Diverse TB drugs and natural products	24	42	Forward feed MLR gave a cross-validated $R = 0.72$ with dielectric energy and heats of formation contributing to activity	Dwivedi et al. (39)

HQSAR hologram quantitative structure–activity relationship, *MLR* multiple linear regression, *PCA* principal component analysis, *PLS-QSAR* partial least-squares quantitative structure–activity relationship, *TAACF* Tuberculosis Antimicrobial Acquisition and Coordinating Facility

[a]STATOO Consulting (Berne, Switzerland)

Table 4
Three-dimensional quantitative structure–activity relationship models

Compound types	Number of molecules in training set	Algorithm used	Statistics	References
1,4-Dihydropyridines	35	CoMFA and CoMSIA	Cross-validated ($R^2 = 0.56$ and 0.62, respectively) as well as external validation ($R^2 = 0.74$ and 0.69, respectively)	Manvar et al. (40)
Diaryloxymethano-phenanthrene derivatives	37	CoMFA and CoMSIA	CoMFA ($q^2 = 0.625$) and CoMSIA ($q^2 = 0.486$) models and 7-compound external test set with very good predictive value	Shagufta et al. (41)
Deoxythymidine monophosphate derivatives that inhibit thymidine monophosphate kinase	36	MFA	Alignments performed with least-squares (predictive $R^2 = 0.70$), pharmacophore (0.56), or docked conformations (0.72). Receptor-based alignment performed best	Aparna et al. (42)
Nitrofuranyl derivatives	95	CoMFA and CoMSIA	Tested with a set of 15 molecules. CoMFA ($R^2 = 0.78$) outperformed CoMSIA. c Log P and polar surface area or steric bulk did not improve the models	Hevener et al. (43)
4-Adamantan-1-yl-quinoline-2-carboxylic acid alkylidene hydrazides	30	CoMFA and CoMSIA	Models tested with 14 molecules CoMFA ($R^2 = 0.49$) and CoMSIA ($R^2 = 0.49$)	Nayyar et al. (44)
Ring-substituted quinolines	70	CoMFA and CoMSIA	Tested with 24 molecules. With the CoMFA model $R^2 = 0.42$. 18 molecules were suggested for synthesis based on the CoMFA predictions	Nayyar et al. (45)
Nitroimidazoles	21	Catalyst pharmacophore	Tested with 22 molecules. No test set correlation value reported but correlation looked similar to the training set ($R = 0.96$)	Kim et al. (46)
1,5-Diarylpyrrole derivatives		Catalyst pharmacophore	Had difficulty predicting *N*-methylpiperazine and thiomorpholine derivatives. No numerical prediction data were presented	Biava et al. (47)

CoMFA comparative molecular field analysis, *CoMSIA* comparative molecular similarity indices analysis methods, in general, are dependent on the molecule conformation, force fields, and the active compounds selected to build the model (48).

Machine learning and classification methods have been used sparingly for TB drug discovery (49). Such methods are valuable in the rapid virtual screening of compound libraries for novel active compounds (50, 51) and generally use much larger training sets. Classification methods can also be used as local models for lead optimization with smaller datasets (52). The value of such general classification models has been illustrated with several recent studies using much larger datasets, such as the 3,770 compounds used to build Bayesian classification models with extended class fingerprints. This model was tested on a dataset of 2,880 compounds (with activity against *Mtb*) with accuracy >70% and was used to screen the ZINC database, ultimately suggesting that 4 compounds be prioritized for future testing (24). To date we have not seen a publication that clarifies whether these predicted compounds were active.

Inspired by this work (24), we used Bayesian methods (24) with molecular function class fingerprints of maximum diameter 6 (53) to identify substructures that were shown to be important in recent TB screening datasets (21–23). Bayesian models were built with the previously described Molecular Libraries Small Molecule Repository (MLSMR) 220,463 library (4,096 active compounds) (15) and dose–response data using 2,273 molecules (475 active compounds). In addition, these models were tested (15) with the National Institute of Allergy and Infectious Diseases (NIAID) data and GVK Biosciences (Hyderabad, India) datasets used by Prathipati et al. (24). We have validated the models with compounds left out of the original models, in some cases showing up to tenfold enrichments in finding active compounds in the top-ranked 600 molecules (22).

We have also used a set of compounds from the Novartis (Basel, Singapore) aerobic assay as a test set for these models (22, 23) to provide further validation of the approach, using the published data from a different group. More than 35% of the total hits were found with the dose–response model in the top 8% of molecules (enrichment greater than fourfold) (21). A further validation used data from a study containing a set of 1,514 known drugs (some US FDA approved and some approved by other governments) screened against Mtb; the minimum inhibitory concentration (MIC) values were determined using the Alamar blue susceptibility assay (54). Twenty compounds were identified that had known antitubercular activity, and 18 novel hits were found. Twenty-one of these active compounds were seeded in a larger set of 2,108 FDA-approved molecules and used as a test set. Both Bayesian models used initially had ~ tenfold enrichments with >60% of the active compounds in the top 14% (21).

Recent work has also used these models as part of a strategy selecting several Mtb targets, metabolites, and their pathways using a combined cheminformatics and bioinformatics approach in CDD (55, 56). This approach leverages an important effort to classify genes (and by extension, in many cases, the corresponding proteins and enzyme substrates/products) as essential or nonessential under a range of in vitro and in vivo conditions that model various aspects of human infection (57–59). Under a set of in vitro conditions, transposon mutagenesis data have been harnessed to afford a set of essential metabolites for Mtb (56). The question was posed as to how to best use this set of essential metabolites to afford novel chemical probes (60) and drug leads. The hypothesis was put forth that essential metabolite mimics could bind to the respective essential enzyme in place of the essential metabolite (substrate/product) and afford competitive inhibition of catalytic activity and Mtb stasis or death. Cheminformatics has been the key methodology used thus far in attempting translation from essential metabolite to useful small-molecule tool.

For example, pharmacophore models for a subset of essential substrates in TB were used to screen more than 80,000 commercial compounds. The combined 842 molecules retrieved using pharmacophores to search vendor libraries were run through the SMARTS filters and Bayesian models for whole-cell TB activity in Discovery Studio (22, 61, 62), and 234 were flagged as failing the SMARTS filters. All compounds were imported into the CDD database. The molecules were then sorted to focus on those passing SMARTS with molecular weight 280–430 Da, log P 3–5, polar surface area (PSA) 50–100 $Å^2$, Bayesian score in the single-point model >0.3, Bayesian score in the dose–response model >1.37, and Bayesian score in the Novartis model >1.11, signified predicted activity. A set of 60 molecules was then sorted based on these Bayesian score dose–response cutoffs (including compounds with data from public datasets from Southern Research Institute) and exported to Excel before further filtering to manually exclude those already tested in public data. We also included three examples of compounds that had poor physicochemical properties to further illustrate the importance of hydrophobicity on permeability and TB activity. Twenty-three compounds for this study were imported into the CDD database private vault (Bayesian score dose–response model, range 1.6–11.8). Two suggested mimics of 2 D-fructose 1,6 bisphosphate had MIC values of 20 and 40 μg/mL, respectively, representing an ~10% hit rate, which is also higher than HTS hit rates (frequently <1%) (63, 64).

These results indicate that classification methods could be very effective computational filters if used prior to experimental testing. However, what is apparent from many of the published studies is that prospective use and follow-up testing of compounds suggested is very limited to date.

3 Structure-Based Methods

There are many papers that use computer-generated images of *Mtb* protein binding sites to highlight interactions with the ligand/inhibitor, while there is little published that suggests that such efforts have been used to help design compounds prior to testing (65). Recently we have seen that docking can be used to filter potential compounds for testing against Mtb proteins, as it has been useful outside the TB field (66).

Analyzing recent publications from 2007 to 2012, we found that docking has been used as a virtual screening strategy to identify small molecules with potency against a given *Mtb* target as well as other interpretive uses. Many use docking methods preceded by some form of computational filtering of screening libraries (Table 5) (67–89). Several of these studies have suggested compounds for testing that have been validated, although in several cases we have yet to see predictions verified (90).

A recent example includes the use of FRIGATE docking software to dock more than 2 million compounds from ZINC against the Mtb protein Ag85C. After filtering for favorable solubility properties, the investigators reduced the database by half and selected the 60 top-ranking compounds, ultimately resulting in 23 that were tested and one compound with an MIC of 20 μg/mL against a multidrug-resistant strain of Mtb (91). More often a homology model is used to dock a lead compound to rationalize binding and the SAR. For example, the induced-fit docking protocol from Schrödinger was used to dock a lead compound into an Mt-GuaB2-IMP homology model (92).

Docking combined with other virtual screening and hybrid-type approaches has given promising results (Table 5) yet is still not widely used by all research groups that are involved in HTS efforts in the TB field. We see these approaches as inextricably linked and suggest their use as a complementary strategy that could help inform or enrich whole-cell screening.

4 Discussion

The large-scale whole-cell screening efforts have created large training sets for TB models such as the Bayesian models described by us and others. The caveat with these models is they are likely limited to predicting compound activity under the exact in vitro conditions used (15). However, by considering data from different groups, we have shown that this may not be a major issue (21). Hence such models can be used to filter vendor libraries of molecules and natural products (21).

Analysis of several datasets representing thousands of active compounds suggested that the mean value for various simple

Table 5
Docking and other virtual screening methods

Method	Results	References
Homology models of DevR and pharmacophore model used to screen 2.5 million compounds, followed by docking with MOE[a] and GOLD[b]	Resulted in 11 compounds screened and four hits, including a phenylcoumarin derivative	Gupta et al. (67)
37 Enoyl-acyl carrier protein reductase carboxamide inhibitors were used to build CoMFA model (tested with ten compounds, $R^2=0.88$), followed by the de novo molecule design software LEAPFROG[c]	Suggested 13 molecules with improved binding energy values; however, these have not been synthesized or tested	Kumar et al. (68)
29 Enoyl-acyl carrier protein reductase arylamide inhibitors were used to build CoMFA and CoMSIA models (tested with eight molecules, $R^2>0.87$). A pharmacophore was also used to screen the Maybridge database to retrieve 996 hits, which were then docked with FlexX[d]	The CoMFA and CoMSIA scores were used to suggest 20 molecules for future testing	Kumar et al. (69)
Docking and pharmacophore approach used to suggest type II dehydroquinase inhibitors, starting from 45 published inhibitors used to test docking approach and generate GA-MLR QSAR model (35 train, ten test) using MOE[a] QuaSAR Evolution (q^2 test and train >0.95). The most active was used for FlexX[d] pharmacophore generation. Also looked at interaction fingerprints	Predicted 42 active compounds—no test data	Kumar et al. (70)
Combined experimental and computational approach with 12 new imidazoles and triazole derivatives using AutoDock[e] to dock molecules in sterol 14α-demethylase followed by free energy of binding calculations	Good agreement between calculated ΔG_{bind} and experimental data for MIC	Banfi et al. (71)
Thirty 5′-thiourea-substituted α-thymidine analogues used to develop receptor-independent 4D-QSAR models ($q^2=0.83$) for thymidine monophosphate kinase inhibitors. The model was also put into the context of reported crystallographically characterized inhibitor/enzyme interactions	The model was tested with four compounds and three were predicted within the SD of the assay. Activity also increased with log P	Andrade et al. (72)
Thirty-one 5′-O-(*N*-(salicyl)sulfamoyl)adenosine inhibitors of MbtA (a salicyl AMP ligase) used with molecular dynamics simulations in a homology model to calculate linear interaction energy ($R^2=0.70$)	A single validation molecule was predicted with the LIE models to have a K_i of 1.6 nM, and the actual value was 0.7 nM	Labello et al. (73)

Docking and molecular dynamics were used to study the binding of the isoniazid metabolite INH-NAD to the enoyl-acyl carrier protein reductase	Suggested the role of a water molecule in binding. The modeling supported the role of KatG prior to InhA binding.	Wahab et al. (74)
FlexX[d] and GOLD[b] were used to virtually screen the ChemBridge and NCI databases (covering over half a million compounds) against the ATP phosphoribosyl transferase (HisG). Filtering for drug "likeness" was also used	50 compounds were tested in vitro, and seven were active at 10 μM. Nitrobenzothiazoles were identified as active and co-crystallized and 19 follow-up compounds found in the ChemBridge database (two of which showed inhibition in the target and whole-cell assays)	Cho et al. (25)
UNITY pharmacophore[c], FlexX[d] docking, and structure interaction fingerprint approaches were used to identify compounds in the Maybridge database (59,275 compounds) as potential thymidine monophosphate kinase inhibitors	Ten compounds were ultimately selected, and five compounds showed MIC < 12.5 μg/mL in whole-cell assays with no cytotoxicity, while the binding of these compounds to enzyme remained to be demonstrated	Kumar et al. (75)
CDOCKER[f] used to dock tripeptides into the TB DHFR crystal structure. Molecular dynamics simulation was also performed	WYY was predicted as potent and selective versus human DHFR. This prediction has yet to be verified	Kumar et al. (76)
FlexX[d] used for docking a library of >19,000 ViChem[g] compounds and Tripos[c] Leadquest compounds into NAD synthetase PknB	Nine submicromolar inhibitors were found. Additional further docking for NAD kinase inhibitors found that 22 showed activity versus NAD synthetase and one against NAD kinase out of 100 compounds tested	Hegymegi-Barakonyi et al. (77)
Catalyst HypoGen[f] pharmacophore and GOLD[b] docking were used to develop the composite model for screening potential thymidine monophosphate kinase inhibitors	Screened an in-house database of ~500,000 compounds, subsequently providing 186 virtual hits that do not appear to have been tested in vitro	Gopalakrishnan et al. (78)
ICM and DOCK[h] were used to virtually screen the University of California, Irvine, ChemDB database and NCI databases to identify AccD5 inhibitors	One ligand, NCI-65828, was found to inhibit AccD5 (an essential acyl-CoA carboxylase carboxyltransferase domain) competitively with an experimental K_i of 13.1 μM	Lin et al. (79)
AutoDock[e] was used for docking inhibitors to MshB (a GlcNAc-Ins deacetylase)	Docking used to explain mode of binding for inhibitors only	Metaferia et al. (80)

(continued)

Table 5 (continued)

Method	Results	References
AutoDock[e] and GOLD[b] were used to find inhibitors for the adenylation domain of the NAD^+-dependent ligase with bound AMP (LigA)	A novel class of inhibitors, glycosyl ureides, were identified to compete with the NAD^+. Five compounds with docking scores were tested in vitro versus LigA; no assessment of correlation	Srivastava et al. (81)
AutoDock[e] used for docking in GlmU	Compounds with 60% similarity to the GlmU in PubChem were docked as well as additional anti-infectives. Hybrid QSAR models were also created. 40 compounds were suggested for testing, but no experimental data were provided	Singla et al. (82)
Glide[i] docking in PknG	26 withaferin and 14 withanolide derivatives from PubChem along with commercially available drugs were docked. Withanolide D, E, and F predicted as binding but not validated in vitro	Santhi and Aishwarya (83)
CDOCKER[f] docking of isoniazid in AccD6	Activated isoniazid docked in homology model of AccD6. No experimental verification	Unissa et al. (84)
CDOCKER[f] docking of azoles in CYP121	357 analogues of azole drugs were docked in CYP121 structures. Five of the top 12 out of 53 compounds were ranked by two different scoring functions. No experimental verification was provided	Sundaramurthi et al. (85)
AutoDock[e] of eight methoxy fluoroquinolones against GyrA mutants	Docking confirmed earlier in vitro studies demonstrating that increased lipophilicity in turn increased binding affinity	Anand et al. (86)
Glide docking of *N*-methyl-2-alkenyl-4-quinolones with MurE ligase	The quinolones were predicted as weak binders, which corresponded to their IC_{50} 36–72 μM (phosphate method) and 95–207 μM (HPLC)	Guzman et al. (87)

AutoDock[c] of 251D in DNA polymerase III α-subunit	A homology model was used to dock a potent inhibitor of bacterial replicative DNA polymerase 251D, which suggested active site residues involved in the interaction	Chhabra et al. (88)
FRIGATE docking into Ag85C	Over two million compounds docked from ZINC with some molecule property filters; found one compound with MDR-Mtb MIC 20 μg/mL activity	Scheich et al. (91)
Induced-fit docking protocol from Schrödinger[i] used to dock into Mt-GuaB2-IMP homology model	Lead compound 7,759,844 K_i = 0.603 μM docked to show binding orientation and rationalize SAR	Usha et al. (92)
AutoDock4 and Dock6 used with a homology model of thiamin phosphate synthetase	NCI diversity set II docked in homology model. 39 compounds tested, 25 showed activity, and seven have >20% inhibition at 100 μg/mL. One compound also had an MIC_{99} of 6 μg/mL	Khare et al. (89)
Glide docking into L-aspartate α-decarboxylase	333,761 compounds including Maybridge, ZINC, NCI, and FDA drugs were docked in the crystal structure and then narrowed to 703 hits and further limited to 28 and then eight compounds. No experimental validation was reported	Sharma et al. (90)

AccD5 acyl-CoA carboxylase domain 5, *CoMFA* comparative molecular field analysis, *CoMSIA* comparative molecular similarity indices analysis, *DevR* dormancy regulon, *DHFR* dihydrofolate reductase, *GA-MLR* genetic algorithm–multiple linear regression, *HPLC* high-performance liquid chromatography; *KatG* peroxidase-peroxynitritase, *InhA* enoyl reductase from Mycobacterium tuberculosis, *INH* isoniazid, *LIE* linear interaction energy, *MDR* multidrug resistance, *MIC* minimum inhibitory concentration, *MOE* Molecular Operating Environment, *NCI* National Cancer Institute, *WYY* H–tryptophan–tyrosine–tyrosine–OH

[a]Molecular Operating Environment (Chemical Computing Group, Montreal, Canada)
[b]CCDC (Cambridge, UK)
[c]Tripos, Inc. (St. Louis, MO)
[d]BioSolveIT GmbH (Sankt Augustin, Germany)
[e]Scripps Research Institute (La Jolla, CA)
[f]Accelrys (San Diego, CA)
[g]ViChem (Budapest, Hungary)
[h]MolSoft (La Jolla, CA)
[i]Schrödinger (Portland, OR)

molecular descriptors, e.g., PSA, is statistically different from that of FDA-approved drugs (23). This analysis follows studies on molecular property values for antibiotics in general (93), including those that have evaluated and calculated/predicted hydrophobicity (c log P) and molecular mass (6), as well as earlier studies on antitubercular compounds (20). Generally, FDA-approved TB drugs are more like inhaled drugs (molecular weight mean 370, PSA 89.2 Å^2, c log P 1.7) (94). An initial analysis of the largest public screening sets (>300,000 compounds) to date using the MLSMR dataset (16) and the TAACF (Tuberculosis Antimicrobial Acquisition and Coordinating Facility)–NIAID-CB2 dataset (15) suggests that the molecular weight, log P, and rule-of-five alerts were statistically significantly higher in the most active compounds in the MLSMR screening data, whereas the PSA was slightly lower compared with the inactive compounds. The active compounds in the TAACF–NIAID–CB2 set have statistically higher mean log P and rule-of-five alerts (95) and also have lower hydrogen bond donor count, atom count, and PSA than inactive compounds (22). These types of physicochemical property and virtual screening insights help define the "*Mtb*-active compound" and can be used to design or select small-molecule libraries for whole-cell phenotypic screens and to efficiently guide medicinal chemistry optimization efforts to find compounds with activity in vitro.

Current studies ongoing with different research groups are using the Bayesian models to prioritize compounds prospectively for testing in whole-cell assays. Other collaborations are using docking into homology models and crystal structures of Mtb proteins to select vendor compounds for testing against the target itself. Combined, such cheminformatics approaches offer great promise to limit continued random HTS, such that selecting a fraction of a percent of the best scoring compounds will be more than enough to afford interesting leads in the majority of cases. Given limited financial resources and an increased urgency to discover novel antibiotics with new mechanisms of action, we should ensure that all TB groups have access to cheminformatics tools and databases (96, 97) to ensure that they test the best compounds. We should also apply such approaches to other neglected diseases, which have similarly limited budgets. Learning from prior data is possible and also cost-effective.

Acknowledgments

The CDD TB database was developed with funding from the Bill and Melinda Gates Foundation (grant #49852, "Collaborative Drug Discovery for TB through a Novel Database of SAR Data Optimized to Promote Data Archiving and Sharing"). We acknowledge our many collaborators.

References

1. Infectious Diseases Society of America (2012) Infectious Diseases Society of America's (IDSA) statement promoting anti-infective development and antimicrobial stewardship through the U.S. Food and Drug Administration Prescription Drug User Fee Act (PDUFA) reauthorization before the House Committee on Energy and Commerce Subcommittee on Health, 8 Mar 2012. http://www.idsociety.org/uploadedfiles/idsa/policy_and_advocacy/current_topics_and_issues/advancing_product_research_and_development/bad_bugs_no_drugs/statements/idsa%20pdufa%20gain%20testimony%20030812%20final.pdf
2. Infectious Diseases Society of America (2010) The 10 x '20 Initiative: pursuing a global commitment to develop 10 new antibacterial drugs by 2020. Clin Infect Dis 50: 1081–1083
3. Boucher HW, Talbot GH, Bradley JS et al (2009) Bad bugs, no drugs: no ESKAPE! An update from the Infectious Diseases Society of America. Clin Infect Dis 48:1–12
4. Nordberg, P., Monnet, D.L., Cars, O. (2005) Priority medicines for Europe and the world: a public health approach to innovation. Antibacterial drug resistance. Background document for the WHO project. World Health Organization. http://apps.who.int/medicinedocs/en/m/abstract/Js16368e/
5. Nathan C, Goldberg FM (2005) Outlook: the profit problem in antibiotic R&D. Nat Rev Drug Discov 4:887–891
6. Payne DJ, Gwynn MN, Holmes DJ, Pompliano DL (2007) Drugs for bad bugs: confronting the challenges of antibacterial discovery. Nat Rev Drug Discov 6:29–40
7. Liu N, Cummings JE, England K et al (2011) Mechanism and inhibition of the FabI enoyl-ACP reductase from Burkholderia pseudomallei. J Antimicrob Chemother 66:564–573
8. England K, am Ende C, Lu H et al (2009) Substituted diphenyl ethers as a broad-spectrum platform for the development of chemotherapeutics for the treatment of tularaemia. J Antimicrob Chemother 64:1052–1061
9. Xu H, Sullivan TJ, Sekiguchi J et al (2008) Mechanism and inhibition of saFabI, the enoyl reductase from Staphylococcus aureus. Biochemistry 47:4228–4236
10. Tipparaju SK, Mulhearn DC, Klein GM et al (2008) Design and synthesis of aryl ether inhibitors of the Bacillus anthracis enoyl-ACP reductase. ChemMedChem 3:1250–1268
11. Fischbach MA, Walsh CT (2009) Antibiotics for emerging pathogens. Science 325:1089–1093
12. Balganesh TS, Alzari PM, Cole ST (2008) Rising standards for tuberculosis drug development. Trends Pharmacol Sci 29:576–581
13. Zhang Y (2005) The magic bullets and tuberculosis drug targets. Annu Rev Pharmacol Toxicol 45:529–564
14. Ballel L, Field RA, Duncan K, Young RJ (2005) New small-molecule synthetic antimycobacterials. Antimicrob Agents Chemother 49:2153–2163
15. Ananthan S, Faaleolea ER, Goldman RC et al (2009) High-throughput screening for inhibitors of Mycobacterium tuberculosis H37Rv. Tuberculosis (Edinb) 89:334–353
16. Maddry JA, Ananthan S, Goldman RC et al (2009) Antituberculosis activity of the molecular libraries screening center network library. Tuberculosis (Edinb) 89:354–363
17. Reynolds RC, Ananthan S, Faaleolea E et al (2012) High throughput screening of a library based on kinase inhibitor scaffolds against Mycobacterium tuberculosis H37Rv. Tuberculosis (Edinb) 92:72–83
18. Macarrón R (2010) Contributions of HTS to drug discovery: a historical perspective. In: Meeting 4th Annual CDD Community Meeting, San Francisco
19. Schneider G (2010) Virtual screening: an endless staircase? Nat Rev Drug Discov 9:273–276
20. Barry CE 3rd, Slayden RA, Sampson AE, Lee RE (2000) Use of genomics and combinatorial chemistry in the development of new antimycobacterial drugs. Biochem Pharmacol 59:221–231
21. Ekins S, Freundlich JS (2011) Validating new tuberculosis computational models with public whole cell screening aerobic activity datasets. Pharm Res 28:1859–1869
22. Ekins S, Kaneko T, Lipinksi CA et al (2010) Analysis and hit filtering of a very large library of compounds screened against Mycobacterium tuberculosis. Mol Biosyst 6:2316–2324
23. Ekins S, Bradford J, Dole K et al (2010) A collaborative database and computational models for tuberculosis drug discovery. Mol Biosyst 6:840–851
24. Prathipati P, Ma NL, Keller TH (2008) Global Bayesian models for the prioritization of antitubercular agents. J Chem Inf Model 48:2362–2370
25. Cho Y, Ioerger TR, Sacchettini JC (2008) Discovery of novel nitrobenzothiazole inhibitors for Mycobacterium tuberculosis ATP phosphoribosyl transferase (HisG) through virtual screening. J Med Chem 51:5984–5992
26. Oprea TI, Davis AM, Teague SJ, Leeson PD (2001) Is there a difference between leads and

drugs? A historical perspective. J Chem Inf Comput Sci 41:1308–1315
27. Sundaramurthi JC, Brindha S, Reddy TB, Hanna LE (2012) Informatics resources for tuberculosis–towards drug discovery. Tuberculosis (Edinb) 92:133–138
28. Ekins S, Freundlich JS, Choi I (2011) Computational databases, pathway and cheminformatics tools for tuberculosis drug discovery. Trends Microbiol 19:65–74
29. Fernandes JP, Pasqualoto KF, Felli VM et al (2010) QSAR modeling of a set of pyrazinoate esters as antituberculosis prodrugs. Arch Pharm (Weinheim) 343:91–97
30. Dolezal R, Waisser K, Petrlikova E et al (2009) N-benzylsalicylthioamides: highly active potential antituberculotics. Arch Pharm (Weinheim) 342:113–119
31. Nayyar A, Malde A, Coutinho E, Jain R (2006) Synthesis, anti-tuberculosis activity, and 3D-QSAR study of ring-substituted-2/4-quinolinecarbaldehyde derivatives. Bioorg Med Chem 14:7302–7310
32. Macaev F, Rusu G, Pogrebnoi S et al (2005) Synthesis of novel 5-aryl-2-thio-1,3,4-oxadiazoles and the study of their structure-antimycobacterial activities. Bioorg Med Chem 13:4842–4850
33. Ventura C, Martins F (2008) Application of quantitative structure-activity relationships to the modeling of antitubercular compounds. 1. The hydrazide family. J Med Chem 51:612–624
34. Andrade CH, Salum Lde B, Castilho MS et al (2008) Fragment-based and classical quantitative structure-activity relationships for a series of hydrazides as antituberculosis agents. Mol Divers 12:47–59
35. Sivakumar PM, Geetha Babu SK, Mukesh D (2007) QSAR studies on chalcones and flavonoids as anti-tuberculosis agents using genetic function approximation (GFA) method. Chem Pharm Bull(Tokyo) 55:44–49
36. Periwal V, Rajappan JK, Jaleel AU, Scaria V (2011) Predictive models for anti-tubercular molecules using machine learning on high-throughput biological screening datasets. BMC Res Notes 4:504
37. Periwal V, Kishtapuram S, Open Source Drug Discovery Consortium, Scaria V (2012) Computational models for in-vitro anti-tubercular activity of molecules based on high-throughput chemical biology screening datasets. BMC Pharmacol 12:1
38. Pytela O, Klimesova V (2011) Effect of substitution on the antimycobacterial activity of 2-(substituted benzyl)sulfanyl benzimidazoles, benzoxazoles, and benzothiazoles–a quantitative structure-activity relationship study. Chem Pharm Bull 59:179–184
39. Dwivedi N, Mishra BN, Katoch VM (2011) 2D-QSAR model development and analysis on variant groups of anti-tuberculosis drugs. Bioinformation 7:82–90
40. Manvar AT, Pissurlenkar RR, Virsodia VR et al (2010) Synthesis, in vitro antitubercular activity and 3D-QSAR study of 1,4-dihydropyridines. Mol Divers 14:285–305
41. Shagufta, Kumar A, Panda G, Siddiqi MI (2007) CoMFA and CoMSIA 3D-QSAR analysis of diaryloxy-methano-phenanthrene derivatives as anti-tubercular agents. J Mol Model 13:99–109
42. Aparna V, Jeevan J, Ravi M et al (2006) 3D-QSAR studies on antitubercular thymidine monophosphate kinase inhibitors based on different alignment methods. Bioorg Med Chem Lett 16:1014–1020
43. Hevener KE, Ball DM, Buolamwini JK, Lee RE (2008) Quantitative structure-activity relationship studies on nitrofuranyl anti-tubercular agents. Bioorg Med Chem 16:8042–8053
44. Nayyar A, Monga V, Malde A et al (2007) Synthesis, anti-tuberculosis activity, and 3D-QSAR study of 4-(adamantan-1-yl)-2-substituted quinolines. Bioorg Med Chem 15:626–640
45. Nayyar A, Malde A, Jain R, Coutinho E (2006) 3D-QSAR study of ring-substituted quinoline class of anti-tuberculosis agents. Bioorg Med Chem 14:847–856
46. Kim P, Kang S, Boshoff HI et al (2009) Structure-activity relationships of antitubercular nitroimidazoles. 2. Determinants of aerobic activity and quantitative structure-activity relationships. J Med Chem 52:1329–1344
47. Biava M, Porretta GC, Poce G et al (2006) Antimycobacterial agents. Novel diarylpyrrole derivatives of BM212 endowed with high activity toward Mycobacterium tuberculosis and low cytotoxicity. J Med Chem 49:4946–4952
48. Kortagere S, Ekins S (2010) Troubleshooting computational methods in drug discovery. J Pharmacol Toxicol Methods 61:67–75
49. Prakash O, Ghosh I (2006) Developing an antituberculosis compounds database and data mining in the search of a motif responsible for the activity of a diverse class of antituberculosis agents. J Chem Inf Model 46:17–23
50. Garcia-Garcia A, Galvez J, de Julian-Ortiz JV, Guna R et al (2005) Search of chemical scaffolds for novel antituberculosis agents. J Biomol Screen 10:206–214
51. Planche AS, Scotti MT, López AG et al (2009) Design of novel antituberculosis compounds using graph-theoretical and substructural approaches. Mol Divers 13:445–458
52. Saquib M, Gupta MK, Sagar R et al (2007) C-3 alkyl/arylalkyl-2,3-dideoxy hex-2-

enopyranosides as antitubercular agents: synthesis, biological evaluation, and QSAR study. J Med Chem 50:2942–2950
53. Jones DR, Ekins S, Li L, Hall SD (2007) Computational approaches that predict metabolic intermediate complex formation with CYP3A4 (+b5). Drug Metab Dispos 35:1466–1475
54. Lougheed KE, Taylor DL, Osborne SA et al (2009) New anti-tuberculosis agents amongst known drugs. Tuberculosis (Edinb) 89:364–370
55. Sarker M, Talcott C, Madrid P et al (2012) Combining cheminformatics methods and pathway analysis to identify molecules with whole cell activity against Mycobacterium tuberculosis. Pharm Res 29:2115–2127
56. Lamichhane G, Freundlich JS, Ekins S et al (2011) Essential metabolites of M. tuberculosis and their mimics. MBio 2:e00301–e00310
57. Sassetti CM, Rubin EJ (2003) Genetic requirements for mycobacterial survival during infection. Proc Natl Acad Sci USA 100:12989–12994
58. Sassetti CM, Boyd DH, Rubin EJ (2003) Genes required for mycobacterial growth defined by high density mutagenesis. Mol Microbiol 48:77–84
59. Lamichhane G, Tyagi S, Bishai WR (2005) Designer arrays for defined mutant analysis to detect genes essential for survival of Mycobacterium tuberculosis in mouse lungs. Infect Immun 73:2533–2540
60. Workman P, Collins I (2010) Probing the probes: fitness factors for small molecule tools. Chem Biol 17:561–577
61. Ekins S, Williams AJ (2010) Meta-analysis of molecular property patterns and filtering of public datasets of antimalarial "hits" and drugs. Medchemcomm 1:325–330
62. Ekins S, Williams AJ (2010) When pharmaceutical companies publish large datasets: an abundance of riches or fool's gold? Drug Discov Today 15:812–815
63. Polgar T, Baki A, Szendrei GI, Keseru GM (2005) Comparative virtual and experimental high-throughput screening for glycogen synthase kinase-3beta inhibitors. J Med Chem 48:7946–7959
64. Doman TN, McGovern SL, Witherbee BJ et al (2002) Molecular docking and high throughput screening for novel inhibitors of protein tyrosine phosphatase-1B. J Med Chem 45:2213–2221
65. Willand N, Dirie B, Carette X et al (2009) Synthetic EthR inhibitors boost antituberculous activity of ethionamide. Nat Med 15:537–544
66. Kolb P, Ferreira RS, Irwin JJ, Shoichet BK (2009) Docking and chemoinformatic screens for new ligands and targets. Curr Opin Biotechnol 20:429–436
67. Gupta RK, Thakur TS, Desiraju GR, Tyagi JS (2009) Structure-based design of DevR inhibitor active against nonreplicating Mycobacterium tuberculosis. J Med Chem 52:6324–6334
68. Kumar A, Siddiqi MI (2008) CoMFA based de novo design of pyrrolidine carboxamides as inhibitors of enoyl acyl carrier protein reductase from Mycobacterium tuberculosis. J Mol Model 14:923–935
69. Kumar A, Siddiqi MI (2010) Receptor based 3D-QSAR to identify putative binders of Mycobacterium tuberculosis Enoyl acyl carrier protein reductase. J Mol Model 16:877–893
70. Kumar A, Siddiqi MI, Miertus S (2010) New molecular scaffolds for the design of Mycobacterium tuberculosis type II dehydroquinase inhibitors identified using ligand and receptor based virtual screening. J Mol Model 16:693–712
71. Banfi E, Scialino G, Zampieri D et al (2006) Antifungal and antimycobacterial activity of new imidazole and triazole derivatives. A combined experimental and computational approach. J Antimicrob Chemother 58:76–84
72. Andrade CH, Pasqualoto KF, Ferreira EI, Hopfinger AJ (2009) Rational design and 3D-pharmacophore mapping of 5′-thiourea-substituted alpha-thymidine analogues as mycobacterial TMPK inhibitors. J Chem Inf Model 49:1070–1078
73. Labello NP, Bennett EM, Ferguson DM, Aldrich CC (2008) Quantitative three dimensional structure linear interaction energy model of 5′-O-[N-(salicyl)sulfamoyl]adenosine and the aryl acid adenylating enzyme MbtA. J Med Chem 51:7154–7160
74. Wahab HA, Choong YS, Ibrahim P et al (2009) Elucidating isoniazid resistance using molecular modeling. J Chem Inf Model 49:97–107
75. Kumar A, Chaturvedi V, Bhatnagar S et al (2009) Knowledge based identification of potent antitubercular compounds using structure based virtual screening and structure interaction fingerprints. J Chem Inf Model 49:35–42
76. Kumar M, Vijayakrishnan R, Subba Rao G (2010) In silico structure-based design of a novel class of potent and selective small peptide inhibitor of Mycobacterium tuberculosis Dihydrofolate reductase, a potential target for anti-TB drug discovery. Mol Divers 14(3):595–604
77. Hegymegi-Barakonyi B, Szekely R, Varga Z et al (2008) Signalling inhibitors against Mycobacterium tuberculosis–early days of a new therapeutic concept in tuberculosis. Curr Med Chem 15:2760–2770
78. Gopalakrishnan B, Aparna V, Jeevan J et al (2005) A virtual screening approach for thymidine monophosphate kinase inhibitors as

antitubercular agents based on docking and pharmacophore models. J Chem Inf Model 45:1101–1108
79. Lin TW, Melgar MM, Kurth D et al (2006) Structure-based inhibitor design of AccD5, an essential acyl-CoA carboxylase carboxyltransferase domain of Mycobacterium tuberculosis. Proc Natl Acad Sci USA 103:3072–3077
80. Metaferia BB, Fetterolf BJ, Shazad-Ul-Hussan S et al (2007) Synthesis of natural product-inspired inhibitors of Mycobacterium tuberculosis mycothiol-associated enzymes: the first inhibitors of GlcNAc-Ins deacetylase. J Med Chem 50:6326–6336
81. Srivastava SK, Tripathi RP, Ramachandran R (2005) NAD+-dependent DNA Ligase (Rv3014c) from Mycobacterium tuberculosis. Crystal structure of the adenylation domain and identification of novel inhibitors. J Biol Chem 280:30273–30281
82. Singla D, Anurag M, Dash D, Raghava GP (2011) A web server for predicting inhibitors against bacterial target GlmU protein. BMC Pharmacol 11:5
83. Santhi N, Aishwarya S (2011) Insights from the molecular docking of withanolide derivatives to the target protein PknG from Mycobacterium tuberculosis. Bioinformation 7:1–4
84. Unissa AN, Sudha S, Selvakumar N, Hassan S (2011) Binding of activated isoniazid with acetyl-CoA carboxylase from Mycobacterium tuberculosis. Bioinformation 7:107–111
85. Sundaramurthi JC, Kumar S, Silambuchelvi K, Hanna LE (2011) Molecular docking of azole drugs and their analogs on CYP121 of Mycobacterium tuberculosis. Bioinformation 7:130–133
86. Anand R, Somasundaram S, Doble M, Paramasivan C (2011) Docking studies on novel analogues of 8 methoxy fluoroquinolones against GyrA mutants of Mycobacterium tuberculosis. BMC Struct Biol 11:47
87. Guzman JD, Wube A, Evangelopoulos D et al (2011) Interaction of N-methyl-2-alkenyl-4-quinolones with ATP-dependent MurE ligase of Mycobacterium tuberculosis: antibacterial activity, molecular docking and inhibition kinetics. J Antimicrob Chemother 66:1766–1772
88. Chhabra G, Dixit A, Garg LC (2011) DNA polymerase III alpha subunit from Mycobacterium tuberculosis H37Rv: homology modeling and molecular docking of its inhibitor. Bioinformation 6:69–73
89. Khare G, Kar R, Tyagi AK (2011) Identification of inhibitors against Mycobacterium tuberculosis thiamin phosphate synthase, an important target for the development of anti-TB drugs. PLoS One 6:e22441
90. Sharma R, Kothapalli R, Van Dongen AM, Swaminathan K (2012) Chemoinformatic identification of novel inhibitors against Mycobacterium tuberculosis L-aspartate alpha-decarboxylase. PLoS One 7:e33521
91. Scheich C, Szabadka Z, Vertessy B et al (2011) Discovery of novel MDR-Mycobacterium tuberculosis inhibitor by new FRIGATE computational screen. PLoS One 6:e28428
92. Usha V, Hobrath JV, Gurcha SS et al (2012) Identification of Novel Mt-Guab2 inhibitor series active against M. tuberculosis. PLoS One 7:e33886
93. O'Shea R, Moser HE (2008) Physicochemical properties of antibacterial compounds: implications for drug discovery. J Med Chem 51:2871–2878
94. Ritchie TJ, Luscombe CN, Macdonald SJ (2009) Analysis of the calculated physicochemical properties of respiratory drugs: can we design for inhaled drugs yet? J Chem Inf Model 49:1025–1032
95. Lipinski CA, Lombardo F, Dominy BW, Feeney PJ (1997) Experimental and computational approaches to estimate solubility and permeability in drug discovery and development settings. Adv Drug Deliv Rev 23:3–25
96. Hohman M, Gregory K, Chibale K et al (2009) Novel web-based tools combining chemistry informatics, biology and social networks for drug discovery. Drug Discov Today 14:261–270
97. Ekins S, Hohman M, Bunin BA (2011) Pioneering use of the cloud for development of the collaborative drug discovery (CDD) database. In: Ekins S, Hupcey MAZ, Williams AJ (eds) Collaborative computational technologies for biomedical research. Wiley, Hoboken, NJ

Index

A

B

C

D

E

F

Sandhya Kortagere (ed.), *In Silico Models for Drug Discovery*, Methods in Molecular Biology, vol. 993, DOI 10.1007/978-1-62703-342-8, © Springer Science+Business Media, LLC 2013

Q

R

S

T

V

MIX
Papier aus verantwortungsvollen Quellen
Paper from responsible sources
FSC® C105338

If you have any concerns about our products,
you can contact us on
ProductSafety@springernature.com

In case Publisher is established outside the EU,
the EU authorized representative is:
Springer Nature Customer Service Center GmbH
Europaplatz 3, 69115 Heidelberg, Germany

Printed by Libri Plureos GmbH
in Hamburg, Germany